E. Grandjean · Fitting the task to the man

E. Grandjean

Director of the Institute for Industrial Hygiene and Work Physiology
Swiss Federal Institute of Technology, Zürich

Fitting the task to the Man

An ergonomic approach

Taylor & Francis Ltd, London
1980

Originally published 1963 by Ott Verlag Thun under the title
Physiologische Arbeitsgestaltung
2nd Edition published 1969
3rd Edition published 1979 by Ott Verlag Thun. All rights reserved.
© 1979 Ott Verlag Thun.
English edition translated by Harold Oldroyd and published 1980
by Taylor & Francis Ltd., 10/14 Macklin Street, London WC2B 5NF.
© 1980 Taylor & Francis Ltd.

Printed and bound in Great Britain by
Taylor & Francis (Printers) Ltd.,
Rankine Road, Basingstoke, Hampshire.
ISBN 0 85066 191 9 (Cloth)
ISBN 0 85066 192 7 (Paper)

Contents

Acknowledgments

My thanks are due to the Directors of the Swiss Federal College of Technology for a sabbatical term during which I was able to carry out most of the work for this third edition.

Special thanks are due to:

- the late Harold Oldroyd for his translation of the text
- Herr Ing. grad. Wilhelm Hünting for the art origination
- Fräulein Esther Knabenhans and Frau Beatrice Brändli for producing a clean copy and corrections.

Etienne Grandjean

Foreword to the third edition

Ergonomics is a study of man's behaviour in relation to his work. The object of this research is man at work in relation to his spacial environment.
Ergonomic research is used in the adaptation of work conditions to the physical and psychological nature of man, and this results in the most important principle of ergonomics:

Fitting the task to the man.

Ergonomics is interdisciplinarian: it bases its theories on physiology, psychology, anthropometry, and various aspects of engineering.
The problems of ergonomics are not new; man has always aimed at alleviating the strains of work by designing practical instruments. Of course yesterday's tools are today's machines, electronic devices, chemical plants, or conveyor belts. Hands have few remaining creative functions; instead, we make considerable demands on the innumerable functions of the brain. This change in work conditions has demanded the presentation of usable knowledge in the form of readily available data; this has resulted in the science of ergonomics.
As a science ergonomics is neither 'good' nor 'bad': the crucial factor is the way in which it is put to use. In the past it served mainly to increase efficiency, and thereby productivity. This is no longer the prime goal. Indeed now the following objectives more closely define the benefits to be gained by ergonomic research:

- Fitting the demands of work to the efficiency of man in order to reduce stress.
- Designing machines, equipment, and installations so that they can be operated with great efficiency, accurately and safely.
- Working out proportions and conditions of the work place to ensure correct body posture.
- Adapting lighting, air-conditioning, noise, etc., to suit man's physical requirements.

This list of individual objectives could easily be lengthened. They all share a prime purpose: to contribute to human needs in a work environment, including promotion of health and wellbeing. In brief, they should heighten the quality of life in work conditions. The second edition of this book goes back to twelve years ago. Meanwhile the work environment has changed even more. A few examples serve to illustrate this change: shiftwork has increased; the man–machine systems are to some extent more monotonous, but also entail greater complications and responsibilities; mental and physical stresses have increased. These factors have been incorporated into this third edition. Each chapter has been revised and up-dated, and in addition the following new topics have been included: physical measurements as structural measurements, visual and accoustic perception, mental attainments, perceptual duration, monotony, and finally vibration in the chapter on noise.

The aim and purpose of this book have remained the same: to impart the rudiments of ergonomics in a simple and lucid form to the engineer, the technical specialist in work organisation, the works manager, the design engineer, and the production manager, so that those responsible can successfully put into practice the principles of ergonomics. References to further reading have been included to allow for special industrial problems that might occur and could not be dealt with in detail here. The interested reader can therefore refer to other works relevant to the subject.

Zurich, 1 May 1979

Etienne Grandjean

1
Muscular work

Physiological principles

Structure of a muscle

The human body is able to move because it has a widely distributed system of muscles, which together make up approximately 40 % of the total body weight. Each muscle is made up of a large number of muscle fibres, which can be between 0.5 cm and 14 cm long, according to the size of the muscle. The diameter of a muscle fibre oscillates about 0.1 mm. A muscle contains between 100 000 and one million such fibres, each of which is drawn out into a sinew at each end. The fibres of long muscles are sometimes bound together in bundles. At each end of the muscle the sinews are combined into a tough, non-elastic tendon, which in turn is firmly attached to the bony skeleton.

Muscular contraction

The most important characteristic of a muscle is its ability to shrink to half its normal length, a phenomenon we call muscular contraction. The work done by a muscle in such a complete contraction increases with its length: for this reason athletes try to make their muscles longer by stretching exercises.

Each muscle fibre contains proteins, among which actin and myosin assume special importance, since this system is contractile, and contributes most to the muscular contraction. Actin and myosin are present as fibres which can slide over each other. During the process of contraction the actin fibres are believed to ensheath themselves in between the myosin fibres, as illustrated in *Figure 1*.

Muscle power

Each muscle fibre contracts with a certain force, and the strength of the whole muscle is the sum of these muscle fibres. The maximum strength of a human muscle lies *between 3 and 4 kg/cm² of the cross-section* [141]: thus a muscle of 1 cm² cross-section can support a weight of 3–4 kg.

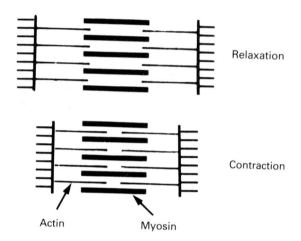

Relaxation

Contraction

Actin   Myosin

Fig. 1 Model of muscular contraction. The actin fibres slide between the myosin fibres and the two ends of the muscle are drawn closer together.

Hence a person's inherent muscular strength depends, in the first instance, on the cross-section of his or her muscles. Given equal training, a woman, with her narrower muscles, can exert about 30 % less force than a man (*Scherrer*, [277]). A muscle produces its greatest force at the beginning of its contraction, when it is still at its relaxed length. As the muscle shortens, its power declines. Many of the recommendations in *Chapter 3* for improving working efficiency are based on this relationship.

Regulation of muscular effort

The number of actively contracting muscle fibres determines how power is developed during the period of contraction. As we shall see later, a muscle fibre is made to contract by incoming nervous impulses; hence the amount of muscle power produced is determined by the number of nervous impulses, that is by the number of motor nerve cells in the brain that have been excited. The speed of a muscular contraction depends upon how quickly power is developed during a given interval of time, so the rapidity of a movement is governed by the number of actively contracting muscle fibres.

When muscular contraction is slow, or maintained for a long time (static muscular work), muscle fibres are brought into active contraction in succession, alternately. Each individual fibre has resting periods, which permit a certain amount of "recuperation" to take place.

Sources of energy

During contraction, mechanical energy is developed at the expense of the reserves of chemical energy in the muscle. Muscular work involves the transformation of chemical into mechanical energy. The energy released by chemical reaction acts upon the protein molecules of the actin and myosin fibres, causing them to change position, and so bring about contraction. The immediate sources of energy for contraction are energy-rich phosphate compounds which change from a high-energy to a low-energy state in the course of chemical reactions. The source of energy most widely used by living organisms is *adenosine triphosphate* (ATP), which releases considerable amounts of energy when it is broken down into adenosine diphosphate. Moreover, ATP is present not only in muscles, but in nearly every kind of tissue, where it acts as a reservoir of readily available energy. Another source of chemical energy in muscle fibres is phosphagen (phosphocreatine) which releases an equally significant amount of energy when it is broken down into phosphoric acid and creatine.

The low-energy phosphate compounds are continuously converted back to the high-energy state in the muscles, so that the reserves of energy remain undiminished. This is one of the wonders of nature: it is as if the exhaust gases of a car could be re-converted into petrol!

The role of glucose, fat and proteins

This regeneration of high-energy phosphate compounds itself consumes energy, which is obtained from glucose and from components of fat and proteins. Glucose, the most important of the sugars circulating in the blood, is the main energy supply in intensive physical work. Under conditions of rest or moderate physical work the components of fat (fatty acids) and of proteins (amino acids) are the dominant energy supplies. *These nutritive substances, glucose, fat and protein, are, therefore, the indirect energy sources for the continuous replenishment of energy reserves in the form of ATP or other energy-rich phosphate compounds.* The glucose passes out of the bloodstream into the cells, where it is converted by various stages into *pyruvic (pyro-racemic) acid*. Further breakdown can take two directions, depending on whether oxygen is available (aerobic glycolysis), or whether the oxygen supply is deficient (anaerobic glycolysis).

The role of the oxygen

If oxygen is present, then the pyruvic acid is further broken down by oxidation (i.e. under continuous oxygen consumption), the end-products being water and carbon dioxide. This releases enough energy to reconstitute a large amount of ATP.

If oxygen is lacking, then the normal breakdown of the pyruvic acid does not take place. Instead it is converted into *lactic acid*, a form of metabolic waste product which plays a vital part in symptoms of muscle fatigue and "muscular hangover". This process releases a lesser amount of energy for the reconstitution of energy-rich phosphate compounds, but it allows a higher muscular performance under low-oxygen conditions, at least for a short time.

Oxygen-debt

After heavy muscular effort a person is said to be "out of breath". This means that he is making up for a shortage of oxygen by breathing more heavily; *he is paying off his oxygen debt*. This oxygen debt arises from previous consumption of energy; the extra oxygen is needed to convert lactic acid back into pyruvic acid, and to reconstitute energy-rich phosphate compounds. After that, energy can again be obtained by oxidative breakdown of pyruvic acid.

Figure 2 shows a much simplified diagram of the energy supply of a muscle.

Protein and fat

As already mentioned, fats and proteins are also involved in these metabolic changes. When breakdown of these substances has reached a certain stage, a *common metabolic pool* comes into being. The remaining fragments of fatty acids (from breakdown of fats) and amino-acids (from breakdown of proteins) undergo the same final breakdown as the pyruvic acid, and end up as water and carbon dioxide. This last stage therefore makes its own contribution of energy to the muscular effort.

The blood-supply

These substances that are so important for energy production – glucose and oxygen – are stored only in small amounts in the muscles themselves. Both of them must therefore be continuously transported to the muscles by the blood. For this reason, in the final analysis it is the blood supply that is the limiting factor in the efficiency of the muscular machinery. During effort, a muscle increases its need for blood several fold, and to supply this the most important adaptations of the blood system are more active pumping by the heart, raising of the blood pressure, and enlargement of the blood vessels that lead to the muscles.

According to *Scherrer* [277] the following increases in circulation can be expected:

Muscle at rest:	4 ml*/min/100 g muscle
Moderate work:	80 ml/min/100 g muscle
Heavy work:	150 ml/min/100 g muscle
After a restriction of the blood circulation:	50–100 ml/min/100 g muscle.

Heat-production

According to the first law of thermodynamics, a muscle must be supplied with the same amount of energy that it uses. In practice the incoming energy is transformed into the following:
(a) work performed;
(b) heat;
(c) energy-rich chemical compounds.

Energy stored in the form of phosphate compounds is the smallest

---

* 1 ml = 1 cm^3

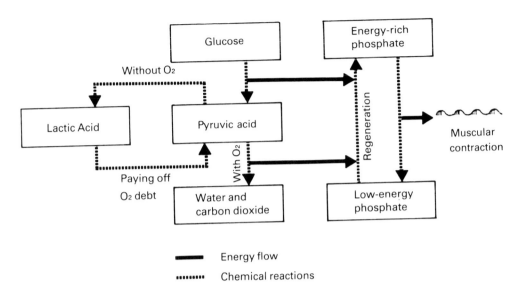

Fig. 2   Diagram of metabolic processes which take place during muscular work.

component; in contrast, generation of heat is by far the biggest. If heat-generation within the muscles is measured with delicate instruments, the following constituent parts are recognisable:

(a) *resting heat production* amounts to 0.3 kcal*/min for a man weighing 70 kg. According to *Scherrer* [277] this serves to maintain the molecular structure and the electrical potentials in the muscle fibre;

(b) *initial heat* is considerably greater than the resting heat. This is the heat produced during the whole course of contraction of the muscle, and is proportional to the work done;

(c) *recovery heat* is produced after the contraction is finished and over a longer period, up to 30 min. This is obviously the output from the oxidative processes of the recovery phase, and is of about the same order of magnitude as the initial heat.

Associated electrical phenomena

It has been known for a long time that muscular contraction is accompanied by electrical phenomena, which are very similar to the processes of transmission of impulses along a nerve. In recent decades these electrical processes have been studied in more detail, using very delicate electrophysiological methods.[1]

---

[1] Several details of such electrical processes will be dealt with in *Chapter 2*: "Nervous control of movements".

*Note: the calorie as a unit of energy, heat or work is now obsolete and has been superseded by the joule (1 calorie = 4.2 joules). However, since the majority of readers will not be familiar with the joule, the calorie has been retained throughout this text. (1 kcal = 4.2 kJ).

In simple terms we can make the following points:

(a) the resting muscle fibre exhibits an electrical potential – the so-called *resting membrane potential* – of about 90 mV. The interior of the fibre is negatively charged in relation to the exterior;

(b) the beginning of contraction is associated with a collapse of the resting potential, and an overriding positive charge in the interior. This reversal of potential is called the *action potential*, so-called because it arises "during the action" of the nerve. The action potential in the muscle lasts about 2–4 ms, and spreads along the muscle fibre at a speed of about 5 m/s;

(c) the action potential involves depolarisation and repolarisation of the membrane of the muscle fibre. During this period the muscle fibre is no longer capable of being excited; this period is called the absolute refractory period, and lasts 1–3 ms. By analogy with the processes that go on in nerves, in the muscle fibre also de- and repolarisation are manifestations of reciprocal streams of potassium and sodium ions through the membrane of the muscle fibre.

Electromyography

The electrical activity of a muscle can be recorded with the help of an amplifier, a technique known as *electromyography*.

The electric current can be tapped by attaching suitable electrodes to the surface of the skin immediately over the muscles to be studied. It is necessary to make sure that the electromyogram is recording the electrical activity of individual motor units. Another technique, less often used, is to insert needle-electrodes into the muscles and thereby to be able to monitor individual muscle fibres. An electromyogram registering the output of surface electrodes records the total electrical activity of the entire muscle. For this purpose two electrodes, each about 1 cm² in area are applied a few cm apart. Nowadays the output of the electrodes is usually integrated and amplified electronically. Results by this method have so far had only a relative significance because they are valid only for one particular set of electrodes in one particular experiment.

Nevertheless, electromyography has shown that electrical activity increases with the level of muscular force developed.

Electromyography is specially useful for investigating the amount of muscular effort expended in different bodily attitudes.

Static muscular effort

There are two kinds of muscular effort:

(a) *dynamic (rhythmic) effort;*

(b) *static (postural) effort.*

Dynamic and static forms of work

Figure 3 illustrates the two kinds of muscular activity. The dynamic example is cranking a wheel, and the static example is supporting a weight at arm's length.

The two forms of muscular effort can be described as follows:

(a) *the dynamic effort is characterised by a rhythmic alternation of contraction and extension, tension and relaxation;*

(b) *the static effort, in contrast, is characterised by a prolonged state of contraction of the muscles, which usually implies a postural stance.*

In a dynamic situation the effort can be expressed as the product of the shortening of the muscle and the force developed (work = weight × height it is raised). During static effort the muscle is not allowed to extend, but remains in a state of heightened tension, with force exerted over an extended period. During static effort no useful work is externally visible, nor can this be defined by a formula such as weight × distance. It rather resembles an electro-magnet, which has a steady consumption of energy while it is supporting a given weight, but does not appear to be doing useful work.

Blood supply

There are certain basic differences between static and dynamic muscular effort.

During static effort the blood vessels are compressed by the internal pressure of the muscle tissue, so that blood no longer flows through the muscle. During dynamic effort, on the other

Fig. 3   Diagram of dynamic and static muscular effort.

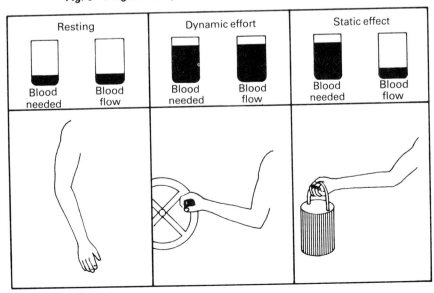

hand, as when walking, the muscle acts as a pump in the blood system. Compression squeezes blood out of the muscle, and the subsequent relaxation releases a fresh flow of blood into it. By this means the blood supply becomes several times greater than normal: in fact the muscle may receive between 10 and 20 times as much blood as when it is resting. A muscle performing dynamic work is therefore flushed out with blood and retains the energy-rich sugar and oxygen contained in it, while at the same time waste products are removed.

In contrast, a muscle that is performing heavy static work is receiving no sugar or oxygen from the blood, and must depend upon its own reserves. Moreover – and this is by far the most serious disadvantage – waste products are not being excreted. Quite the reverse: these waste products are accumulating and produce the acute pain of muscular fatigue.

For this reason we cannot continue a static muscular effort for very long; the pain will compel us to desist. On the other hand a dynamic effort can be carried on for a very long time without fatigue, provided that we choose a suitable rhythm for it. There is one muscle that is able to work dynamically throughout our lives, without interruption, and without tiring: the muscle of the heart. *Figure 4* shows how the two kinds of muscular effort affect the blood supply to the working muscle.

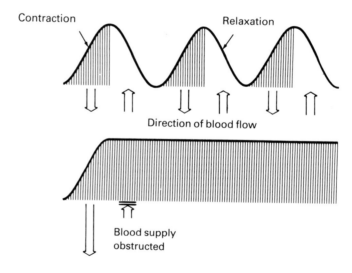

Fig. 4 Flow of blood through muscles during dynamic and static effort.
The curves show the variation of muscular tension (internal pressure).
Upper: dynamic effort operates like a pumping action, which ensures a flow of blood through the muscle.
Lower: static effort obstructs the flow of blood.

*Examples of
static effort*

Our bodies must often perform static effort during everyday life. Thus when we stand up, a whole series of muscle groups in the legs, the hips, in the back and neck are stressed for long periods. It is thanks to these static efforts that we can hold selected parts of our bodies in any desired attitude. When we sit down, the static effort of the legs is relieved, and the total muscular strain of the body is reduced. When we lie down, nearly all static muscular effort is avoided; that is why a recumbent posture is the most restful. There is no sharp line between dynamic and static effort. Often one particular task is partly static and partly dynamic. Since static effort is much more arduous than dynamic, the static component of mixed effort assumes the greater importance.

In general terms, static effort can be said to be considerable under the following circumstances:

(a) if a high level of effort is maintained for 10 seconds or more;
(b) if moderate effort persists for 1 minute or more;
(c) if slight effort (about one third of maximum force) lasts for 4 minutes or more.

There is a static component in nearly every form of factory work or any other occupation. The following are some of the commonest examples:

(a) jobs which involve bending the back either forwards or sideways;
(b) holding things in the arms;
(c) manipulations which require the arms to be held out horizontally;
(d) putting the weight on one leg while the other works a pedal;
(e) standing in one place for long periods;
(f) pushing and pulling heavy objects.

Figure 5 shows examples of static loads.

*Effects of
static work*

During static effort the flow of blood is constricted in proportion to the force exerted. If the effort is 60 % of the maximum, the flow

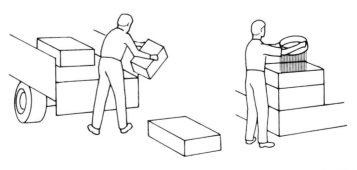

Fig. 5 Examples of static muscular effort. Left: loading parcels. Right: sieving sand into a mould in a foundry. In both examples there are high static loads in the muscles of back, shoulders and arms.

is almost completely interrupted, but a certain amount of circulation of blood is possible during lesser efforts, because the tension in the muscles is less. When the effort is less than 15–20% of the maximum, blood flow should be normal.

Obviously, therefore, the onset of muscular fatigue from static effort will be more rapid the greater the force exerted, that is the greater the muscular tension. This can be expressed in terms of the relation between the maximal duration of a muscular contraction, and the force expended. This was systematically studied by *Monod* [226] and later confirmed by *Rohmert* [265]. *Figure 6* shows the results obtained by *Monod* for four muscles.

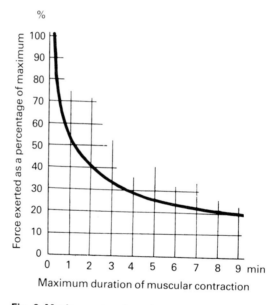

Fig. 6 Maximum duration of static muscular effort in relation to the force exerted. After *Monod* [226].

It appears from this work that a static effort which exerts 50% of maximum force can last no more than one minute, whereas, if the force expended is less than 20% of maximum, the muscular contraction can continue for very long periods.

Rohmert [265], who obtained his results from 21 men and women carrying out 12 grasping operations, was able to show that the maximum period of retention was independent of the maximum possible force of the group of muscles, or of the research subject. He found the following maximum periods of retention:

(a) at 100% of maximum force: 0.1 min
(b) at 75% of maximum force: 0.35 min
(c) at 50% of maximum force: 1.0 min
(d) at 25% of maximum force: 3.4 min.

Static muscular effort is strenuous

Under roughly similar conditions a static muscular effort, compared with dynamic work, leads to:
(a) a higher energy consumption;
(b) raised heart rate;
(c) longer rest periods needed.

This is easy to understand if we bear in mind that, on the one hand, the metabolism of sugar with an inadequate supply of oxygen releases less energy for regeneration of energy-rich phosphates, and on the other hand produces a large amount of lactic acid, which interferes with muscular effort. Oxygen deficiency, which is unavoidable during static muscular effort, inevitability lowers the effective working level of the muscle.

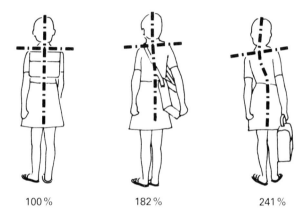

100%   182%   241%

Oxygen consumption

Fig. 7   Effect of static effort on energy consumption (measured by oxygen consumption) for three ways of carrying a school satchel. After *Malhotra* and *Sengupta* [213].

Figure 7 shows a good example of this, repeating the research results of *Malhotra* and *Sengupta* [213]. These authors showed that schoolchildren who carried their satchel in one hand needed more than twice as much energy as when they carried the satchel on their back. This increased energy consumption must be attributed to the high static loads on the arms, shoulders and trunk. *Figure 8* shows another example from *Hettinger* [142], involving potato planting. In one case the basket of potatoes was carried in one hand; in the other the basket hung in a harness in front of the body. Carrying the basket by hand raised the heart rate by 45 pulses per minute, compared with only 31 pulses per minute when the potato-basket was carried in a harness. Carrying the basket required a muscular effort in the left arm amounting to 38% of its maximum. *Hettinger* drew the conclusion that the increased strain on the heart was entirely caused by the heavy static effort needed to carry the basket in one hand.

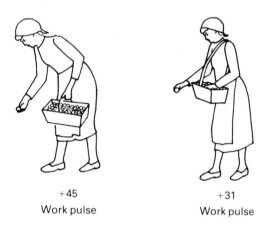

+45
Work pulse

+31
Work pulse

Fig. 8 Static muscular effort in the left arm during potato-planting; the use of a sling has avoided static effort in the left arm. Over a period of 30 minutes the heart rate rose up to 45 pulses per minute (left) or 31 pulses per minute (right). After *Hettinger* [142].

Complaints and illnesses

Static muscular effort produces a troublesome fatigue in the muscles concerned, which can build up into an intolerable pain. If the static load is repeated daily over a long period, more or less permanent aches will appear in the limbs concerned, and may involve not only the muscles but also the joints, tendons and other tissues. *A long-continued and excessive static work load thereby leads to deterioration of joints, ligaments and tendons.* Many practical observations lead to the conclusion that increased static loads lead to increases in:

(a) inflammation of the joints;
(b) inflammation of the tendon sheaths;
(c) inflammation of the attachment points of tendons;
(d) symptoms of chronic degeneration of the joints, in the form of arthritis;
(e) disc trouble.

These overstrain problems can be separated into two groups:

Group 1: The troubles are short-lived. The pains are mostly localised in the muscles and tendons, and pass off as soon as the static load is relieved. The work is made substantially harder, but when the job is finished, the pains vanish. So the first group of troubles is reversible: *they are the pains of weariness.*

Group 2: These troubles are also in the muscles and tendons, but they affect the joints as well. The pains do not disappear when the work stops, but go on afterwards. They are usually preceded by pains belonging to Group 1. Most of all, pains of Group 2 are associated with particular movements, or particular postures. *These persistent pains are attributable to conditions of inflammation and degeneration in the overloaded tissues.* Ailments of this second group appear particularly among older operatives; according to *van Wely* [325] they are commonly seen among operatives who work all the year round at the same machine at which the components are supplied at either too high or too low a level. These pains may become very much more acute, ending in deformations of the muscles and tendons, muscular cramps, or sudden attacks of inflammation in the tendon sheaths, attachments, or joints.

The troubles that may be expected to follow from certain forms of static load are set out in *Table 1*:

Table 1   Static load and bodily pains.

Work-posture	Possible consequences affecting:
Standing in one place	Feet and legs, possibly varicose veins
Sitting erect without back support	Extensor muscles of the back
Seat too high	Knee; calf of leg; foot
Seat too low	Shoulders and neck
Trunk curved forward, when sitting or standing	Lumbar region; deterioration of intervertebral discs
Arm outstretched, sideways, forwards or upwards	Shoulders and upper arm; possibly periarthritis of shoulders
Head excessively inclined backwards or forwards	Neck; deterioration of intervertebral discs
Unnatural grasp of hand grip or tools	Forearm; possibly inflammation of tendons

An example of increased pains from a bad working posture is reproduced as *Figures 70, 72* in *Chapter 6.*

A work of *Corlett* and *Bishop* [58] may also be mentioned in this context. Pains brought on by bodily posture were investigated, using special recording equipment, and, as an example, they were able to demonstrate quantitatively a distinct reduction in postural pains in a study of spot-welders, after their work places had been ergonomically redesigned.

Examples of morbid symptoms

During a training course lasting 12 weeks, *Tichauer* [306] compared the effects of using a wire cutter that was shaped to the hand with those of a normal pair of pliers. The normal pliers called for an unnatural grip, with the hand rotated round the ulnar, and hence a pronounced static muscular strain. During those 12 weeks, 25 out of the 40 workers who had the unnatural grip showed morbid symptoms, in the sense of inflammation of the tendon attachments or of the tendon sheaths. In contrast, there were only four cases of inflamed tendon sheaths among an equal number of workers who used the curved wire-cutters which permitted a natural grip. The two forms of pliers are shown in *Figure 9*.

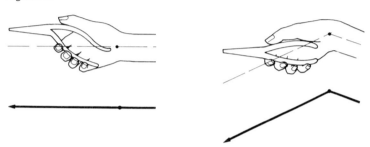

Fig. 9 Left: the curved pliers are shaped to fit the hand, which remains in line with the forearm. Right: pliers of the traditional shape require sustained static effort, with the wrist turned outwards; the hand is no longer in line with the forearm. Modified from *Tichauer* [306].

Van Wely [325] reported a study of the work places of 50 employees of the Philips works at Eindhoven, who were being treated in the works clinic for movement problems such as those described above. In 39 out of 50 cases he showed a clear link with a bad working posture. At 40 out of the 50 work places there was a bad layout, which led to unnatural postures: in 19 of these the machine was at fault, in 21 it was bad seating.

Figure 10 shows two work places at a machine which could lead to a danger of aches and other unhealthy symptoms.

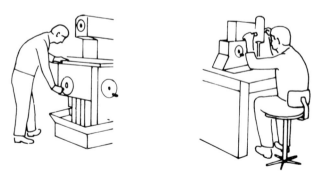

Fig. 10 Unnatural postures while at work, which involve static loads, can lead to physical disabilities. Left: risk of back troubles. Right: risk of aches in shoulders and upper arms.

<table>
<tr><td>Standing in
one place</td><td>"Standing in one place" involves static effort, through prolonged immobility of the joints of the feet, knees and hips. The force involved is not great, and falls below the critical level of 15 % of the maximum. All the same, "standing in one place" for a long time is wearisome and painful. This is not entirely the effect of static muscular effort; pain is also partly caused by increased hydrostatic pressure of the blood in the veins of the legs and general restriction of lymphal circulation in the lower extremities. In practice the hydrostatic pressure in the veins, when standing motionless, is increased as follows:</td></tr>
</table>

(a) at the level of the feet by 80 mm Hg;
(b) at the level of the thighs by 40 mm Hg.

During walking the muscles of the legs act as a pump, which compensates for the hydrostatic pressure of the veins by actively propelling blood back towards the heart.
Thus prolonged "standing in one place" causes not only fatigue of the muscles which are under static load, but also discomfort, which is attributable to insufficient return flow of the venous blood.
This unhealthy state of the blood circulation is the cause of cumulative ailments of the lower extremities in occupations which call for prolonged standing without movement. Such occupations involve increased liability to:

(a) dilatation of the veins of the legs (varicose veins);
(b) swelling of the tissues in the calves and feet (oedema of the ankle; dropsy);
(c) inflammation of the veins of the legs, with the formation of blood-clots (thromboses);
(d) ulceration of the oedematous skin.

<table>
<tr><td>Bodily postures
of saleswomen</td><td>In one of our own investigations among saleswomen in a departmental store we studied on the one hand their working postures, and on the other their bodily aches and pains, by means of a questionnaire. Working postures were studied through 5280 observations on 24 employees during 24 working days. The first thing we did was to make allowances for the different working attitudes involved: see Figure 11. It emerges from this that out of a total working day of 8¼ hours, 5 hr 25 min were spent standing in one place.</td></tr>
</table>

The saleswomen concerned were therefore exposed to considerable static strain.
200 saleswomen from the same store, with approximately similar situations, were questioned about their aches and pains. 79 (about 40 %) said that they did have ailments, and the various forms of ailment are analysed in *Table 2*.

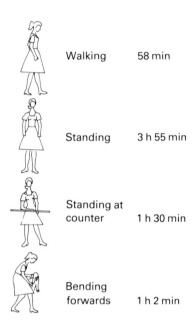

	Walking	58 min
	Standing	3 h 55 min
	Standing at counter	1 h 30 min
	Bending forwards	1 h 2 min

Fig. 11 Working attitudes of saleswomen, with the time spent in each attitude during a working day. 5280 observations on 24 saleswomen, by a multi moment observation procedure. After *Grandjean* and others [112].

Table 2 Distribution of complaints among 79 saleswomen (spontaneous information). 100 % = 79 individuals.

Complaints	% of those questioned
Pains in legs and feet	20
Pains in back	19
Headache	19
Digestive organs and liver	9
Rheumatism; arthritis; neuralgia	7
Nervous troubles	6
Heart trouble	5
Kidney and bladder troubles	5
Various other complaints	10

Comparison of these results with other investigations, in which the same questions were asked, showed that the saleswomen had a significantly higher incidence of ailments affecting the legs and feet. This is all the more grave a situation because more than half of them were still under 22 years of age.

The conclusion is therefore obvious that prolonged "standing in one place" is a common cause of ailments affecting the legs and feet among saleswomen.

2
Nervous control of movements

Physiological principles

Structure of the nervous system

The central nervous system consists of the brain and the spinal cord. The peripheral nerves either run outwards from the spinal cord to end in the muscles (motor nerves), or inwards from the skin, the muscles, or the sense organs to the spinal cord or the brain (sensory nerves). The sensory and motor nerves, together with their associated tracts and centres in the spinal cord and brain comprise the somatic nervous system, which links the organism with the outside world through perception, awareness and reaction.

Complementary to this is the visceral or autonomic nervous system, which controls the activities of all the internal organs: blood circulation, breathing organs, digestive organs, glands, and so on. The visceral nervous system therefore governs those internal mechanisms that are essential to the life of the body.

The complete nervous system is made up of millions and millions of nerve cells, or neurones, each of which has basically a cell body and a comparatively long nerve fibre. The cell body is a few thousandths of a millimetre thick, whereas the nerve fibre can be more than a metre long. *Figure 12* shows a diagrammatic picture of a neurone.

The function of nerves

The nervous system is essentially a control system, which regulates external and internal movements, as well as monitoring a variety of sensations. The working of a neurone depends upon its being sensitive to stimuli, and being able to transmit a stimulus along the length of the nerve fibre. When a nerve cell is stimulated, the resulting impulse travels along the nerve fibre to the operative organ, which may be a muscle fibre, among others.

Nervous impulses are of an electrochemical nature. Nerves are not just "telephone wires", transmitting impulses passively. A nervous

18

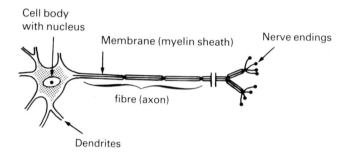

Cell body
with nucleus

Membrane (myelin sheath)

Nerve endings

fibre (axon)

Dendrites

Fig. 12 Diagram of a motor neurone, consisting of cell-body, dendrites, a nerve fibre and nerve endings.

impulse is an active process, self-generating and energy-consuming, more like a length of fuse, or slow match. Unlike a fuse, however, a nerve fibre is not dead once it has been used, but is regenerated in a fraction of a second, becoming receptive once again after the so-called refractory period. The nerve fibre cannot transmit a continuous "direct current", but only single impulses, with brief interruptions between them.

The speed of transmission is very different in different types of nerve: motor fibres transmit at 70–120 m/s; other fibres in the region of 12–70 m/s.

The nature of nervous impulses

What are these nervous impulses? Like muscle fibres, nerve fibres have a resting membrane potential. *In the resting phase the membrane of the neurone is polarised; positive charges predominate on the outer surface, while negative charges predominate internally. Depolarisation of the membrane produces the nervous impulse.* The resting potential, −70 mV, collapses completely, and depolarisation continues until a reverse peak of +35 mV is reached. Then comes a repolarisation, with a quick return to the resting potential of −70 mV. *Figure 13* shows the trace of an action potential as a nervous impulse passes along a nerve fibre.

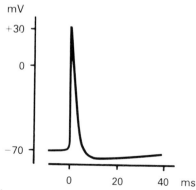

mV

+30

0

−70

0   20   40   ms

Fig. 13 The action potential in a nerve fibre, which indicates the passage of a nervous impulse.

As mentioned in Chapter 1, this electrical oscillation between de- and repolarisation of the membrane is called the action potential. This is what nervous impulses are all about.

Causes of the action potential

The action potential is therefore the electrical manifestation of a wave of de- and repolarisation, which travels along the nerve fibre at a speed of between 12 and 120 m/s. Depolarisation, the breakdown of the resting potential, is brought about by a sudden change in the permeability of the membrane, allowing a stream of positively charged sodium ions to penetrate to the interior of the fibre. Almost simultaneously, positively charged potassium ions move from inside to outside of the membrane, but these are fewer than the sodium ions moving inwards, so the net result is a sharp increase in the total positive charge in the interior. *It is this displacement of electrical charges which produces depolarisation and the associated action potential.*

In the subsequent repolarisation, these charged ions move in the reverse directions: sodium outwards, potassium inwards. We are back at the starting-point, with the membrane potential restored. This mechanism is called the *"sodium-potassium pump"* for short. It is not confined to nerve fibres but occurs in nearly all living cells. The energy for this pumping mechanism is derived from adenosine triphosphate (ATP). The "sodium-potassium pump" is an essential requirement before a nerve can react to stimuli and transmit impulses.

The nervous system requires a supply of energy, principally to maintain the membrane potential, and derives this energy from ATP. The metabolic activity of a nerve approximately doubles when it is active; a small increase compared with that of skeletal muscle, which increases its metabolism one hundredfold when it is working.

Innervation of muscles

Every muscle is connected to the brain, the overriding control centre, by nerves of two kinds: efferent, or motor nerves, and afferent, or sensory nerves.

Motor nerves carry impulses, in this case movement orders, from the brain to the skeletal muscles, where they bring about a contraction, and collectively control muscular activity. Within a muscle the nerve divides into its constituent fibres, each nerve fibre serving to innervate several muscle fibres. Each single motor neurone, together with the muscle fibres that it innervates, constitutes a *motor-unit*. In muscles which carry out fine and precise movements, as in skilled work, there are only three to six muscle fibres per motor-unit, whereas muscles which do heavy work may have 100 muscle fibres innervated by one neurone.

The bundle of motor nerve fibres shown above in *Figure 12* ends in so-called "*motor end plates*", where the cell membrane of the nerve fibre is thickened. This is where the motor impulse leaps across from the nerve fibre to the muscle fibre and where the action potential finally provokes muscular contraction.

Sensory nerves conduct impulses from the muscles into the central nervous system, either the spinal cord or the brain. *Sensory impulses are bearers of "signals"*, which will be utilised in the central nervous system in part to direct muscular work and in part to store as information.

Receptor organs of a special kind are the *muscle spindles*, parallel to the muscle fibres, and ending in both sides of the tendons. The muscle spindles are sensitive to stretching of the muscle, and send signals about this to the spinal cord (proprioceptor system).

Further organs of sensory response are the *Organs of Golgi* (Golgi–Mazzoni corpuscles), which are composed of a network of nodular nerve endings, and are embedded in the tendons. These organs, too, transmit sensory impulses to the nerve cord whenever the tendon is under tension.

Within the spinal cord the sensory impulses pass by means of an intermediate neurone over to the motor nerve, so that new impulses flow back to the muscles. This system of an afferent sensory nerve and an efferent motor nerve running back to the same muscle is called a *reflex arc*. Such a reflex arc keeps muscle tension and muscle length continually adjusted to each other, *the muscle spindle and the Organ of Golgi being the "detectors" in this "regulatory system"*.

Other sensory nerves conduct impulses from the muscles: over a first intermediate neurone in the spinal cord, and over a second in the medulla of the brain (brain-stem), up to the cerebral cortex *where the incoming pulses are finally experienced as a sensation*. This is how pains, which arise in the muscles, are felt. The innervation of the skeletal muscles with sensory and motor nerve fibres is demonstrated in *Figure 14*. At the same time the sensory paths leading to the spinal cord and into the brain, as well as the motor paths running back from the brain to the muscles are drawn in very simply.

Reflexes and skills

One special way in which movement and activity are controlled is by means of reflexes. Because these are not consciously directed, *they are "automatic" in a physiological sense*. A reflex consists of the initial stimulus: an impulse which travels along a sensory nerve, carrying the information to the spinal cord or to the brain; intermediate neurones, which pass the impulse across to a motor nerve; and a final impulse along the motor nerve which activates

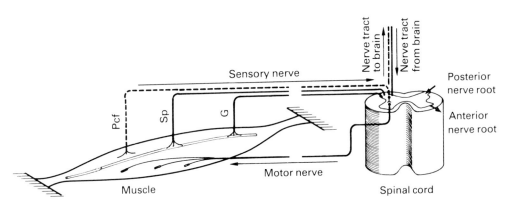

Fig. 14 Innervation of a muscle. The sensory nerve comprises pain-conducting fibres (pcf) as well as fibres from the muscle spindle (Sp) and from the Golgi receptors of the tendon (G). The fibres of the motor nerve terminate in the motor end plates in the membranes of the muscle fibres.

the appropriate muscle. An example is reflex blinking of the eyelids. When anything unexpected moves close to either eye, both eyes close automatically. The unexpected movement in the visual field provides the initial stimulus; the sensory impulse travels to a particular centre in the brain, which itself acts as intermediate neurone, and passes on the message to a motor nerve; the motor nerve in turn operates the muscles of the eyelid. Reflex blinking is thus an automatic protective mechanism, safeguarding the eyes against damage. The body makes use of thousands of such reflexes, not only for protection, but as part of normal control functions.

Reflexes also play an essential part in muscular activity. One example has been described above, in connection with skeletal muscle, and the reflex arc involved is shown diagrammatically in *Figure 14*. Another example of a more complicated and important reflex is the antagonistic control of a muscular movement. When the lower arm is bent the bending muscles are caused to contract by stimulation of their motor nerves, and if this is to proceed smoothly the opposing muscles behind the arm must be simultaneously relaxed by exactly the right amount. This is a reflex phenomenon, i.e. automatic direction, by which a trouble-free movement is carried out.

Skilled work

To give an idea of the complexity of nervous control, the most important stages in a piece of skilled work are shown graphically in *Figure 15*. During a simple grasping operation such as that illustrated, the first step is to make use of visual information to direct the movements of the arms, hands and fingers towards the object to be grasped. For this purpose nervous impulses travel along the optic nerve from the retina of the eye to the brain and are there integrated into a picture; hand–finger–object to be grasped are perceived. These impulses are then passed over to other centres, in the medulla and cerebellum, which control muscular activity. On the strength of the visual signals it has received, the brain thus decides what the next move will be. When the object has been grasped, pressure-sensitive nerves in the skin send new signals to the brain, and the operator can adjust his finger pressure accordingly.

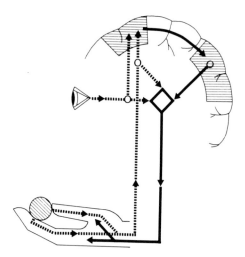

Fig. 15 Diagram illustrating nervous controls of skilled work.

Conditioned reflexes

Becoming skilled is largely a matter of developing new reflexes, which go on without conscious control, and are called conditioned reflexes. *Figure 15* indicates new reflex pathways as arrows, passing directly from the synapses (intermediate neurones) of the sensory pathways, into the muscular control centres,[1] where *combinations of movements are retained as a sort of template.* In other words, whenever a sequence of movements is practised for a long time, *the complete movement-pattern becomes "engraved" in the brain.* Coordination and delicate adjustment of individual muscular movements are achieved when a continuous stream of sensory information reaches these motor coordination centres. Skill reaches a maximum when learning has eliminated conscious control and movements have become automatic. The task of the conscious mind is to concentrate all nervous activity upon the job in hand and to give overriding "orders" to the control centres.

An example: writing

The whole process of automatism can be illustrated by the example of learning to write. First the child learns to control the

---

[1] According to current theory the centres responsible for precise regulation and coordination lie in the basal ganglion of the medulla and in the cerebellum; movement patterns are "localised" in these areas.

movements of hand and fingers so delicately that the correct shapes appear on the paper. Then he begins to make the necessary movements at will. After a very long period of practice the sequence of movements necessary to write each letter has become "engraved" as a pattern in the motor-control centres of his brain. The writing process itself becomes more and more automatic, and eventually the conscious mind concerns itself only with finding the right words and making them into sentences.

3
Improving working efficiency

Optimal use of muscle-power

General principles   When any form of bodily activity calls for a considerable expenditure of effort the necessary movements must be organised in such a way that the muscles are developing as much power as possible. In this way the muscles will be at their most efficient and most skilful.

Where the work involves holding something statically, a posture must be taken up which will allow as many strong muscles as possible to contribute. This is the quickest way to bring the loading of each muscle down below 15% of its maximum.

Since a muscle is most powerful at the beginning of its contraction it is a good idea, in principle, to start from a posture in which the muscle is fully extended. There are so many exceptions to this general rule, however, that it has more theoretical than practical value. One must also take into account the leverage effect of the bones, and, if several muscles join forces, their total effect. In the last case the force exerted is usually at its greatest when as many muscles as possible contract simultaneously. The maximum force of which a muscle, or a group of muscles, is capable depends upon:

(a) age;
(b) sex;
(c) constitution;
(d) state of training;
(e) momentary motivation.

Age and sex   *Figure 16* sets out the effects of age and sex on muscle power, according to dates obtained by *Hettinger* [141]. The peak of muscle power for both men and women is reached between the ages of 25 and 35 years old. Older workers aged between 50 and 60 can produce only about 75–85% as much muscular power.

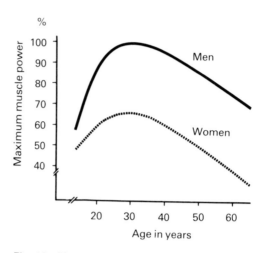

Fig. 16 Muscle power in relation to age and sex. After *Hettinger* [1].

As has already been pointed out in *Chapter 1*, it can be assumed that the average woman will be only about two-thirds as powerful as the average man [277].
Hettinger [142] has studied the maximum force developed by both men and women when using three groups of muscles which he considered have special significance in the assessment of human muscular power. His results are summarised in *Table 3*.

Table 3 Maximum muscular power for men and women, according to *Hettinger* [142]. s = standard deviation of the individual values.

Functions	Max. force men (kp)*	s	Max. force women (kp)	s
Hand-clasp	46	12	28	7
Kicking (with knee bent at 90°)	40	6	32	5
Stretching the back	109	16	74	16

Maximum power when sitting at work

The following rules, deduced from studies made by *Caldwell* [45], apply to test subjects who are sitting down, with their backs against a back rest:

(a) the hand is significantly more powerful (18 kp) when it turns inwards (pronation) than when it turns outwards (supination) (11 kp);

(b) this rotation force is greatest if the hand is grasping 30 cm in front of the axis of the body;

(c) the hand is significantly more powerful when it is pulling downwards (37 kp) than when pulling upwards (16 kp);

* kp = kilopond.

(d) the hand is more powerful when pushing (60 kp) than when pulling (36 kp);

(e) pushing power is greatest when the hand is grasping 50 cm in front of the axis of the body;

(f) pulling power is greatest at a grasping distance of 70 cm.

Maximum foot pressure

When a control mechanism calls for very great strength a pedal is particularly suitable for a seated operator, since the muscles of the human leg are capable of great pushing power. As long ago as 1936 *E. A. Müller* [231] had carried out a systematic study of foot pressures in relation to the place of the pedal, in which he showed that *the greatest pushing power was obtained with knee angles in the range 140–160°, and a small downwards tilt of 20–30°*. From this he drew the practical conclusion that the height of the seat above the pedal should be as little as possible, and that the pedal distance should be around 90% of the full reach of the operator's foot.

A selection of *E. A. Müller's* results are shown in *Figure 17*.

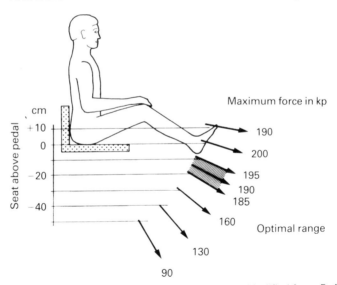

Fig. 17 Maximum foot pressure when sitting. Modified from *E. A. Müller* [231]. The optimal combination is a seat height of 20 cm above the pedal while pushing at an angle of 75° to the vertical.

Later, *Rohmert* and *Hettinger* [269] testing 60 students, obtained somewhat lower absolute values for foot pressures, though the relation between pedal position and maximum foot pressure was confirmed. *Rohmert* and *Jenik* [268] carried out the same tests on 10 women. Under optimal conditions maximum foot pressures varied over the range 90–100 kg, but the effects of seat height and knee angle were of the same order of magnitude as they had been for the men.

The maximum force exerted by the muscles which bend the elbow (biceps muscles) seems to be particularly dependent upon leverage in the arm. The results of studies by *Clarke* [54] and *Wakim* [318] and their co-workers, as set out in *Figure 18*, show that *the greatest bending moment is achieved at angles between 90–120°.*

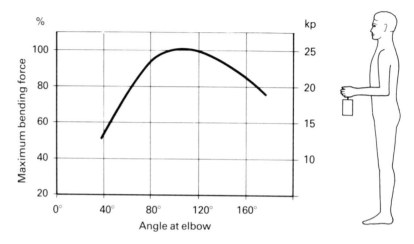

Fig. 18 Maximum bending strength in the elbow joint in relation to the angle at the elbow. After *Clarke* [54] and *Wakim* [318].

Maximum force for standing work

The maximum pulling and pushing powers of the hand during standing work have been thoroughly studied by *Rohmert* [266] and by *Rohmert* and *Jenik* [268]. A selection of their results for men are set out in *Figure 19*.

The following conclusions may be drawn from *Rohmert's* [266] studies:

(a) at most positions of the arms, while standing up, the pushing power is greater than the pulling power;

(b) both pulling and pushing power are greatest in the vertical plane, and lowest in the horizontal plane;

(c) pulling and pushing power are of the same order of magnitude whether the arms are held out sideways or forwards (in the sagittal plane);

(d) pushing power in the horizontal plane may amount to:
    (i) for men: 16–17 kg;
    (ii) for women: 8–9 kg.

Practical guidelines for work layout

Most important principles

As has already been fully emphasised, static loads on the muscles lead to painful fatigue; they are wasteful and exhausting. For this reason, a major objective in the design and layout of jobs, work

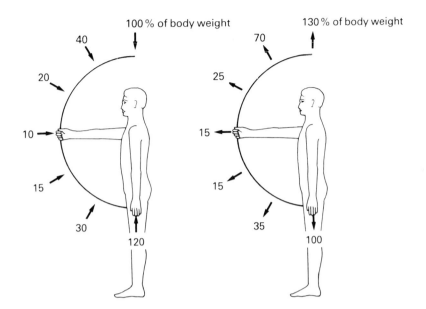

Fig. 19 Maximum power of pulling (left) and pushing (right) for a man.
Feet 30 cm apart. The values at various angles are given as a percentage of
the body weight. Simplified, after *Rohmert* [266].

places, machines, instruments and tools should be *to minimise or abolish altogether the need to grasp and hold things.*
Unavoidable static effort should be reduced to not more than 15 % of the maximum.
According to *van Wely* [325], dynamic effort of a repetitive nature should not exceed 30 % of the maximum, although it may rise to 50 % as long as the effort is not prolonged for more than 5 minutes.

7 guidelines   1. *Avoid any kind of bent or unnatural posture* (see *Figure 20*).
Bending the trunk or the head sideways is more harmful than bending forwards.

Fig. 20 Strongly bent posture while cleaning a casting in a foundry. If the casting stood up vertically it could be cleaned with sub-stantially less static load on the muscles of the back.

2. *Avoid keeping an arm outstretched either forwards or sideways*. Such postures not only lead to rapid fatigue, but also markedly reduce the precision and general level of skill of operations using the hands and arms (see *Figure 21*).

Fig. 21 Fitting components on to a machine with the arm continuously outstretched. The static loads on the right arm and on the shoulder muscles are tiring, and reduce skill. The machine should be redesigned so that the operator could work with her elbow lowered and bent at right angles.

3. *Work sitting down as much as possible*. Work places at which the operator can either stand or sit are to be recommended (see *Figure 34*).

4. *Arm movements should be either in opposition to each other, or otherwise symmetrical*. Moving one arm by itself sets up static loads on the trunk muscles. Furthermore, symmetrical movements of the arms facilitate nervous control of the operation.

5. *The working field (either the work itself or the table on which it stands) should be at such a height that it is at the best distance from the eyes of the operator*. The shorter his sight, the higher should be the working field (see *Figure 22*).

Fig. 22 The work level should be at such a height that the body takes up a natural posture, slightly inclined forwards, with the eyes at the best viewing distance from the work. This work bench is a model of its kind, with the elbows supported in a natural position, without static effort.

6. Hand grips, operating levers, tools and materials should be arranged around the work place in such a way that the most frequent movements are carried out with the elbows bent and near to the body. *The best attitude for both strength and skill in the hands is to have them 25–30 cm from the eyes, with the elbows lowered and bent at right-angles.*

7. *Hand-work can be raised up by using supports under the elbows, forearms or hands.* These supports should be padded with felt or some other soft, warm material, and should be adjustable to suit people of different sizes (see *Figure 23*).

Fig. 23 The forearms and elbows can be supported by adjustable, padded rests. Restful support is also given to the legs by a footrest that is adjustable for height, and which allows all normal leg movements to be made. For women the bars of the footrest should be closer together, to prevent their shoe heels becoming trapped.

4
Problems of body size

Basics

Since natural postures – attitudes of the trunk, arms and legs which do not involve static effort – and natural movements are a necessary part of efficient work, it is also essential that *the work place should be suited to the body size of the operator.*

Variation

Here we are soon up against a problem; the enormous variation in body size between individuals, the two sexes, and different races. It is not enough, as a rule, to design a work place to suit an average person. Most often it is necessary to take account of the tallest persons (e.g. to decide leg room under a table), or of the shortest persons (e.g. to make sure they can reach high enough). To take an extreme example: if the heights of doorways were fixed to suit the average person, many people using them would strike their heads on the lintel.

Figure 24 illustrates the extent of individual variation in body size, as measured by height, from a survey undertaken in Switzerland. The results from 508 women and 500 men are displayed as a frequency-distribution curve which shows that, besides a significant difference between the sexes, there is also wide variation within each sex. It should be further noted that the male workers and employees studied included 358 Swiss, 85 Italians and 57 of other nationalities, while the women included 402 Swiss, 84 Italians and 22 others.

Confidence interval

Since it is not usually possible to design work places to suit the very biggest or the very smallest workers, we must be content with meeting the requirements of the majority. A selection is therefore made of *either 95% or 90% respectively.* A confidence interval of 95% means that the smallest 2.5% and the largest 2.5% are excluded from consideration. Any particular percentage is called a

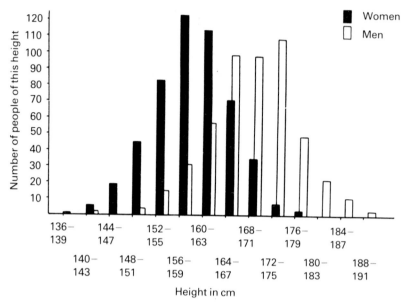

Fig. 24 **Body length of factory workers**. The measurements of 508 women and 500 men show a wide range of variation, because they included people of several nationalities.

percentile: thus in the present example only the percentiles lying between 2.5% and 97.5% are being considered. If we know the mean ($\bar{x}$) and the standard deviation (s) of any group of measurements, then:

Ci 95% = $\bar{x} \pm 1.95\ s$
Ci 90% = $\bar{x} \pm 1.65\ s$.

Figure 25 shows the frequency distribution of a body of American recruits.

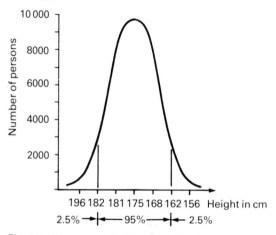

Fig. 25 **Frequency distribution of body-length among American recruits. 95% confidence interval is indicated.** From measurements made by *Hertzberg* and others [139].

Relevant literature　　There are many accounts of human measurements in the anthropometrical literature of different countries. The following may be mentioned here:

(a) Germany: *Kroemer* [184], *Jürgens* [167]
(b) England: *Murrell* [237]
(c) France: *Bouisset* & *Monod* [32], *Wisner* & *Rebiffé* [333], *Bouisset* [30]
(d) Sweden: *S. Thiberg* [303]
(e) Switzerland: *Grandjean* & *Burandt* [106]
(f) United States: *Hertzberg, Daniels* & *Churchill* [139], *McFarland* & *Stoudt* [222], *Morgan and others* [228], *U.S. Dept of Health, Education and Welfare* [313], *Henry Dreyfus Associates* [66].

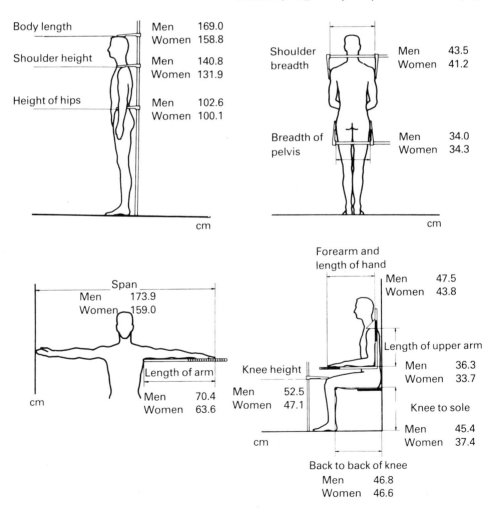

Fig. 26　Average bodily dimensions of industrial workers (508 women and 500 men) in Switzerland [106].

Effects of age

Several of these results came from surveys of recruits [139], [167]. Since with increasing age people become shorter, but heavier and more corpulent, work places should be so designed that they allow for the bodily dimensions of people of all ages between 20 and 65 years. The American survey [313] showed that people in the 45–65-years age group had changed by the following amounts compared with the 20-year age group:

(a) body length (men and women):   − 4 cm
(b) body weight (men):   + 6 kg
(c) body weight (women):   + 10 kg.

Some of the results of our own survey [106] are set out below, along with the results of studies by *Kroemer* [184].

Body size in Switzerland

The body sizes and mean values for men and women, recorded in our survey [106] are set out in *Figure 26*. As we mentioned above, our survey differs from others by including a considerable proportion of Southern Europeans in its numbers. For the same reason the mean figures are not quite 2% lower than those of comparable measurements in other countries.

Body size according to Kroemer

Figure 27 shows the criteria used by *Kroemer* [184] in his survey, and *Table 4* summarises the bodily dimensions that he found.

The question arises whether the figures given in *Table 4* are applicable to other industrial countries. To answer this question, *Table 5* compares the mean values for men in five different countries.

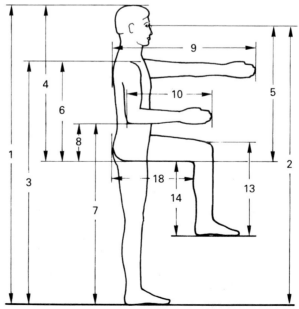

Fig. 27 **Indication of measurements listed in Table 4.** After *Kroemer* [184].

Table 4 External bodily dimensions (in cm) of men and women. Means ($\bar{x}$) and 90% confidence intervals (Ci) for persons aged 20–65 years from the Federal Republic of Germany. No. of men 15 700; no. of women 17 700. Measurements of the unclothed body: add 2.5 cm to men's heights and 4 cm to women's, to allow for height when wearing shoes. After *Kroemer* [184].

Measure- ment no.	Part of body measured	Men $\bar{x}$	Ci 90%	Women $\bar{x}$	Ci 90%
1	Standing height	172	160–184	161	150–172
2	Eye-level, standing	161	150–172	150	138–162
3	Shoulder height, standing	142	131–153	131	120–142
4	Top of head, above seat	90	84– 96	85	79– 91
5	Eye-level, above seat	79	73– 85	74	68– 80
6	Shoulder height, above seat	59	54– 64	54	49– 59
7	Elbow height, standing	106	98–114	97	89–105
8	Elbow height, above seat	24	20– 28	24	20– 28
9	Forward reach (fingers in grasping position)	82	75– 87	70	63– 77
10	Underarm length (elbow to fingertip)	47	43– 51	42	38– 46
11	Span of arms	175	159–191	155	139–171
12	Buttocks to knee	59	54– 64	57	52– 62
13	Sole of foot to knee	55	51– 59	50	46– 54
14	Sole of foot to hollow of knee	45	42– 48	43	40– 46
15	Shoulder breadth	45	41– 49	41	37– 45
16	Breadth of hips	35	31– 39	37	33– 41
17	Thickness of thigh	14	12– 17	14	12– 17
18	Back to hollow of knee	50	46– 54	46	43– 50

The comparison of Table 5 shows that the anthropometric data are very similar for the Anglo-Saxon, the German and the French people. On the other hand these mean values might not be valid for people of Scandinavia or Southern Europe, and of course even less for some people of Asia.

Hand size

Hand size is particularly important when designing controls. *Figure 28* and *Table 6* summarise relevant information according to *Jürgens* [167].

Table 5 Comparison of several bodily measurements in five different countries. $\bar{x}$ = Mean values for males of all age groups. In brackets, reference number of authority. [1] inclusive of 20% Southern Europeans.

Bodily measurement	W. Germany [184] $\bar{x}$	France [30] $\bar{x}$	Britain [237] $\bar{x}$	U.S.A. [313] $\bar{x}$	Switzer- land[1] [106] $\bar{x}$
Body length	172	170	171	173	169
Body length, above seat	90	88	85	86	–
Back to front of knee	59	60	60	59	–
Elbow to ground	106	105	107	106	104
Top of knee to ground	55	54	–	55	52
Shoulder breadth	45	–	46	45	44
Breadth of hips	35	35	–	35	34

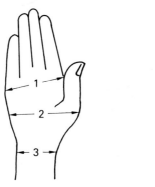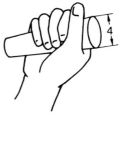

Fig. 28 Indication of measurements listed in Table 6. After *Jürgens* [167].

Table 6 Dimensions of one hand. Mean values ($\bar{x}$) and 90% confidence interval (Ci) 90% for 8000 20-year-old men, with additional sample checks on women in the Federal Republic of Germany. After *Jürgens* [167].

Measure- ment no.	Bodily dimension (in cm)	$\bar{x}$	Men Ci 90%	$\bar{x}$	Women Ci 90%
1	Circumference of hand	21.1	19.3–23.0	18.7	17.5–20.1
2	Breadth of hand	10.6	9.8–11.3	–	–
3	Circumference of wrist	17.1	15.5–18.8	16.1	14.3–17.9
4	Maximum grasp (circumference of thumb and forefinger)	13.4	12.0–15.3	–	–

Angle of rotation of joints

The operating space of a limb is a product of its length and the angle of rotation of its joint. *Figure 29* shows some operating spaces available to the hand.

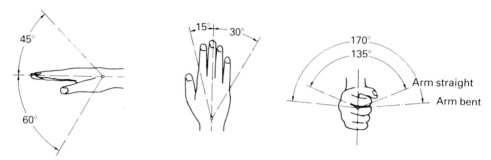

Fig. 29   Arcs of movement of wrist and hand. After *Stier* and *Meyer* [297].

The arm can rotate through an angle of 250° about its axis, in the sagittal plane, of which a half-circle (180°) lies in front of the body, and a further 70° or thereabouts, backwards [228]. The angle of rotation of the ankle is shown in *Figure 30*.

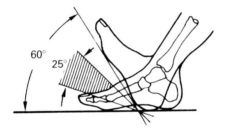

Fig. 30   Bending and stretching at the ankle. Maximum arc 60°; most convenient arc 25–30°. Modified after *Stier* and *Meyer* [297].

5
The design of work places

Working heights

*Working heights
when standing*

Working height is of critical importance in the design of work
places. If the work is raised too high the shoulders must frequently
be lifted up to compensate, which may lead to painful cramps at
the level of the shoulder blades, and in the neck and shoulders (see
later in this Chapter). If the working height is too low the back must
be excessively bowed, which again often causes backache. Hence
the work table must be of such a height that it suits the height of
the operator, whether he stands or sits at his work.

*The most favourable working height for handwork while standing
is 5–10 cm below elbow level.* The average elbow height (distance
from floor to underside of elbow when it is bent at right angles
with the upper arm vertical) is 105 cm for men and 98 cm for
women.

*Hence we may conclude that on average working heights of
95–100 cm will be convenient for men, and 88–93 cm for women.*
Besides these anthropometric considerations, we must also allow
for the nature of the work:

(a) for delicate work (e.g. drawing) it is desirable to support the
    elbow to help to reduce static loads in the muscles of the back.
    A good working height is about 5–10 cm above elbow height;

(b) during manual work the operator often needs space for tools,
    materials, and containers of various kinds, and a suitable
    height for these is 10–15 cm below elbow height;

(c) during standing work, if this involves much effort and makes
    use of the weight of the upper part of the body (e.g. wood-
    working, or heavy assembly work), the working surface needs
    to be lower; 15–40 cm below elbow height is adequate.

Recommended working heights when standing are set out in
Figure 31.

The dimensions recommended in *Figure 31* have merely the force

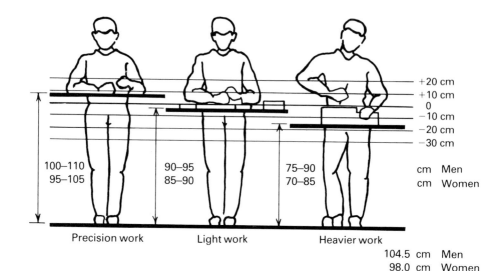

Precision work   Light work   Heavier work

104.5 cm Men
98.0 cm Women

Fig. 31 Recommended heights of bench for standing work. The reference line (± 0) is the height of the elbows above the floor, which averages 105 cm for men and 98 cm for women.

of general guidelines, since they are based on average body measurements and make no allowances for individual variation. The table heights quoted are too high for small persons, who will need to use wooden foot rests, or some similar support. On the other hand tall persons will need to bend over the work table, which will be a cause of backache.

Fully adjustable work tables

Ergonomically speaking, therefore, it is often desirable to be able to adjust the working height to suit the individual. In addition to improvisations such as foot supports, or lengthening the legs of the work table, a fully adjustable bench is to be recommended. *Figure 32* shows recommended working heights for light standing work, in relation to the body length of the operators.

If a firm is unable, for administrative or practical reasons, to provide fully adjustable benches, or if the operating level at a machine cannot be varied, then in principle *working heights should be set to suit the tallest operators*: smaller persons can be accommodated by giving them something to stand on.

Work places for delicate sedentary work

For sedentary work, a few cm below elbow height is equally suitable.

Since the work may also require fine or precise manipulation, working heights must take note of the optimum visual distance, as *Figure 22* has already shown. In such cases the plane of the work must be raised until the operator can see clearly while holding his back in a natural posture.

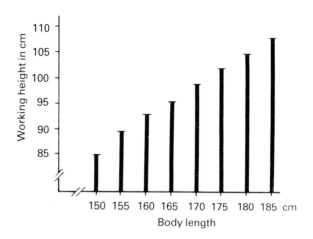

Fig. 32 Working heights for light work while standing, in relation to body length.

Typing

The opposite, i.e. lowering of the working surface, is necessary when handwork calls for great force, or much freedom of movement. Typing requires a low table, since the height of keyboards defines the working surface: the table should be little above elbow height. However, the need for a low table conflicts with the necessity for enough knee room under the table, and this can be a limiting factor. If we extract the measurement ground-to-upper surface of knee, for large individuals, from *Table 4*, and if we make certain additions to allow for heels, and for a minimum amount of movement, we arrive at *the following figures for free knee room:*
men: (59 cm + 5 cm) = 64 cm
women: (54 cm + 7 cm) = 61 cm.
If we allow 4 cm for the thickness of the table top, assuming it has no rim, *this brings the lowest table level up to*:
men: 68 cm
women: 65 cm.
This lowest working surface is recommended for typing, and for prolonged use of a calculating machine, as well as for heavy assembly work, for working over any kind of container, and for preparatory work in the kitchen. It should also be noted here that the maximum distance from seating surface to the underside of the table should be 17 cm (95 % of the thickness of the thighs).

*Ordinary
office work*

It is less difficult to choose a suitable working height when we are dealing with everyday work which does not involve problems of visual distance, machine height, or the need to apply sufficient force. The most widely used work table in this category is the ordinary office desk, as used for reading and writing. A slight forward stoop, with arms on the desk is only minimally tiring, but to make this relaxed back possible *the distance from seating surface to desk top must be 27–30 cm.*

In one of our own investigations [107], which will be reported later on, we were able to show that the great majority of the office workers studied had chosen to set their fully-adjustable office chairs to a height that was 27– 30 cm below that of the desk top. It emerged from this investigation that a person sitting at his work looked first of all for a comfortable and relaxed position for his back, and would often accept a seat height that was bad for his legs, rather than sacrifice a comfortable back.

Table heights for normal reading and writing (but not typing!) follow the same rule as for standing work: it is more practical to choose a height to suit the tall person rather than the short person, because the latter can always be given some support, or a raised seat. On the other hand, a tall person given a table that is too low for him can do nothing about it, except to set the seat so low that it gives him aches and pains in the legs.

We therefore conclude that *office desks to be used without a typewriter should be at heights of*:

men: 74–78 cm,
women: 70–74 cm,

assuming that the chairs are fully adjustable, and footrests are available for short people.

It is important the office desks should allow plenty of room for leg movement, and it is an advantage if the legs can be crossed. For this reason there should be no drawers above the knees, and no thick edge to the desk top.

Recommended table heights for seated work are summarised in *Figure 33* and *Table 7.*

Table 7   Table heights (in cm) for seated work.

Kind of work	Men	Women
Precision work, at close visual range	90–110	80–100
Reading and writing	74– 78	70– 74
Typing; manual work requiring strength	68	65

Alternate sitting and standing

A work place which allows the operator to sit or stand, as he wishes, is highly to be recommended from a physiological and orthopaedic point of view. There is less need to hold things when one is sitting down than when standing up; on the other hand sitting is the cause of many aches and pains, which can be relieved by standing. Standing and sitting impose stresses upon different muscles, so that each changeover relaxes some muscles and

For typing on office machine   For writing

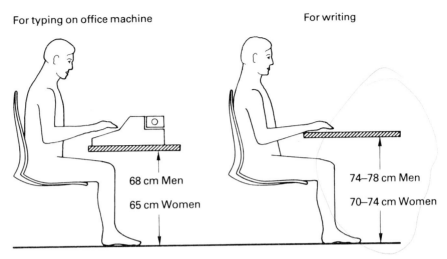

Fig. 33   Recommended table heights for sedentary work.

stresses others. Furthermore we have good grounds for believing
that each change from standing to sitting (and vice versa) is
accompanied by variations in the supply of nutrients to the intervertebral discs [183], so that the change is beneficial in this respect,
too.

Figure 34 shows a machine which permits alternate sitting and
standing, and the relevant measurements are as follows:

(a) knee room: 30 × 65 cm
(b) height of working field above seat: 30– 60 cm
(c) height of working field above floor: 100–120 cm
(d) range of adjustment of seat: 80–100 cm.

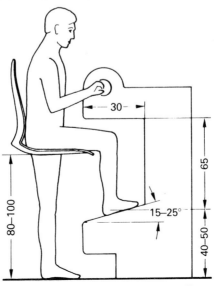

Fig. 34   Alternately sitting and standing at work.

Comfortable head position

Whether standing or sitting, the area of the working field that needs to be kept in constant view must be so placed that the operator's head remains comfortable. Too much tilt either up or down will give rise in time to aches in the neck muscles. *Lehmann* and *Stier* [195] investigated this problem experimentally and found that for seated workers the most comfortable attitude of the head was when the angle between the line of sight and the horizontal was 32–44°; for standing workers this was 23–37°. These lines of sight were achieved by bending the head and by lowering the eyes, in about equal amounts. Recommendations based on these findings are set out in *Figure 35*, and indicate where the most important visual tasks should be located.

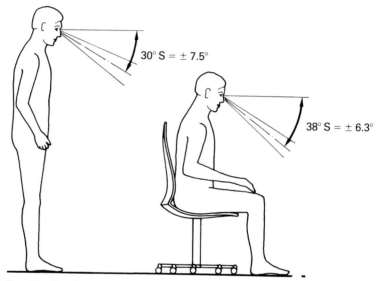

30° S = ± 7.5°

38° S = ± 6.3°

Fig. 35 Comfortable inclination of vision when standing (left) and when sitting down (right). S = standard deviation. After *Lehmann* and *Stier* [195].

The tilt of the head (angle between the axes of head and trunk) associated with such a line of sight is:
(a) standing: 8–22°
(b) sitting: 17–29°.

Inclined work surface

Recommended inclinations of the line of sight of 30° when standing and 38° when seated are often impractical: for example, writing, reading and drawing on a flat table may require an excessive forward tilt of the head.

In our own investigation of 10 tables sold commercially for technical drawing we were able to show that height and inclination of the table top had a considerable effect on how the head and trunk of the operator were held. *Figure 36* illustrates the postures that were observed at the 10 tables tested.

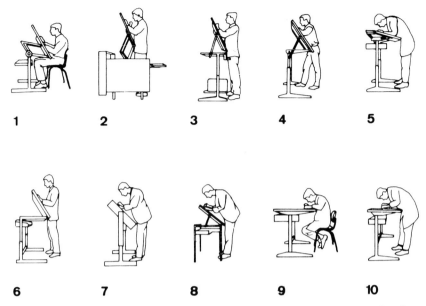

Fig. 36 Body postures at each of ten commercially produced drawing-boards. After *Grandjean, Kahlcke* and *Wotzka* [111].

Photographic records were used to measure the angle between the axes of head and trunk, and between the axis of the trunk and the vertical plane, for short, average and tall persons. The following results were obtained:

(a) bending of the trunk at the 4 best tables: 7–9°
(b) bending of the trunk at the 4 worst tables: 19–42°
(c) bending of the head at the 4 best tables: 29–33°
(d) bending of the head at the 4 worst tables: 30–36°.

It appears, therefore, that if the work table is horizontal, or placed too low, the operator compensates for this by excessive bending of his body, since the head clearly cannot tilt much more than 30°. On the evidence of this research we can make the following recommendations for work tables for technical drawing:

(a) the table top must be adjustable both for height and inclination;
(b) the height must be adjustable between 65 and 130 cm (from front edge to floor);
(c) inclination must be adjustable between zero and 75° to the horizontal.

The question of work surfaces for special purposes also arises in relation to school desks and to tables where much reading and writing is carried on. *Eastman* and *Kamon* [77] recently made a contribution to this problem. Six male subjects carried out writing and reading exercises for 2½ hours each at a flat table, and at ones which were tilted to 12° and 24° respectively. Inclinations of the body (measured as the angle between the horizontal plane and a straight line from the twelfth thoracic vertebra to the eyes), were as follows:

(a) flat table: 35–45°
(b) table tilted 12°: 37–48°
(c) table tilted 24°: 40–50°.

So tilting of the table top leads to a more erect posture, as well as to fewer pulses on the electromyogram. Subjectively, there was an impression of fewer aches and pains. As a general rule, it seems that a tilted work table is an improvement, posturally as well as visually. These advantages have certain drawbacks, however, notably that it is difficult to put things down on an inclined surface.

Room to grasp and move things

Vertical grasp

An understanding of how much room the hands and arms need to take hold of things and to move them about is an important factor in the planning of controls, tools, accessories of various kinds, and places on which to put these down. Reaching too far to pick things up leads to excessive movements of the trunk, making the operation itself less accurate, and increasing the risks of pains in the back and shoulders.

Vertical grasp means the radius of action of the arms, with the hands in a grasping attitude. A decisive factor is the height of the axis of the shoulder joint, and the distance from this joint to the closed hand. This is a case in which we need to consider the 5 percentiles of short people, whose measurements are as follows:

Shoulder height when standing:
(a) small men   130 cm
(b) small women   120 cm
Shoulder height above seat:
(a) small men   54 cm
(b) small women   49 cm
Arm length (closed hand):
(a) small men   65 cm
(b) small women   58 cm.

The vertical grasp in the sagittal plane of the body is in practice an arc of radius 65 cm (men) and 58 cm (women) above shoulder height. If we also allow for lateral movements of the arm the grasping space becomes a semi-circular shell of comparable radius.

Figure 37 shows vertical grasp in the sagittal plane:

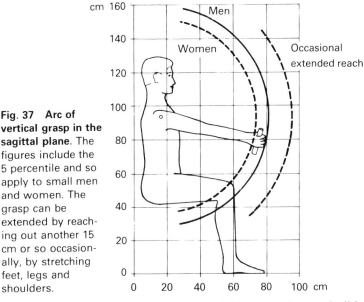

Fig. 37 Arc of vertical grasp in the sagittal plane. The figures include the 5 percentile and so apply to small men and women. The grasp can be extended by reaching out another 15 cm or so occasionally, by stretching feet, legs and shoulders.

An occasional stretch to reach beyond this range is permissible since the momentary effect on trunk and shoulders is transient.

Height of reach

For the positioning of shelves, storage surfaces, special hand grips, and other controls it is necessary to know what is the maximum possible reach upwards.

Thiberg [303] has calculated regression lines and correlation coefficients for the ratio between body length and height of reach in both men and women. His results when a hand is extended towards a shelf are illustrated in *Figure 38*.

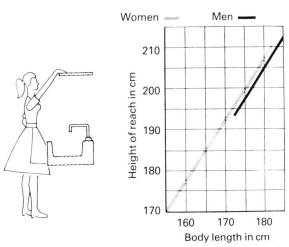

Fig. 38 Height to which a free-standing person can reach and place a hand flat on a shelf. Drawn from data given by *Thiberg* [303].

The following is the ratio between height of reach and body length:

maximum height of reach = 1.24 × body length.

Table 8 summarises maximum reach and vertical grasp for men and women.

Table 8 Maximum height of reach for men and women. Based on the measurements quoted in *Table 4*.

	Percentile	To fingertip (cm)	Grasping height (cm)
Men:			
Large	95	228	216
Average	50	213	201
Small	5	198	186
Women:			
Large	95	213	201
Average	50	200	188
Small	5	186	174

It is often necessary to be able to look to the back of the shelf, in order to see the object to be grasped: a bottle, a box or some other container. Under these conditions the highest shelf should be not more than:

(a) *men:*   *150–160 cm*
(b) *women: 140–150 cm.*

At these heights we can allow a further 60 cm for the depth of the shelf.

Horizontal grasp at table top level

Figure 39 illustrates grasping and working space over a table top.

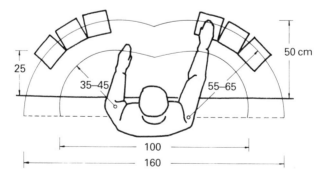

Fig. 39 Horizontal arc of grasp, and working area at table top height. The grasping distance takes account of the distance from shoulder to hand; the working distance only elbow to hand. The values include the 5 percentile, and so apply to men and women of less than average size.

All materials, tools, controls and containers should be deployed within this space, remembering that an occasional stretch up to distances of 70–80 cm can be undertaken without ill-effects.

Operating space
for the legs

Room to operate the feet is essential for all forms of pedal control. This is illustrated in *Figure 40*.

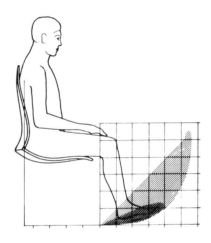

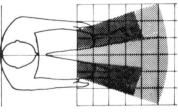

Fig. 40   **Range of operation of the feet.** The optimal range for delicately controlled pedal work, where only slight force is needed, is double-shaded. After *Kroemer* [185].

10 cm squares

Kroemer [185] carried out a very thorough investigation into the optimal placing and construction of pedals, to which we shall return in *Chapter 9*.

Seating at work

Historical review

Many primitive peoples still living today have no knowledge of seats of any kinds; they crouch, kneel or squat. So where did seats come from? The anthropological answer is that they originated as status symbols; only the chief had the right to be raised up. Hence the gradual development of ceremonial stools, which indicated status by their size and decoration. The highest point of this development in cultural history was the splendour of a throne. This status function persists to the present day. Anyone who doubts this should look inside a factory making office furniture, and notice that there is a type of chair appropriate to each salary level. Thus there is a wooden chair for typists; for the senior clerk one that is thinly upholstered; a thickly upholstered chair for the office manager, and for directors a swivel armchair upholstered in leather.

At the beginning of the present century the idea gradually emerged that wellbeing and efficiency are improved and fatigue reduced if people can sit at their work. The reason is a physiological one. As long as a person is standing it requires an outlay of static muscular effort to keep the joints of feet, knees and hips in fixed positions; this muscular effort ceases when the person sits down.

This realisation led to a greater application of medical and ergonomic ideas to the design of seats for work. This development gained in importance as more and more people sat down at their work, until today about three quarters of all operatives in industrial countries have sedentary jobs.

Pros and cons

The advantages of sedentary work are:
(a) taking the weight off the legs;
(b) ability to avoid unnatural body postures;
(c) reduced energy consumption;
(d) fewer demands on the blood system.
These advantages must be set against certain drawbacks. Prolonged sitting leads to a slackening of the abdominal muscles ("sedentary tummy"), and to curvature of the spine, which in turn is bad for the organs of digestion and breathing.

Main problem is the back

The most severe problem involves the spine and the muscles of the back, which in many sitting positions are not merely not relaxed, but are positively stressed in various ways.

Intervertebral discs

About 60 % of adults have backache at least once in their lives, and the commonest cause of this is disc trouble.
An intervertebral disc is a sort of cushion, which separates two vertebrae, and collectively gives flexibility to the spine. It consists internally of a viscous fluid, enclosed in a tough, fibrous ring, which encircles the disc.
For reasons that are still unknown, intervertebral discs may degenerate and lose their strength: they become flattened, and in advanced cases the viscous fluid may even be squeezed out. This impairs the working of the vertebral column, and allows tissues and nerves to be strained and pinched, leading to various diseases of the pelvis, to lumbago, and even to paralysis of the legs.
Unnatural postures and bad seating can speed up the deterioration

of the discs, resulting in all the ailments mentioned above. For this reason many orthopaedists in recent years have concerned themselves with the medical aspect of seating posture; they include *Akerblom* [3], *Schoberth* [282], *Yamaguchi* [342], *Keegan* [173], *Nachemson* [241], *Andersson* [7], *Rosemeyer* [272] and *Krämer* [183].

Orthopaedic research

The Japanese *Yamaguchi* [343] and the Swedish investigators *Nachemson* [241] and *Andersson* [7] have used delicate methods to measure the pressure inside an intervertebral disc during a variety of bodily postures and sitting positions. These workers are unanimous in finding that bodily posture had a profound effect on disc pressure. Since increased internal pressure in the discs means that they are being overloaded and will wear out more quickly, the results obtained by these three investigators have considerable medical importance.

Disc pressure

Nachemson [240] found that the pressure in the disc amounted to about 1.5 times the vertical force exerted upon it by the overlying vertebrae. The effects of various bodily postures upon the disc pressure of 9 healthy subjects are shown in *Table 9*.

Table 9 Effects of bodily posture on the pressure inside the disc between the third and fourth lumbar vertebrae, expressed as a percentage of the pressure when standing erect.

Posture	Disc pressure %
Standing erect	100
Lying flat on the back	24
Sitting, trunk erect	140
Sitting, bent forwards	190

This shows that *disc pressure is greater when sitting than when standing.*

The explanation lies in the mechanism of the pelvis and sacrum. During the changeover from standing to sitting down:

(a) the thigh is used as a lever;

(b) the upper edge of the pelvis is rotated backwards;

(c) the sacrum turns upright;

(d) the vertebral column changes over from a lordosis[1] to either a straight or a kyphotic[1] shape.

These effects of sitting posture are illustrated in *Figure 41*.

---

[1] Lordosis means that the spine is curved forwards, as it normally is in the lumbar region when standing up. kyphosis means curving backwards, and is normal in the thoracic region when standing. See *Figure 41*, left.

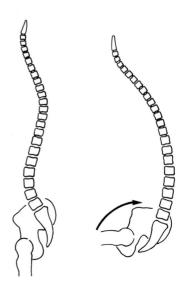

Fig. 41 Rotation of the pelvis when changing from a standing to a sitting posture. Left: standing upright. Right: sitting posture. Sitting down involves a backward rotation of the upper edge of the pelvis (indicated by the arrow), bringing the sacrum to an upright position, and changing the lumbar lordosis into a kyphosis.

The backwards rotation of the pelvis puts the spine into a state of kyphosis at this point, which in turn increases the pressure within the discs.

Erect or relaxed sitting posture

Many orthopaedists still recommend an upright sitting posture, since this holds the spinal column in its natural shape of an elongated S, and in fact disc pressures are lower than if the body is curved forwards (see *Table 9*). This orthopaedic advice, however, conflicts with the fact that a slightly curved attitude relieves the pressure on the back muscles, and makes the whole posture more comfortable. This was established, in part, through electro-myographic studies made by Lundervold [204]. Some of his results are shown in *Figure 42*.

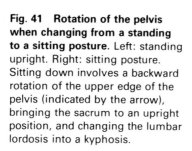

Muscles

——— Right trapezins

——— Right latissimus dorsi

——— Right sacrospinalis

——— Left sacrospinalis

Fig. 42 Electrical activity (EMG) in the back muscles while sitting upright and in a relaxed posture, slightly bent forwards. While sitting upright there is considerable electrical activity, resulting from the static effort imposed upon the muscles of the back. After *Lundervold* [204].

It emerges from this that a relaxed posture, with the trunk slightly bent forwards holds the weight of the trunk in balance with itself. This is the attitude that most people adopt when they are sitting, as far as their work permits, because it is relaxed and puts a minimum of strain on the muscles of the back. Thus we have a "conflict of interests" between the demands of the muscles and those of the intervertebral discs. While the discs prefer an erect posture, the muscles prefer a slight forward curvature. So it is understandable, as we have said above, that the sitting posture poses problems for both the vertebral column and the intervertebral discs.

"Feeding the intervertebral discs"

Here we must again refer to the interesting work of *Krämer* [183], who has made a close study of the nutritional needs of intervertebral discs. The interior of a disc has no blood supply, and must be fed by diffusion through the fibrous outer ring. *Krämer* has produced evidence that pressure on the disc creates a diffusion gradient from interior to exterior, so that tissue fluid leaks out. When the pressure is reduced, this gradient is reversed, and tissue fluid diffuses back, taking nutrients with it. It seems from this that, to keep the discs well nourished and in good condition, they need to be subjected to frequent changes of pressure, as a kind of pump mechanism.

From a medical point of view, therefore, *an occasional change of posture from bent to erect, and vice versa must be beneficial.*

Angle of seat and discs

The investigators mentioned above [7], [240], (343] also studied the effects of seat angle (between backrest and seat), and the shape of the backrest, on disc pressures. They used the electrical activity of the back muscles as a measure of static load, while simultaneously measuring disc pressure [241]. The most important results are displayed in *Figures 43* and *44.*

Yamaguchi and his team [343] studied the effects of seat angle on the electrical activity of the muscles of the back and abdomen, both in healthy subjects and in some who suffered from back ailments. They advised that the best conditions for relaxation of the spine were provided by a seat angle of 10° and an angle between the seat and the backrest of 115–120°. The effect was more marked among the healthy subjects than among those with back troubles, and the latter group found it difficult to relax their back muscles completely. Taking these orthopaedic studies as a whole, the following design features for seats can be deduced:

(a) pressure within the intervertebral discs is at its lowest when the back is tilted backwards and relaxed;

(b) the more the backrest is inclined, the lower the loading on the muscles;

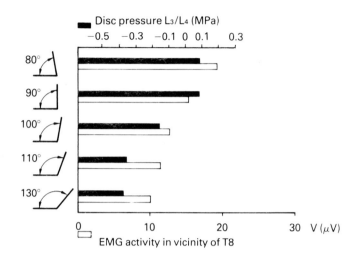

Disc pressure L₃/L₄ (MPa)

Fig. 43 **Effect of seat-angle (i.e. between backrest and seat) on pressure in the intervertebral discs, and on the electrical activity (EMG) in the back muscles at the level of the eighth thoracic vertebra (T$_8$).** MPa = 10.2 kp/cm². Zero (0) on the pressure scale is a relative reference value between the third and fourth lumbar vertebrae, (L$_3$/L$_4$), with a seat angle of 90°. Absolute values were at the reference level zero about 0.5 MPa (= 5 kp/cm²). Drawn from the results obtained by *Nachemson* [241] and *Andersson* [7].

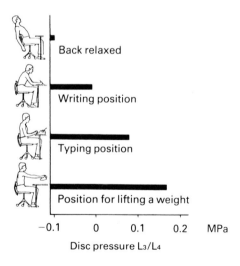

Fig. 44 **Effects of various sitting postures on disc pressure.** MPa = 10.2 kp/cm². L$_3$/L$_4$ = third and fourth lumbar vertebrae. Zero (0) on the pressure scale is a relative reference value for a seat angle of 90°. Absolute values were at the reference level zero about 0.5 MPa (= 5 kp/cm²). Drawn from the results obtained by *Nachemson* [241] and *Andersson* [7].

(c) both disc pressure and muscular activity are lower with a lumbar pad 5 cm thick than with a flat backrest;

(d) optimum conditions, as measured by disc pressure and muscular activity, are present when the backrest is inclined at 120°, and the seat at 14°, to the horizontal, with a lumbar pad 5 cm thick.

These results indicate that resting the back against a sloping backrest transfers a proportion of the weight of the upper part of the body to the backrest and reduces the pressure in the intervertebral discs. *This effect continues as the angle between seat and back is increased, at least up to 110°.*

Sitting habits in offices

In our own study [107], mentioned earlier, we studied the sitting habits of 261 men and 117 women, who were employed on the usual kinds of office work, with the help of time-lapse flash photography. Some of the results are shown in *Figure 45.*

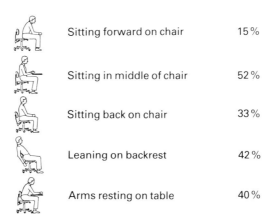

Sitting forward on chair	15%
Sitting in middle of chair	52%
Sitting back on chair	33%
Leaning on backrest	42%
Arms resting on table	40%

Fig. 45 Sitting postures of 378 office workers, as shown by a multimoment observation technique. 4920 observations. The percentages quoted indicate how much of the working period was spent in that posture. After *Grandjean* and *Burandt* [107].

Aches and pains when sitting

Associated with this study was a survey by questionnaire of bodily aches and pains of all kinds which seemed to be brought on by the work. These are shown in *Figure 46.* Finally, the principal body measurements of the employees concerned were recorded.

To test whether there was any significant relationship between the observations, the frequency of aches and pains, and the dimensions of the work places, we calculated correlations between these quantities, using the Chi^2 test. From a large number of results and the calculated correlations, the following conclusions emerged:

(a) seat heights between 38–54 cm were preferred for a comfortable posture of the upper part of the body;

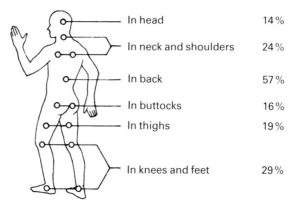

In head	14%
In neck and shoulders	24%
In back	57%
In buttocks	16%
In thighs	19%
In knees and feet	29%

Fig. 46 Frequency of bodily aches among sedentary workers. 246 persons questioned = 100%: multiple answers were possible. After *Grandjean* and *Burandt* [107].

(b) aches in the thighs were mostly caused by putting the weight on the thighs while working, and only to a lesser extent by the height of the seat;

(c) a desk top 74–78 cm high gave the employees most scope to adapt it to suit themselves, provided that a fully adjustable seat and footrests were available;

(d) 57% of those questioned complained of back troubles while they were sitting;

(e) 24% reported aches in neck and shoulders, and 15% in arms and hands, which most of them, especially the typists, blamed on too high a desk top (cramped hunching of the shoulders);

(f) 29% reported aches in knees and feet, mostly among small people who had to sit forward on their chairs because they had no footrests;

(g) regardless of their body length, the great majority of the workers liked the seat to be 27–30 cm below the desk top. This permitted a natural posture of the trunk, obviously a point of first priority with these workers (see earlier under "ordinary office work");

(h) the frequency of back complaints (57%) and the common use of a back rest (42% of the time) showed the need to relax the back muscles from time to time, and may be quoted as evidence of the importance of a well-constructed backrest.

Ergonomic research A "sitting-machine" [105], and a variety of moulded seat shells [110] of different profiles, were tested upon a large number of people, including a group of 68 who complained of back ailments. They were asked to give their subjective impressions of the different seat profiles, and their effects upon various parts of the body. The profile of a multipurpose seat, and one of an easy-chair, which the test subjects voted to produce the fewest aches and pains, are shown in *Figure 47*.

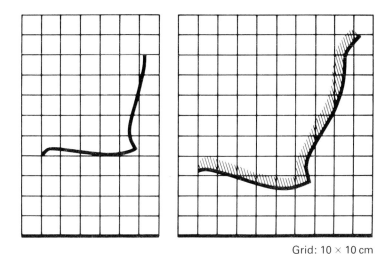

Grid: 10 × 10 cm

Fig. 47 Seat profiles of a multi purpose chair (left) and an easy-chair (right) both of which caused a minimum of subjective complaints. Grid 10 × 10 cm. After *Grandjean* and others [105, 110].

The best chair for relaxing

The result for the easy-chair agrees almost exactly with what the orthopaedists say. *A seat profile which produces only a low pressure in the intervertebral discs and requires very little static muscular effort, is also the one that causes the fewest aches and pains. When more discomfort is experienced, it is evidently associated with stresses falling upon the discs and fatigue symptoms in the muscles.* This link is also pointed out by *Rosemeyer* [272], who noted that opening out the angle between seat and backrest to 110° resulted in less electrical activity in the muscles and greater comfort. Moreover our profile for an easy-chair is in good agreement with the recommendations of *Ridder* [260] and *Branton* [34].

Taking these researches as a whole, the following recommendations for an easy-chair have both orthopaedic/medical and ergonomic backing:

(a) *the seat should be tilted backwards* so that the buttocks will not slide forwards. A tilt of 14–24° to the horizontal is recommended;

(b) the backrest should be inclined at the following angles:
   (i) *to the seat:*   *105–110°*
   (ii) *to the horizontal:*   *110–130°;*

(c) *the backrest should be provided with a lumbar pad*, as the Swede *Akerblom* said in 1948, when he pioneered the study of seating. The apex of this pad should meet the spine between the third and fifth lumbar vertebrae. This means that its vertical height above the back of the seat should be 10–18 cm.

The pad should reduce kyphosis of the lumbar region, and hold the spine in as natural a position as possible.

For further details on the easy-chair, see Reference [98].

The work seat

As far as work seats are concerned, our own researches indicate that a high backrest, slightly concave to the front at its top end, and distinctly convex in the lumbar region, is good both medically and ergonomically. A seat profile such as that shown on the left of *Figure 47* gives support to the lumbar region when the occupant is leaning forwards (in a working attitude) yet relaxes the back muscles thoroughly when leaning backwards, because it then holds the spine in a natural position.

Hünting [151, 152] studied a fixed chair with high backrest, both in the laboratory and under practical conditions, recording sitting habits and reports of aches and pains. He compared tilting chairs with similar fixed ones, as well as a traditional type of office chair with an adjustable backrest. The tilting and adjustable chairs are illustrated in *Figure 48*.

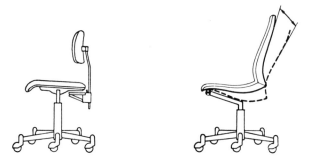

Fig. 48 Tilting chair with high backrest (right) and a commercial type of office chair with adjustable backrest (left).

The tilting and non-tilting chairs were not very different. The clearest results came when people were asked to say which they preferred after trying each type for two weeks. These results are shown in *Figure 49*.

This survey showed very clearly that the office workers favoured both types of chair because they had a high backrest, and their opinions confirm the view we expressed earlier, that a high backrest is good for sedentary work. *We therefore consider that any work places where it is possible to lean back in one's seat occasionally should be provided with high backrests.*

Recommendations for work seats

As far as we know at present the following recommendations for the design of work seats can be made:

(a) *every work seat should be adjustable for height* (between 38 and 53 cm if it is for office workers);

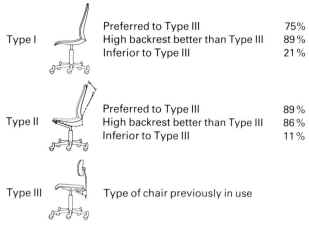

Type I	Preferred to Type III	75%
	High backrest better than Type III	89%
	Inferior to Type III	21%

Type II	Preferred to Type III	89%
	High backrest better than Type III	86%
	Inferior to Type III	11%

| Type III | Type of chair previously in use |

Fig. 49 Comparative assessment of three experimental chairs, which were used by 66 office workers over a period of two weeks. Type I = fixed moulded chair with high backrest. Type II = moulded chair which tilts 2° forwards and 14° backwards, freely movable with high backrest. Type III = standard office chair with adjustable backrest. After *Hünting* and *Grandjean* [152].

(b) *the seat should be safe against tipping or slipping.* It should have five feet, set in a circle at least as big as the seat itself (40–45 cm diameter);

(c) *the seat should allow the occupant sufficient freedom of movement.* If the occupant needs to get up often, or to move the seat sideways, then the five feet should have casters;

(d) hollow spaces under the desk top help to reduce aches in the legs, because these can be moved about.

(e) *the seat surface* should be 40–45 cm across and 38–42 cm from back to front. A slight hollow in the seat, with the front edge turned upwards about 4–6° will prevent the buttocks from sliding forwards (see *Figure 47*). Of course the front edge should be rounded off;

(f) *a light padding* with 2 cm of latex, covered with non-slip, permeable material (wool or Dralon) is a great aid to comfort;

(g) *a work seat that is either fixed or tilting, but with a high, non-adjustable backrest* give good opportunities to relax the back muscles occasionally. Such a backrest should extend 48–50 cm vertically above the seat surface[1] and have a breadth of 32–36 cm. The backrest should have a lumbar pad, which should give good support to the spine and the sacrum at a point 10–20 cm above the lowest point of the seat surface. The upper part of the backrest should be concave forwards. The backrest may, with advantage, be concave in all horizontal planes, with a radius of 40–50 cm;

[1] The minimum of 32 cm, approved in the German Norm DIN4552 (67), must certainly be rejected since it does not give nearly enough support to the back.

(h) *a workseat with adjustable backrest* should give as much support as possible to the lumbar vertebrae. The backrest should be 30 cm high and 38 cm broad, and be adjustable in both horizontal and vertical planes;

(i) *foot rests are important so that small people can avoid sitting badly,* or to meet similar problems in people of other sizes. They should be an integral part of each work table;

(j) *a work seat should be designed in conjunction with the work place at which it is to be used. The most important consideration is the distance from seat height to work height.* Assuming that the elbows are held downwards and the arms bent at 90°, this distance should be between 27 and 30 cm.

Figure 50 shows the most important dimensions for work seats both at normal desks and at desks for typists.

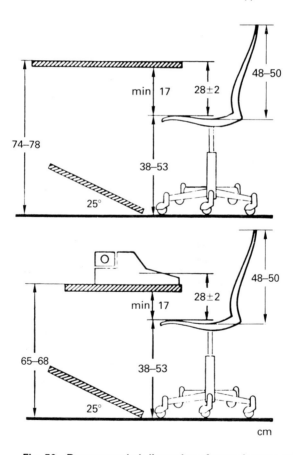

Fig. 50 Recommended dimensions for work seats, tables and footrests. All measurements in cm.

6
Heavy work

Physiological principles

Heavy work is any activity that calls for great physical exertion, and is characterised by a high energy consumption, and severe stresses on the heart and lungs. Energy consumption and cardiac capacity set limits to the performance of heavy work, and these two functions are often used to assess the degree of severity of a physical task.

Mechanisation has reduced the demands for strength and energy of the operator at many kinds of work, nevertheless in many industries there are still jobs which rank as heavy work, and which not infrequently lead to overstrain. Thus, for example, *Hettinger* [142] reports that studies that he made between 1961 and 1969 in the German Iron and Steel Industry showed excessive physical demands being made on the operators at about one quarter of the 337 work places he investigated (a work pulse of more than 40). Heavy work is also common in mining, building, haulage, agriculture and forestry.

Metabolism

A fundamental biological process is to take in nutrients and to convert their chemical energy into mechanical energy and heat. Food is progressively broken down in the intestines until its constituents can pass through the gut wall and be absorbed into the blood. Most of the nutrients then pass to the liver, where they are stored as glycogen, an energy reserve. When needed, they again pass into the blood stream as readily usable compounds, mainly sugars. Only a small proportion of the food is used for building up body tissues, or reaches the adipose tissues as fat.

The blood carries nutrients to all the cells of the body, where they are broken down still further by precisely controlled stages, ending up as water, carbon dioxide and urea. The collective term for these processes is *metabolism*, which can be compared to a slow, self-regulated combustion. The comparison is the more apt, since metabolism, like combustion, needs a supply of oxygen, which it obtains via the lungs and the blood stream. These metabolic processes liberate heat, as well as mechanical energy, depending on what muscular activity is going on. *Figure 51* shows these processes diagrammatically.

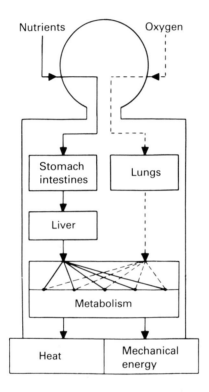

Fig. 51 Diagram illustrating the conversion of nutrients into heat and mechanical energy in the human body.

Energy consumption

Energy consumption in man is measured in kilocalories (kcal):[1] 1 kcal is the amount of heat required to raise the temperature of one litre[2] of water from 14.5°C to 15.5°C. Energy consumption can be measured indirectly by measuring the consumption of oxygen, since the two are directly related. *When one litre of oxygen is consumed in the human body, there is, on average, a turnover of 4.8 kcal of energy. This is called the calorific value of oxygen: to obtain the energy consumption, the consumption of oxygen must be multiplied by 4.8.*

---

[1] See *Chapter 1 footnote*, p. 5, regarding use of the calorie. [2] 1 litre = 1 dm³.

Measurement of *energy consumption*	Oxygen consumption is usually measured by means of a gas meter, carried on the back of the person being tested. A mask with valves directs the exhaled air through the instrument. Part of this air is stored in a container and later taken to the laboratory where its oxygen content is assessed. *The difference between the oxygen content of the inhaled and exhaled air multiplied by the volume of air passing through the gas meter gives a measure of the total amount of oxygen consumed. This can be easily converted into energy consumption either during a given time, or for the performance of a specified task.*
Basal metabolism	Measurements show that a resting person has a steady consumption of energy, depending on size, weight and sex. When the person is lying down, with the stomach empty, this quantity is known as *the basal metabolism.* For a man weighing 70 kg it amounts to about 1700 kcal per 24 hours, and for a woman weighing 60 kg, about 1400 kcal. Under these conditions of basal metabolism nearly all the chemical energy from nutrients is converted into heat.
Work calories	*Energy consumption at work*

As soon as physical work is performed, energy consumption rises sharply. The greater the demands made on the muscles by one's occupation, the more the energy consumed.

The increased consumption associated with a particular activity is expressed in *work calories*, and is obtained by measuring the energy consumption while working and subtracting from this the resting consumption, or basal metabolism. *Lehmann* [194] called these occupational work calories.

These work calories indicate the level of bodily stress, and in relation to heavy work they can be used to assess the level of effort, to work out necessary rest periods, and to compare the efficiency of different tools and different ways of arranging the work. In this context it should be clearly understood that energy consumption measures only the level of physical effort; it tells us nothing about mental stress, about demands the work may make in alertness, concentration or skill, nor about any special physical problems such as excessive heat or static loads from awkward postures.

Hence energy consumption should be used as a measure of comparison only for strenuous physical effort and never for studying mental activities or skilled work.

Leisure calories
Everyday activities also consume energy: what we may call *leisure calories*. A fair average daily consumption would be 600 kcal for a man and 500–550 kcal for a woman.

Thus the total energy consumption is made up as follows:

(a) basal metabolism;

(b) leisure calories;

(c) work calories.

The four categories of work calories, according to *Hettinger* [142] are summarised diagrammatically in *Figure 52*.

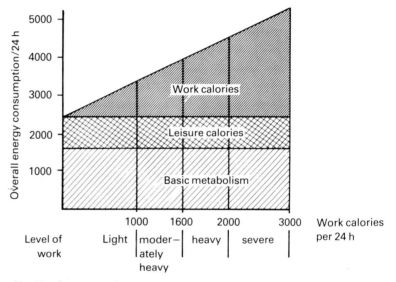

Fig. 52 Summary of overall energy consumption of a man compared with his working consumption. After *Hettinger* [142].

Energy consumption in different occupations
Lehmann and others [194], and *Durnin* and *Passmore* [76] are among the research workers who have studied energy consumption in a variety of occupations. *Table 10* summarises a selection of results from *Lehmann* [194].

Energy consumption in individual activities
Work physiologists have studied the energy consumption involved in a great many separate forms of bodily activity, both at work and at home, and used the mean values obtained as a basis for assessing the severity of the task. *Table 11* summarises the effort involved.

Table 10 Energy demands of various occupations. The calorific values are the annual mean of daily consumption. After *Lehmann* [194].

Men kcal/day	Women kcal/day	Type of work	Example of occupation
2400	2000	Light manual work, sitting	Bookkeeper
2700	2250	Light manual work, sitting	Shorthand typist; watchmaker
		Light manual work, standing	Hairdresser
		Walking	Lowland shepherd
3000	2500	Heavy manual work, sitting	Weaver; basket worker
		Heavy arm work, sitting	Bus driver
		Light bodily work, standing	Mechanic
		Light manual work, walking	Fitter; general practitioner; meter reader
3300	2750	Heavy manual work, sitting	Shoemaker
		Light bodily work, walking	Electrical fitter
		Light bodily work, climbing stairs	Postman (flats)
3600	3000	Heavy arm work, sitting	Stonemason
		Moderate bodily work, standing	Locksmith; masseur
		Moderate bodily work, walking	Butcher
		Moderate bodily work, with heavy arm work	Chimney-sweep
3900	3250	Very heavy bodily work, standing	Sawing firewood
		Heavy bodily work, walking	Ballet dancer; shunter
		Moderate bodily work, climbing	Carpenter on building site
4200	–	Extreme bodily effort, standing	Coal miner (if lucky!)
		Very heavy bodily work, walking	Agricultural labourer
		Heavy bodily work, climbing	Worker in hillside vineyard
4500	–	Extreme bodily effort, standing	Tree feller; lumber-jack
		Very heavy bodily work, walking	Coal cutter; carrying sacks of flour
4800	–	Extreme bodily effort in worst position	Coal miner, lying down
5100	–	Extreme bodily effort, walking	Harvesting by hand

Table 11 Energy consumption in work calories during various forms of physical activity. Work calories = total energy consumption during the work, minus consumption at rest. After *E. A. Müller* and *Spitzer* [234], *Lehmann* and others [194].

Activity	Conditions of work	kcal/min
Walking, empty-handed	Level, smooth surface 4 km/h	2.1
	Metalled road, heavy shoes	
	4 km/h	3.1
Walking, with load on back	Level, metalled road	
	10 kg load 4 km/h	3.6
	30 kg load 4 km/h	5.3
Climbing	16% gradient climbing speed	
	11.5 m/min	
	without load	8.3
	20 kg load	10.5
Climbing stairs	30.5° gradient climbing speed	
	17.2 m/min	
	without load	13.7
	with 20 kg load	18.4
Cycling	Speed 16 km/h	5.2
Pulling hand cart	3.6 km/h, level, hard surface	
	tractive force 11.6 kg	8.5
Working with axe	Two-handed strokes 35	
	strokes/min	9.5–11.5
Working with hammer	Weight of hammer 4.4 kg	
	vertical strokes, 15 per min	7.3
Filing iron	60 strokes/min, 2.82 kcal/g of	
	filings	2.5
Shovelling	10 shovels per min, throwing	
	2 m horizontally and 1 m high	7.8
Sawing wood	Two-handed saw, 60 double	
	strokes/min	9.0
Bricklaying	Normal rate 0.0041 m³/min	3.0
Screwdriving	Screw horizontal	0.5
	Screw vertical	0.7–1.6
Digging	Garden spade in clay soil	7.5–8.7
Mowing	Clover	8.3
Household work	Cooking	1.0–2.0
	Light cleaning; ironing	2.0–3.0
	Making beds; beating carpets;	
	washing floors	4.0–5.0
	Heavy washing	4.0–6.0

Most of the figures quoted in *Table 11* are those obtained by *Lehmann* and *Müller* and their co-workers [234] [194]. The calorific values refer to the net working time, and when they are converted to hourly or daily rates of consumption, at least 10% must be deducted for short rests.

Energy consumption is affected by bodily posture: *Figure 53* summarises the relevant data.

Sitting	Standing	Stooping	Kneeling
3–5%	8–10%	50–60%	30–40%

Fig. 53 Percentage increase in energy consumption in different bodily postures. 100% = energy consumption lying down. The relative increase, as a percentage of this, is the same for men and women.

Energy consumption and health

Most workers in industrial countries sit down at their work. If we add to this the time they sit while they travel to and from work, and in front of the television screen in the evening, *twentieth-century man is clearly well on the way to becoming a "sedentary animal".* Such a sedentary life leaves many organs of the body underused. Often more chemical energy is taken into the body than is consumed, leading to overweight, with increased risk of cardiac and circulatory disease, as well as metabolic troubles. Research has shown that a healthy occupation should involve a daily energy consumption of 3000–3500 kcal for a man, with 2500–3000 kcal for a woman. This category includes, for example, being a non-motorised postman or a mechanic, a shoemaker, or having one of many jobs in building or agriculture. People who have sedentary jobs can make up some of the deficiency during their leisure time: some examples are given in *Figure 54.*

Limits and norms of energy consumption at work

Grading by work calories

As indicated in *Figure 52* men's work can be graded into *four categories* on a scale of work calories:

work calories per 8-h day, kcal

(a) light: < 1000
(b) moderately heavy: 1000–1600
(c) heavy: 1600–2000
(d) severe: > 2000.

Most experts are undecided whether a similar scale for women is possible. *Hettinger* [142] fixed the rates for women about 30% below these; *Durnin* and *Passmore* [76] only 25%. In any event women should not be concerned with the "severe" category.

From what has gone before, *the average energy consumption, less occupational work calories* – i.e. only the energy consumed by

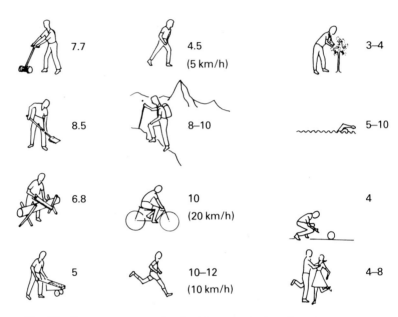

Fig. 54 Energy consumption in leisure activities. The figures are the average overall consumption in kilocalories per minute, for men. Slightly lower figures apply to women, about 10–20% less. Drawn from figures given by *Durnin* and *Passmore* [76].

basal metabolism plus leisure calories – can be estimated as follows:
men: 2300–2400 kcal/24 h
women: 1900–2100 kcal/24 h.
These values also apply to sedentary occupations which do not involve any significant physical effort: e.g. book-keeper, secretary, watchmaker.

Upper limits of energy consumption   When the job involves physical effort the occupational work calories must be added to these values, so that for very heavy work the daily average may reach as much as 5000 kcal. Even this is not the extreme limit, since the human body is capable of putting out much more effort as a temporary exertion. In sport, for example, in cycle racing, long-distance skiing, or during intensive training for endurance tests, energy consumptions of 5000–10 000 kcal per day are possible. Such levels of effort may also be called for during military service, or in strenuous mountaineering, but they cannot be repeated every day. It appears that heavy physical tasks cannot be carried out satisfactorily at an average rate of much more than 5100 kcal per 24 h, over a number of years. These limits are still reached in agriculture, forestry, building trades in hilly districts, or among porters, but they have become rare in modern industry.

*Most work physiologists today consider an energy consumption
of 4800 kcal per working day (averaged over a year) to be a
reasonable maximum for heavy work.*

This corresponds to an average of *2500 occupational kcal* per
working day, and if these are spread over an 8 h work period they
amount to 313 kcal/h or 5.2 kcal/min.

Seasonal workers may exceed these values for a few weeks, or
even a few months, provided slack periods intervene.

This is true in principle for many heavy workers in forestry,
haulage and so on, who can reach levels of 5000–7000 kcal for a
few days without ill effects. This limit of 4800 kcal per working day
is therefore a yearly average which applies only to a heavy worker
in a good state of health. Clearly there are individual variations
both above and below this mean value, depending particularly
upon such factors as constitution, level of training, age and sex.

Age and sex

P. O. *Astrand* [11] and I. *Astrand* [10] measured the maximum
possible oxygen consumption – which is a good indicator of energy
consumption – for a period of 30 minutes, among 350 men and
women of various ages. Their results show clearly the effects of
age and sex, and are set out in *Figure 55*.

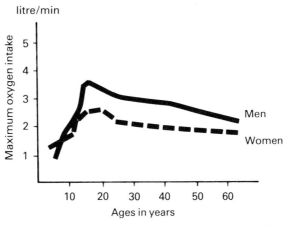

**Fig. 55 Maximum performance, measured by maximum oxygen intake
(ordinate), during a standard task for men and women of various ages.**
After *P. O. Astrand* and *I. Astrand* [10, 11].

If we take performance for either sex as being 100% between the
ages of 20–30 years, then both sexes suffer the following percentage loss of performance with age:

20–30 years: 100%
40 years: 96%
50 years: 90%
60 years: 80%
65 years: 75%.

Normal
performance and
rest pauses

Most scientists [194] [227] [234] [237] are agreed that the norm for energy consumption during heavy work, at an efficiency of 100 %, should be:

work calories   4.0 kcal/min = 240 kcal/h
total energy consumption   5.2 kcal/min = 312 kcal/h.

This working norm can be used as a basis for calculating necessary rest periods. *Murrell* [237] and *Spitzer* [294] devised formulae, that of *Spitzer* [294] being:

$$\text{resting time} = \left(\frac{\text{kcal/min}}{4} - 1 \right) \times 100.$$

Taking shovelling as an example, with a throwing distance of 2 m and a throwing height of 1 m, the energy consumption is 7.8 kcal/min. Applying the formula,

$$\text{resting time} = \left(\frac{7.8}{4} - 1 \right) \times 100 = 95\,\% \text{ of the working time.}$$

This means that the shoveller needs to rest nearly as long as he shovels, so that he can work little more than 30 minutes in every hour.

Spitzer's formula gives very high values for rest periods, and these are certainly valid in theory, within the limits of accuracy possible when measuring energy consumption.

In practice a heavy worker instinctively chooses his own working rhythm. An hourly consumption of about 240 work calories is maintained by varying the tempo of the work, with masked as well as overt pauses for rest. Work-study specialists accept this overall performance as a basis for calculating how much time to allow for the work.

Organisation of heavy work

Efficiency

As far as energy is concerned, a man doing physical work can be compared to an engine. An engine converts the chemical energy of coal or oil into mechanical performance, though with certain losses. Similarly, in the human body, a large part of the energy it receives is wasted by being converted into heat, and only a small proportion becomes useful mechanical energy. In both engine and man the term *efficiency* denotes the ratio between the externally measurable useful effort and the energy consumption that was necessary to produce it.

Under favourable conditions, human physical effort can be 30 % efficient, turning 30 % of the energy consumed into mechanical work and the remaining 70 % into heat. But heat is not the only form in which energy is "wasted". Unproductive static effort is equally important. Hence the highest level of efficiency is possible only by converting as much as possible of the mechanical effort into a useful form, with little or none of it being dissipated in holding or supporting things. *The greater the proportion of mechanical energy to go into static effort, the lower the efficiency.* This is particularly true when work is carried on with a bent back.

In every kind of heavy work it is important to aim at maximum physiological efficiency, not only to use energy economically, but to minimise stresses upon the operator. For this reason work physiologists have made a great many attempts to measure the physiological efficiency of various kinds of working methods, and the use of different tools and other equipment. Their results make it possible to formulate guidelines for the layout of work and the design of equipment, which are of particular importance where strenuous work is concerned.

Some examples are briefly enumerated below:

Table 12 Maximum efficiency in various physical tasks.

$$\% \text{ efficiency} = \frac{\text{useful work}}{\text{energy consumption}} \times 100.$$

Activity	% efficiency
Shovelling in stooped posture	3
Screw driving	5
Shovelling in normal posture	6
Lifting weights	9
Turning a handwheel	13
Using a heavy hammer	15
Carrying a load on the back on the level, returning without load	17
Carrying a load on the back up an incline, returning without load	20
Going up and down ladders, with and without load	19
Turning a handle or crank	21
Going up and down stairs, without load	23
Pulling a cart	24
Cycling	25
Pushing a cart	27
Walking on the level, without load	27
Walking uphill on a 5° slope, without load	30

Shovelling

Shovelling is a common form of manual work, which has been thoroughly studied at the *Max-Planck-Institut für Arbeitsphysiologie in Dortmund, Germany* [194]. The highest level of efficiency was attained when *a load of 8–10 kg was shovelled 12–15 times per minute.* Besides the load, one must take into account the weight of the shovel itelf, so use a big shovel when the material is light, and a small one for heavy materials. For fine-grained materials the shovel should be slightly hollow, spoon-shaped, and with a pointed tip for penetration. For coarse materials, the cutting edge should be straight, and the blade flat, with a rim round the back and sides. For stiff materials such as clay, the cutting edge can be either straight or pointed, but the blade should be flat. The handles of spades and shovels should be 60–65 cm long.

Sawing

In one of our own studies [108] we investigated the physiological efficiency of various types of timber saw. Oxygen consumption was measured before and during the use of five different types of saw, and the number of work calories calculated. This energy consumption related to cutting slices or discs one square metre in area, and the results are shown in *Figure 56*.

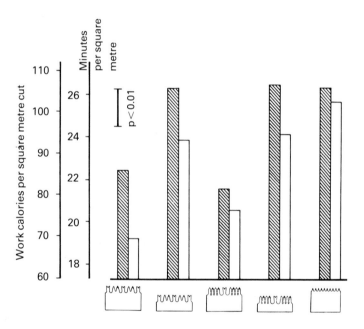

Fig. 56 Energy consumption with five different types of saw. The shaded columns represent the average number of work calories consumed per square metre of cut surface; the white columns show the average time needed to saw one square metre. The vertical marker (p<0.01) indicates a difference in calorie consumption that is statistically significant. After *Grandjean, Egli, Rhiner* and *Steinlin* [108].

The results show that the best saws were the two ripsaws with a broad blade and either two or four cutting teeth in each group, and these were preferred to the others. A more extensive study by *Egli* [78] showed that the best results were obtained from a sawing rate of 42 double strokes per minute, and a vertical pressure of about 10 kg.

Hoeing

Figure 57 shows the results of loosening the soil of a vegetable plot, using hoes of two different types. In soft soil a swivel hoe is much more efficient than the ordinary chopping hoe, but if the soil is hard and dry, the two types are about equally good.

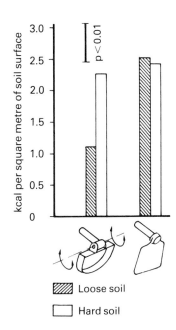

Fig. 57 Energy consumption while using a swivel hoe (left) and a traditional hoe (right), in loose and hard soil. After *Egli, Grandjean* and *Turrian* [78].

Walking

A pleasant and not too strenuous walking pace is 75–110 steps per minute, with a length of pace between 50 and 75 cm, but this is not the most efficient pace when the work performed is compared with the energy consumed. *Figure 58* shows the efficiency of walking, expressed in terms of the energy consumed per mkg of work performed.

From studies by *Hettinger* and *Müller* [144] it appears that *the most efficient walking speed is 4–5 km/h*, reduced to 3–4 km/h in heavy shoes.

Carrying loads

Work physiologists have given special attention to all kinds of jobs which involve carrying heavy loads, since these are rightly considered to be among the most strenuous forms of work.

According to *Lehmann* [194] 50–60 kg is the most efficient load to carry. Smaller ones are more convenient, but require more journeys, and "carrying one's own body weight to and fro" increases the total consumption of energy. If we disregard the "way back", then maximum efficiency, according to *Teeple* [302], is given by:

load: 35% of body weight
speed: 4.5–5 km/h.

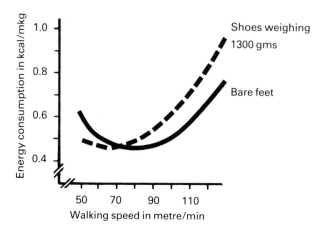

Fig. 58 Best walking speed. Ordinate: efficiency expressed as the energy consumption in kcal per unit of walking effort (metre kilogram). Curves shown for bare feet (solid line) and for shoes weighing 1300 grams (broken line). The heavier the shoes, the lower the best speed.

Nature of the surface

The nature of the surface has a great effect on energy consumption. The following are the distances that can be covered with a load of one tonne on a 4-wheeled truck to use up 250 work calories [194]:

(a) on a light railway track: 850 m
(b) on a good road: 700 m
(c) on a bad road: 400 m
(d) on a dirt road: 150 m.

In this connection it may be said that pushing the truck in front of one requires about 15% less effort than dragging it behind. The handles of the truck should be one metre above the ground and about 4 cm thick. A two-wheeled barrow should have its centre of gravity as low and as close to the axle as possible, for better balance and less weight to support.

Gradients

When a task involves climbing, a gradient of 10° gives the best efficiency. The following heights can be reached for an outlay of 10 work calories:

(a) on a ladder at 90°: 11.5 m
(b) on a ladder at 70°: 14.4 m
(c) on a staircase at 30°: 13.2 m
(d) on a path at 25°: 13.1 m
(e) on a path at 10°: 15.5 m.

Climbing ladders is most efficient when the ladder is at an angle of 70° and the rungs are 26 cm apart, but if heavy loads are being carried 17 cm between rungs is better. If loads are often lifted to considerable height, some sort of lifting gear should be provided. Even a hand-operated crank halves the effort.

Staircases   Climbing stairs is one of the best forms of exercise in everyday life. *From the standpoint of preventive medicine one is strongly advised to avoid using lifts whenever possible and to take every opportunity to climb the stairs as a "gymnastic performance".* At the same time it is sensible to have the stairs designed in such a way that they can be climbed as efficiently as possible. This is particularly so when the stairs are in constant use, or used by infirm and old people.

Lehmann [194] found that least energy was consumed when climbing stairs with a gradient of 25–30°, and he fixed the following empirical norms:

(a) *tread height (riser);*   *17 cm*
(b) *tread depth:*   *29 cm.*

Stairs of these dimensions are not only the most efficient, but also seem to cause fewest accidents. This recommendation can be expressed as a formula:

2 h + d = 63 cm,

where h = height of riser and d = depth of tread, both in cm. *Figure 59* gives optimum dimensions for the design of flights of stairs.

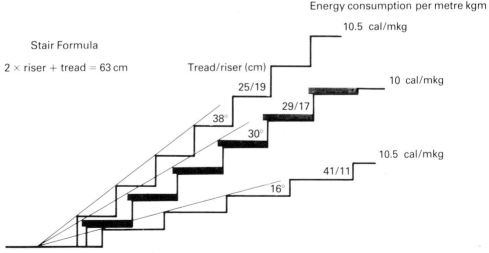

Fig. 59 Physiological recommendations for slope and dimensions of stairs.

Heart rate as a measure of work load

Energy consumption and heart rate   Until recently energy consumption has usually been the means by which the severity of physical stress was estimated, but now it is becoming more and more evident that energy consumption alone is not enough. The degree of physical stress depends not only on the number of calories consumed, but also on the number of muscles involved and on the extent to which they are under static load. A given level of energy consumption is much more strenuous if it is achieved by using only a few muscles than if many others are employed. Similarly, the same energy consumption by static muscular effort is distinctly more tiring than if it is applied to dynamic work.

A further argument against the use of energy consumption as a measure of work load is the heat that is often produced. This may be a trivial part of the energy consumption, yet may cause a sharp rise in the heart rate.

The various ways in which a rise in heart rate is related to work load are shown diagrammatically in *Figure 60*.

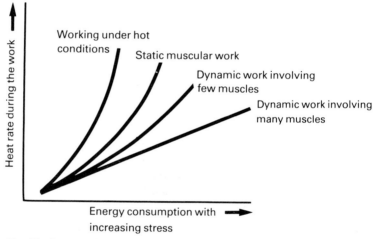

Fig. 60 **Increase in heart rate associated with various types of stress.**

This diagram shows that a given energy consumption can make different demands on the heart according to the circumstances. Summarising, it can be said *the rise in heart rate with increasing work load is the steeper*:
(a) *the higher the ambient temperature;*
(b) *the greater the proportion of static to dynamic effort;*
(c) *the smaller the number of muscles involved.*

For these reasons, heart rate has been used more and more in recent years as an index of work load. Before going into details about heart rate and pulse, we must first consider the relationships between the blood circulation and respiration.

Blood system and respiration

Physical work demands adjustments and adaptations which affect nearly all the organs, tissues and fluids of the body. The most important of these are:

(a) deeper and more rapid breathing;
(b) increased heart rate, accompanied by an initial rise in cardiac capacity and an increased output per minute;
(c) vasomotor adaptations, with dilatation of the blood vessels in the organs involved (muscles and heart), while other blood vessels contract. This diverts blood from the organs not immediately concerned into those which need more oxygen and nutrients;
(d) rise in blood pressure, increasing the pressure-gradient from the main arteries into the dilated vessels of the working organs, and so speeding up the flow of blood;
(e) increased supply of sugar, by release of more sugar into the blood from the liver;
(f) rise in body temperature and increased metabolism. The rise in temperature speeds up the chemical reactions of metabolism, and ensures that more chemical energy is converted into mechanical energy (for this reason athletes "warm up" before a contest).

As work continues, secondary metabolic effects arise, particularly in the chemical composition of the body fluids. There is an accumulation of metabolic waste products, notably lactic acid, and the kidneys have more waste products to excrete. Muscular activity generates more heat in the interior of the body, and to restore the balance more heat must be lost through the skin by increased blood flow and by sweating.

Within certain limits, some of these changes – ventilation of the lungs, heart rate and body temperature – show a linear relationship with the rate of energy consumption, or the work performed. Since these changes can be measured while a person is at work, they can be used to assess the physical effort involved. *Table 13* shows reactions measured at various work loads, after *Christensen* [53].

Table 13 Metabolism, respiration, temperature and heart rate as indications of work load.

Assessment of work load	Oxygen consumption Litres/min	Lung ventilation Litres/min	Rectal temperature °C	Heart rate Pulses/min
"Very low" (resting)	0.25–0.3	6–7	37.5	60–70
"Low"	0.5–1	11–20	37.5	75–100
"Moderate"	1–1.5	20–31	37.5–38	100–125
"High"	1.5–2	31–43	38–38.5	125–150
"Very high"	2–2.5	43–56	38.5–39	150–175
"Extremely high" (e.g. sport)	2.4–4	60–100	over 39	over 175

Measuring heart rate (pulse)

Measuring the heart rate ("taking the pulse") is one of the most useful ways of assessing the work load, because it can be done so easily.

One way is by simply feeling the pulses of the radial artery in the wrist, but to do this while a person is at work is an interruption and may disturb their work, thereby producing a false result. Instruments have been devised which give a continuous record of heart rate, and these have given consistent results during work. The best known of modern methods is the trace shown on an electro-cardiogram, using a small radio transmitter instead of wires from the subject to the tube. This records the action current in the heart muscle, and the heart rate is the number of R-peaks (the strongest action potential) per unit of time.

Heart rate during physical activity

As already mentioned, and within certain limits, the heart rate increases linearly with the work performed, provided that this is dynamic not static, and is performed with a steady rhythm, only the force exerted being variable.

When the work is comparatively light, the heart rate increases quickly to a level appropriate to the effort, then remains constant for the duration of the work. When work ceases, the pulse returns to normal after a few minutes.

With more strenuous work, however, the heart rate goes on increasing until either the work is interrupted, or the operator is forced to stop from exhaustion. *Figure 61* shows diagrammatically the behaviour of the pulse during certain work studies.

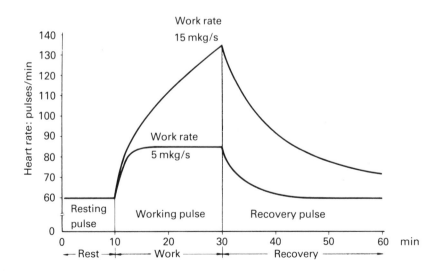

Fig. 61 Heart rate in two different work loads. At the higher level of stress the heart rate goes on increasing as long as the stress lasts, whereas at the lower rate it levels off at a "plateau", or "steady state".

E. A. Müller [232] proposed the following definitions:

(a) resting pulse: average heart rate before the work begins;

(b) working pulse: average heart rate during the work;

(c) work pulse: difference between the resting and working pulses;

(d) total recovery pulse (recovery cost): sum of heart beats from the cessation of work until the pulse returns to its resting level;

(e) total work pulse (cardiac cost): sum of heart beats from the start of the work until resting level is restored.

E. A. Müller [233] considers that the total recovery pulse is a way of measuring fatigue and recovery. Since "fatigue" is a subjective term it would be more correct to regard heart rate, and particularly total recovery pulse, as measuring the physical work load of the individual.

Acceptable limits

Karrasch and *Müller* [172] made use of their studies to lay down an acceptable upper limit of work load as being that within which the working pulse did not continue to rise indefinitely, and when the work stopped, returned to the resting level after about 15 minutes. This limit seems to ensure that energy is being used up at the same rate as it is being replaced, that is to maintain a steady state. *The maximum output under these conditions is the limit of continuous performance throughout an eight-hour working day.*

According to *E. A. Müller* [233] the limit of continuous perform-ance for men is reached when the average working pulse is 30 beats/min above the resting pulse (i.e. work pulse = 30/min), both of these being measured in the same posture (e.g. both standing up), so that the static loads are the same. *Rohmert* and *Hettinger* [270] have made a systematic study of limits of work load, during which the heart rate remained steady, using a cycle-ergometer, and a hand-crank for eight hours at a time. They came to the conclusion that this limit was still valid up to a work pulse of 40/min, provided that *the resting pulse was measured when the operator was lying down*. The authors show that for dynamic work involving a moderate number of muscles *1 work calorie/min = 10 work pulses*.

Several studies undertaken in factories have shown us that it is easier to measure the resting pulse when the subject is sitting than when lying down, so we suggest that, *for men*, we should start from *a resting pulse taken when seated*, and fix:

35 work pulses as the limit for continuous performance.

There are no corresponding *studies of women*, but on physio-logical grounds it seems reasonable to postulate:

30 work pulses as the limit for continuous performance,

again taking the resting pulse in a seated position.

Recovery pulse

In the U.S.A., *Brouha* [39] has made lengthy and detailed studies of heart rate as an index of work load, and the most important of his findings include:

(a) identical exercises on a cycle ergometer,[1] producing between 180 and 1000 mkg/min of work, resulted in 17 women having higher heart rates than the 30 men studied [39].

A performance of 360 mkg/min produced work pulses of:
women: 50
men: 40;

(b) *Brouha* [39] found that the total recovery pulse (recovery cost) was as good as, or even better than, the total work pulse (cardiac cost) as a measure of work load, and that the two quantities were linearly related;

(c) the following procedure seems to be suitable for recording the recovery process: after work has stopped, take the pulse at the wrist during the following 30-second intervals:

from 30 s to 1 min;
from 1½ min to 2 min;
from 2½ min to 3 min.

Take the average of these three readings as being the heart rate during the recovery phase, as well as an indication of the preceding work load. *Figure 62* shows the relation between cardiac cost and average recovery cost;

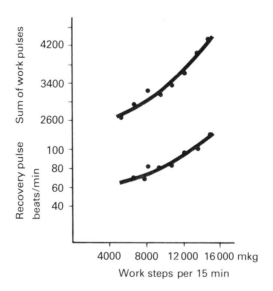

Fig. 62 Work load (abscissa); sum of work pulses (upper ordinate) and number of recovery pulses/min (lower ordinate). The recovery pulse is the mean of three measurements taken at 30 seconds, 1½ minutes and 2½ minutes after the end of the work. After *Brouha* [39].

[1] A cycle ergometer is a sort of stationary bicycle on which the effort of pedalling can be measured either in watts or in metre-kilogrammes/min.

(d) *Brouha* recommended the following criteria for determination of acceptable limits of work load: *the first reading should not exceed 110 pulses/min with a fall of at least 10 pulses between the first and third readings*. Given these conditions the work load could be sustained during an eight-hour working day.

Combined effects of work and heat

As we have already seen, heart rate can be used as a measure of heat load as well as of work load. This is understandable, since the "heart" is merely a "pump" which supplies the muscles with blood, on the one hand, and suffuses the skin with blood, on the other, when it is necessary to eliminate excess heat. So when work is being performed under hot conditions the heart and circulatory system have two functions:

(a) *transporting energy to the muscles;*
(b) *transporting heat from the interior of the body to the skin.*

This double burden upon the heart and circulatory system is a common occurrence in industry and commerce. When heavy work has to be carried on in ambient temperatures of 25°C,[1] the elimination of excess heat throws an additional load upon the heart. An example of such a double loading is drop forging. *Hünting* and his colleagues [153] were able to show that holding the forging during the operation caused considerable loads on the back and arms both static and dynamic. Working conditions were certainly strenuous, and deliberate pauses of up to 60 % of the working time were noted. At the same time the operator was subjected to intensive radiation of heat. *Figure 63* gives a simple sketch of the drop forge, and results from measurements of the work load upon a 47-year-old operator are shown in *Figure 64*.

Fig. 63 Drop-forging. The operator manipulates a piece of glowing metal weighing 18.5 kg with tongs, and receives a stream of radiant heat.

---

[1] Ambient temperature is a rough approximation for the air temperature plus the heat radiated from the forge, and is a suitable way of assessing the heat load.

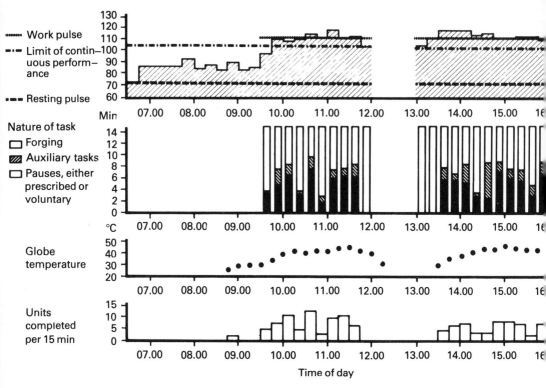

Work pulse

Limit of continuous performance

Resting pulse

Nature of task
- Forging
- Auxiliary tasks
- Pauses, either prescribed or voluntary

Globe temperature

Units completed per 15 min

Time of day

Fig. 64 Pulse measurements, analysis of the task, and globe temperatures of an operator of a drop forge. Above: heart rate (averaged over 15 minutes). Average work pulse during drop forging is 41/min. Middle: distribution in time of the various phases of the job. Pure drop forging occupies 28 % of the total time. Globe temperatures for each phase are given in °C. Below: number of completed units per 15 minute period. After *Hünting, Nemecek* and *Grandjean* [153].

The mean work pulse of 41 beats/min during forging rates as a high level. Figure 64 also indicates the limit of continuous performance set by *E. A. Müller* [233], which is valid in this case because the resting pulse of the operator was measured when he was standing up.

The net working time is naturally related to the number of forgings/min made and to the ambient[1] temperature, and *Figure 64* shows that heart rate follows in step with these two quantities. On days when there is no pause for setting up the forge, this work load must be considered excessive. The management have taken steps to reduce the work load after seeing these results.

Static work and heart rate

Chapter 1 gave an example of "potato planting", where the "static effort" produced a work pulse of 40 beats/min, which fell to 31 beats/min when the need to support the potato basket was eliminated (see *Figure 8*), although the energy consumption remained the same (4.3 kcal/min). This shows that *static effort can cause a rise in heart rate, even though there is no increase in the total consumption of energy. This rise must be interpreted as an increased physical stress*. This phenomenon is also shown in *Figure 60*.

In the laboratory, the effects of static muscular effort have been studied mainly by using loads which had to be either supported or dragged along. *Figure 65* shows the results of experiments by *Lind* and *McNicol* [201].

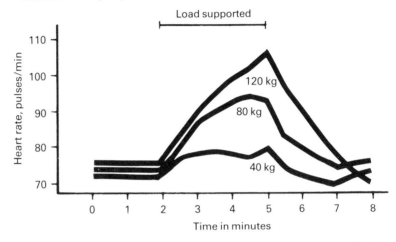

Fig. 65 Heart rate during static effort (carrying a load). The heavier the load, the greater the total of both work pulses and recovery pulses. After *Lind* and *McNicol* [201].

It was found that in spite of considerable static loads the heart rate rose only a little above 100 beats/min, and that the initial resting pulse was quickly restored afterwards.

Monod [227] recorded the heart rate at which static contraction of the muscles of the forearm could no longer be continued because of the severe pain experienced. He found that a weight of 3 kg could be supported for 3 min only, by which time the heart rate had reached 100 beats/min. The less the weight, the longer it could be supported, and the lower the peak pulse. As was explained in *Chapter 1*, the reduced blood flow during a muscular contraction is followed by a very much increased flow (hyperaemia) afterwards, which may be 20–25 times normal. The acids (carbonic and lactic) and other "waste products" that have accumulated during the contraction produce this hyperaemia. This phenomenon gives the muscle the "oxygen debt" back.

Heavy work in the iron and steel industry

Case histories involving heavy work

Hettinger [142], *Scholz* [283], as well as *Steinhausen* and *Bosse* (cited after *Hettinger* [142]) have all carried out intensive studies of work loads in the German iron and steel industry. In *Figure 66* the maximum heart rate of 380 workers, measured over periods of 2 and 4 minutes, is displayed as a frequency distribution curve.

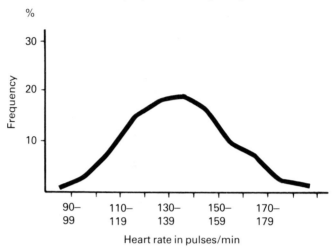

Fig. 66 Frequency distribution of maximal heart rates, measured on 380 factory workers during the course of a working day in the German iron and steel industry, during the years 1961–1969. After *Hettinger* [142].

It is evident from this that the most frequent peak value lies in the range 130–140 beats/min (mean 132.6), and that extreme increases of up to 180 beats/min occur.

The frequency distribution of work calories/min and work pulses/min, which *Hettinger* [142] has compiled from the three most extensive field studies of the other authors, are of particular interest. The results from a total of 552 workers studied over the years 1949–1969 are set out in *Figure 67*.

If we are looking for a common limit of continuous performance, the two frequency distribution curves do not coincide. Values for consumption of work calories are generally lower than values for work pulse, a discrepancy which the authors attribute to the effects of static effort and heat. Furthermore, *Figure 67* indicates that an appreciable percentage (29 % in all) of the work places studied give rise to work pulses above the continuous limit of 40 beats/min.

Table 14 analyses the 215 work places in *Hettinger*'s [142] studies in more detail.

This summary shows that in practice muscular effort and heat combine to raise the heart rate so much that it usually exceeds the limit for continuous performance of 40 work pulses/min.

A work place where castings are polished may be taken as an example of simple steps being sufficient to reduce the work load significantly. *Figure 68* shows the results.

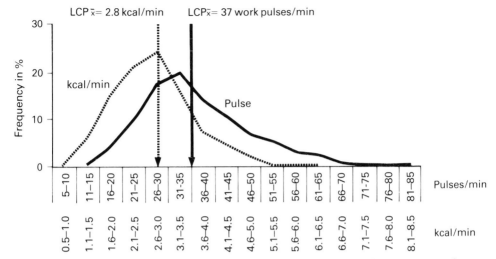

Fig. 67 Frequency distribution of expenditure of work calories (lower pairs of numbers) and of the work pulse frequency (upper pairs of numbers). Studies made at 552 working places in the German iron and steel industry 1949–1969. Dotted curve = kcal/min; solid curve = work pulses/min. LCP = limit of continuous performance. After *Hettinger* [142].

Table 14 Analysis of 215 work places according to the kind of work load most frequently associated with them.

Work place	Number of such places	Average per shift	
		Work calories kcal/min	Work pulse beats/min
Mainly dynamic effort	54	2.7	30
Essentially static work	59	2.5	44
Hot work place	102	2.3	42
Overall	215	2.5	39

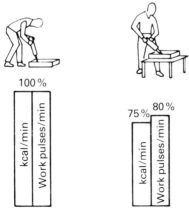

Fig. 68 Stresses when polishing a casting when bending over and when working upright. Avoidance of working in a stooping posture has reduced the work calories per shift by an average of about 25 % and the work pulse by 20 %. After *Scholz* and *Sieber* [284].

*Heavy work
in agriculture*

Despite mechanisation, heavy work still exists in agriculture, and some studies by *Brundke* [43] may be mentioned in this connection.

A fruit grower, a farmer and a dairyman were studied, and their pulse taken throughout a working week. Resting pulse was measured when they were sleeping at night. *Figure 69* shows the results of recording their work pulse, using the data given by *Brundke*.

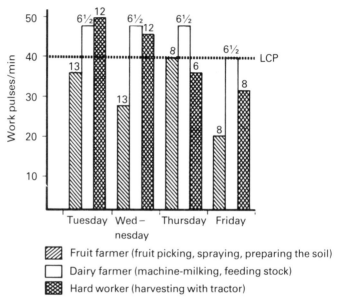

Fig. 69 **Average work pulses for three agricultural occupations**. The resting pulses were measured during nighttime sleep. The figures at the top of each column indicate the approximate length of the working day in hours. LCP = limit of continuous performance for an 8-hour day. Drawn from data given by *Brundke* [43].

It is clear from *Figure 69* that the critical limit of 40 work pulses/min was exceeded on several occasions, especially by the dairyman and by the farmer at harvest time. These studies confirm that *in spite of mechanisation, heavy work is still part of the programme for the agricultural worker.*

*Heavy work by
women in the
textile industry*

Not a few jobs in industry involve heavy work, without appearing to do so at first glance. One such example may be taken from the textile industry, where women have the job of examining spools of artificial fibre and packing them into boxes. The three key moves in this task are illustrated in *Figure 70*.

Reach to the left varies between 50 and 90 cm, while that to the right amounts to about 90 cm. The movement across from left to right requires an effort that is partly static (supporting a weight of 3

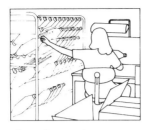  

Grasping motion   Checking and   Packing in the right 2 s
to the left 1–2 s   tying ends 10–12 s

Fig. 70 Inspection of spools for artificial fibre. The three drawings show the three operations involved. Weight of a spool: 3 kg. Performance: 730–960 times 3 kg = 2200–2900 kg per arm per shift. After *Nemecek* and *Grandjean* [242].

kg), and partly dynamic, and each turn is accompanied by slight bending and twisting of the trunk. The factory recently introduced a *"new way"* of knotting the ends of the fibres, and since this rationalisation did not produce the improved performance that was expected, the question arose whether the operators might be physically overstressed.

Because of this situation we selected five workers, and:

(a) took their pulse;

(b) counted how many spools they handled in 15 minutes;

(c) investigated the extent of bodily aches and subjective impressions of fatigue.

To assess the bodily aches, and the fatigue, we used a double-ended questionnaire, both before and after the task. The operators used an agreed system of marking to note their momentary impressions on a scale which was 7 cm long and ran from one extreme impression to its opposite (e.g. wide awake-drowsy). The result of this "self-assessment" was expressed by the difference in mm between the first marking before beginning the task and the second marking after completing it, and this was taken as a measure of the aches and degree of fatigue that had arisen during the task.

The resulting pulse measurements are shown in *Figure 71*, and those from the survey of aches and fatigue in *Figure 72*.

If we first consider the results without reference to the increased performance resulting from *the new method, the peak work pulse and the highest levels of fatigue and of some of the aches come at the end of the working shift*. Assuming that the desirable upper limit for continuous effort for women is 30 work pulses/min, we can see that this has been exceeded in 6 out of 10 cases, since we have twice recorded work pulses of more than 40/min. *The cause is to be found in the increased demand for muscular effort from the arms.* According to *Rohmert* [267] the maximum lifting power of a woman's outstretched arm averages 6 ± 1.5 kg, so that, to lift

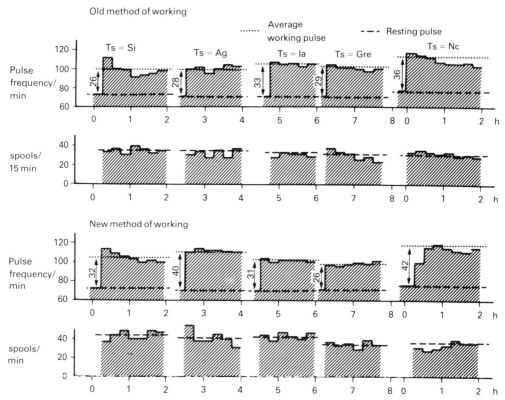

Fig. 71 Pulse frequency and performance of 5 female operatives in checking spools (bobbins) in a textile factory. Ts = operative, designated by an abbreviation of her name. Broken line = resting pulse while sitting. Dotted line = average working pulse. Vertical arrows = work pulse averaged over 2 hours. After *Nemecek* and *Grandjean* [242].

a spool weighing 3 kg, she is almost certainly exerting more than 50 % of her maximum power. This is too high even for dynamic effort (see *van Wely* [325] and *Chapter 1*), and obviously so for static effort, which ought not to exceed 15–20 % of maximum power. Moreover we have not allowed for the weight of the outstretched arm itself, which is well known to be between 19 % and 30 % of its maximum muscular effort. So these studies have shown that *this apparently light form of work involves the operators in too much muscular effort.*

Table 1 (*Chapter 1*) illustrated the risk of bodily aches being caused by bad postures, and the results just given confirm this. *Figure 72* gives direct evidence that during a working shift aches and pains in the right arm, back and neck, in particular, are greatly aggravated. There is no doubt that these are related to the postures that the workers are seen to adopt.

Finally, *Table 15* summarises the effect of the "*new method*", with its demand for increased performance, on the output, the work

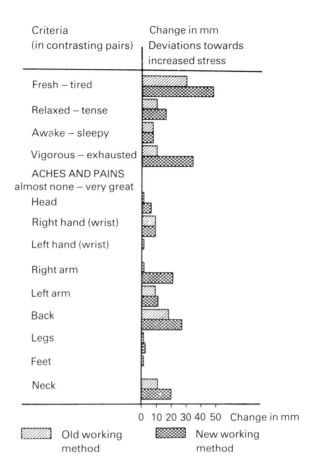

Fig. 72 The self-assessment of fatigue and aches and pains by 5 female textile operatives. The lengths of the blocks indicate the average increase in these complaints during a shift. Maximum possible length of a block = 70 mm. For fuller explanation see text. After *Nemecek* and *Grandjean* [242].

Table 15 Effect of the increased performance of the new method on output, work pulse, fatigue, and aches and pains in five women textile workers.
M = method. After *Nemecek* and *Grandjean* [80].

Worker	No. of spools handled per shift		Work pulses per minute		Increases in daily aches and fatigue (in mm of scale)	
	Old M	New M	Old M	New M	Old M	New M
Ne	730	959	36	42	1	18
Aq	720	912	28	41	22	27
Ge	732	840	29	26	7	19
Ja	732	960	33	31	3	31
Si	852	960	26	32	26	30

pulse, and the level of aches and fatigue. The "self-assessment" scheme made use of an average of the daily variations in replies to the questions involving fatigue, and aches in arms, hands, back and neck.

Although these results are not statistically significant, because of the small number of observations, their tendency is obvious. The work load was already excessive under the "*old method*", and the "*new method*" – by reducing the inspection time, and involving more arm work to left and right – *has increased the work load still further, which tends to raise the work pulse and apparently also to cause more fatigue and aches and pains in the operatives.*

From the evidence of these results, the work could be lightened in the following ways:

(a) shortening the distance to be reached at each side;

(b) lowering the working level;

(c) mechanical aids to reduce the load on the hands: e.g. a swivelling support for the spool;

(d) reorganisation of the work, with a rotation between different operations.

7
Handling loads

Lifting

**Loss of work
through back
troubles**

Lifting, handling and dragging loads involves a good deal of static effort, enough to be classified as heavy work. The main problem of these forms of work, however, is not the heavy loads on the muscles, but much more

the wear and tear on the intervertebral discs,

with the increased risk of back troubles. That is why the handling of loads deserves a chapter to itself.

Disc troubles, and their associated infirmities of the back and legs are essentially a human problem, and one that concerns industrial man. Back troubles are painful and reduce one's mobility and vitality. They lead to long absences from work, and in modern times are among the main causes of early disability. They are comparatively common in the age group 20–40, with certain occupations (labourer, farmer, porter, nursing staff, etc.) being particularly vulnerable to disc troubles. Moreover, we know that workers with physically active jobs suffer more from ailments of this nature, and their work is more affected than is the case with sedentary workers. In Sweden, for instance, *Lundgren* [206] questioned 1200 workers, and the frequency of complaints and periods of absence from work are given in *Table 16*.

Table 16 Absenteeism because of back troubles, in relation to the nature of the work. After *Lundgren* [206]. Each figure is a percentage of the workers questioned.

	Frequency of complaints	Absenteeism		
		1 day	3 weeks	6 months
Light work	52.7	25.5	12.1	2.3
Heavy work	64.4	45.5	25.3	5.4
All	60.0	36	20	4

Comprehensive studies by *Wagenhäuser* [317] of 773 Swiss workers gave the following results:

(a) 73 % of those questioned had rheumatic complaints;

(b) 74 % of those who complained had demonstrable signs of degenerative changes in the spine.

According to *Krämer* [183], disc troubles are the cause of:

(a) 20 % of absenteeism;

(b) 50 % of premature retirements.

These figures show how widespread disc trouble is today, and justify the efforts of ergonomists to reduce the amount of wear and tear on the intervertebral discs by attention to posture.

When the *"work seat"* was discussed in *Chapter 5* we considered the anatomy of intervertebral discs and the nature of disc troubles. Here are a few supplementary remarks.

Disc troubles

It is well known that the vertebral column, or spine, has the shape of an elongate S: at chest level it has a slight backwards curve, called a kyphosis; in contrast the lumbar region is slightly curved forward, the lumbar lordosis. This construction gives the spine elasticity to absorb the shocks of running and jumping.

The loading on the vertebral column increases from above downwards, and is at its greatest in the lowest five lumbar vertebrae. As we have said, each pair of vertebrae are separated by an intervertebral disc.

Degeneration of the discs first affects the margin of the disc, which is normally tough and fibrous. A tissue change is brought about by loss of water, so that the fibrous ring becomes brittle and fragile, and loses its strength. At first the degenerative changes merely make the disc flatter, with the risk of damage to the mechanics of the spine, or even of displacement of the vertebrae. Under these conditions quite small actions such as lifting a weight, or a slight stumble, or similar incidents may precipitate severe backache and lumbago.

When degeneration of the disc has progressed further, any sudden force upon it may squeeze the viscous internal fluid out through the ruptured outer ring, and so exert pressure either on the spinal cord itself or on the nerves running out from it. This is what happens in a case of "slipped disc". Pressure on nerves, narrowing of the spaces between vertebrae, pulling and squeezing at adjoining tissues, ligaments and bursae of the joints are the causes of the variety of aches, muscular cramps and paralyses including lumbago and sciatica which commonly accompany disc degeneration.

Loading of the intervertebral discs

The Swedes *Nachemson* [240] [241] and *Andersson* [7], cited above, have made a thorough study of the effects of bodily posture, and of the handling of loads, on pressures inside the intervertebral discs.

Figure 73 shows the results of handling various weights on these pressures in nine persons, two of whom had back troubles, and the other seven of whom were in good health.

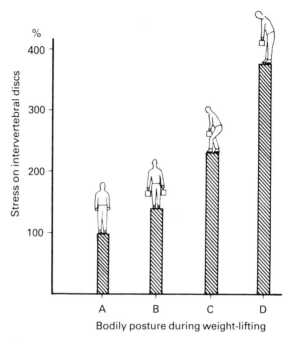

Bodily posture during weight-lifting

Fig. 73 The effect of body posture when lifting weights on the intervertebral disc pressure between the third and fourth lumbar vertebrae. A = upright stance. B = upright stance with 10 kg in each hand. C = lifting a load of 20 kg with knees bent and back straight (correct stance for weight lifting). D = lifting 20 kg with knees straight and back bent. Pressure on discs during upright stance (A) is taken as 100 %. After *Nachemson* and *Elfström* [241].

The figure shows clearly the effect of a bent back on the loading of the discs when a weight was being lifted.

Bending the back, while keeping the knees straight, puts a much greater stress on the discs in the lumbar region than keeping the back as straight as possible and bending the knees.

Figure 74 shows how pressure develops in the discs during the two types of lifting action.

Lifting technique and disc pressure

This pressure curve shows very clearly how lifting a load with bent back brings about a sudden and steep increase of internal pressure in the discs, and quickly overloads them, especially their fibrous rings.

Scientific studies confirm everyday experience that persons who have disc troubles are specially liable to sudden and violent pains,

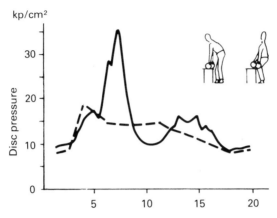

Fig. 74 Curves of pressure within the intervertebral disc between the third and fourth lumbar vertebrae, while lifting a load of 20 kg. A = back rounded, knees straight. B = back straight, knees bent. After *Nachemson* and *Elfström* [241].

and even paralysis. These symptoms are precipitated by sudden heavy loads placed on the discs, a risk which is increased by working methods involving unskilful manipulation.

If a person bends over until the upper part of his body is horizontal, then the leverage effect imposes very heavy pressure on the discs between the lumbar vertebrae. An average weight of the upper part of the body would be about 45 kg and the length of leverage about 35 cm, with a resulting moment of between 100 and 200 kp.* If a weight is lifted at the same time, the force on the discs could rise to 300–400 kp.

The exact pressures in kp per disc for various bodily postures are summarised in *Table 17*.

Table 17 Loading of the disc between the third and fourth lumbar vertebrae in kp per surface, during various postures and tasks. After *Nachemson* and *Elfstrom* [241].

Posture/activity	Loading in kp
Standing upright	86
Walking slowly	92
Bending trunk sideways 20°	114
Rotating trunk about 45°	114
Bending trunk forwards 30°	147
Bending trunk forwards 30°, supporting weight of 20 kg	240
Standing upright holding 20 kg (10 kg in each hand)	122
Lifting 20 kg with back straight and knees bent	210
Lifting 20 kg with bent back and knees straight	327

* kp = kilopond.

Distribution of loads on discs

When a rounded back causes curvature of the lumbar spine, the loads imposed on the intervertebral discs are not only heavy, but asymmetrical, being considerably heavier on the front edge than on the back (see *Figure 75*). *The resultant stresses on the fibrous ring are certainly bad for it, and must be regarded as an important factor in the "wearing out" of the disc.* Further, it must be assumed that the viscous fluid inside the disc will tend to be squeezed towards the side with less pressure, in the back, with the danger that this fluid will leak towards the spinal nerve cord. This is yet another argument for keeping the back as straight as possible when lifting a heavy load.

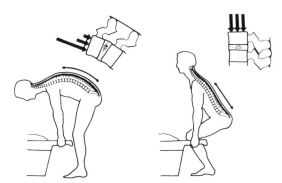

Fig. 75 How the pressures on the intervertebral discs are distributed when a load is being lifted, with bent, and with straight back. Above, effects of the two methods on the discs: the round back (left) leads to heavy pressures on the front edge of the disc, and increases the risk of rupture; the straight back (right) ensures that the loads on the disc are evenly distributed, thus reducing wear and tear on the fibrous ring of the disc.

Recommended ways of lifting loads

The following four rules are based on scientific knowledge as well as on general experience:
- (a) *seize the load and lift it*
  - (i) *with straight back;*
  - (ii) *with the body as erect as possible;*
  - (iii) *with the knees strongly bent,*
    i.e. imitate the stance of a weight-lifter (see *Figure 75*);
- (b) *make sure that your hold on the load is 40–50 cm above ground level.* Design loading ramps with this in mind, and if the load does not have a handle, bring your hand to the correct height by fitting the load with a rope sling, harness or hook;
- (c) *get your body as close to the load as possible* (see *Figure 76*);
- (d) *handle the load in such a way that it can be moved with your back as straight as possible.* A good example is moving a barrel or keg; tip it over so that you can roll it along (see *Figure 77*).

Fig. 76 Lift loads as close to the body as possible.

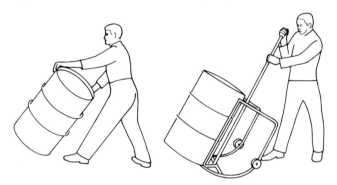

Fig. 77 **Handling casks.** Left: tilting and rolling, with the upper part of the body held upright. Right: lightening the work by using a trolley.

Limiting values
for lifting loads

Since the lifting of heavy loads is dangerous for the intervertebral discs, it is a sound idea to fix limits in industry. The guidelines recommended in *Table 18* owe much to the suggestions of *Münchinger* [236] and to the limits set by the International Labour Office in Geneva.

Table 18 **Upper limits of permissible weight (in kp) to be lifted manually.**

	Adults		Juveniles	
	Men	Women	Male	Female
Occasionally:	50	20	20	15
Frequently:	18	12	11–16	7–11

These limits lie well below the *maximum lifting power of the two arms,* for which the following values apply (after *Rohmert* [267]):
(a) average farmer's wife: 83 kp
(b) powerful young farmer's wife: 108 kp
(c) average foundry worker: 173 kp
(d) muscular foundry worker: 200 kp
The limits recommended in *Table 18* can therefore be taken only as a rough guide, since they do not take into account a number of

relevant factors, such as bodily posture, or the distance of the load from the gravity axis of the body.

An interesting method of estimating maximum permissible load was used by *Davis* and others [62] [64]. They started from the fact that when a heavy load is being lifted the muscles of the trunk, and especially the abdominal muscles, are stressed, thereby raising the pressure in the abdominal cavity, the so-called intra-abdominal pressure.

Intra-abdominal pressure as a limiting factor

The work of *Davis* [63] leads to the conclusion that intra-abdominal pressure can be used as a measure of the load imposed upon the vertebral column. To measure this pressure the person being studied must swallow a capsule containing a pressure-sensitive element and a radio transmitter. Statistical studies carried out in factories where heavy loads are lifted indicate that the risk of back troubles is increased if intra-abdominal pressures exceed 100 mm Hg [65] [297].

Using equipment of this type, the authors measured the intra-abdominal pressures of 200 soldiers, lifting five different weights, with 36 different positions of the arms, and these when standing, sitting and kneeling. From all these readings the authors calculated the weights, or the pulling or pushing forces, which induced an intra-abdominal pressure up to, but not exceeding, 90 mm Hg. The authors [64] rated this as the maximum permissible force to be exerted in jobs which involved lifting, pushing and pulling. Some of these limiting values are summarised in *Table 19*.

Table 19 Maximum permissible force (kp) for young men while lifting, pushing or pulling. These values apply for a movement that is repeated not more than once per minute, and for a maximum intra-abdominal pressure of 90 mm Hg. For more frequent movements, reduce these values to 70%. After *Davis* and *Stubbs* [64].

Conditions	Grasping distance, expressed as a fraction of arm's length			
	¼	½	¾	⁴⁄₄
Standing up:				
Two-handed lift, frontal	35	25	15	10
One-handed lift, frontal	30	22	14	10
One-handed lift, sideways	27	20	13	10
One-handed pushing, frontal	25	23	18	25
One-handed pulling, frontal	45	50	50	50
Seated:				
Two-handed lift, frontal	27	17	12	11
One-handed lift, frontal	35	22	14	10
One-handed lift, sideways	33	21	14	9

Needless to say, these recommendations are valid for young men only. No similar studies have yet been carried out for older men, or for women.

The results obtained by *Davis* and *Stubbs* [64] illustrate the importance of the distance between the load and the body: loads held close to the body produce intra-abdominal pressures about three times as low as when the load is held at arm's length. Another interesting finding of these authors may be mentioned here; the pressures measured when standing may also be exceeded when bending forwards or crouching, *provided the back is kept straight.*

Carrying a burden

Carrying techniques

When a heavy load is being carried, it is advisable to hold it vertically over the centre of gravity of the body, as far as possible. In this way the effort of balancing will be minimised, and unnecessary static muscular work avoided. The best way of carrying loads is by a yoke across the shoulders. It is less good to carry a burden in front of the body, where it places static loads on many muscles of the abdominal wall. *Müller* and *Spitzer* [234] have found that a workman can carry a load of 20 kg the following distances for an energy consumption of 250 kcal:

(a) using a yoke: 4.5 km
(b) on his back: 3.9 km.

Loads are often carried in one hand. According to *Rohmert* [267] the maximum carrying power of one hand held by the side is as follows:

(a) lightly built women: 27 kg
(b) average women: 41 kg
(c) powerfully built women: 54 kg
(d) lightly built men: 63 kg
(e) average men: 81 kg
(f) powerfully built men: 94 kg.

The limits quoted in *Table 18* (above) should be used as guidelines for loads carried in one hand.

It is markedly easier to carry a load in each hand simultaneously, with the hands at the sides. Nevertheless this method is not recommended for very heavy loads, because these interfere with walking and put heavy stress upon the muscles of the shoulders and upper arms.

As we noted above in *Chapter 6*, a load of 50 kg is the most economical from the point of view of energetics, as well as being the most efficient when the number of journeys is taken into account. Hence 50 kg sacks are to be recommended over short distances. In contrast, 100 kg sacks are both bad physiologically and very much more harmful to the spine.

8
Skilled work

Skilled jobs call for a high degree of:
(a) quick and accurate regulation of muscular contraction;
(b) coordination of the movements of the individual muscles;
(c) precision movements;
(d) concentration;
(e) visual control.

In practice, skilled work is mostly a matter for the hands and fingers only. The most important nervous processes that accompany a movement with an element of skill involved were shown in *Figure 15*. It is clear that to perform a delicate movement with speed and precision calls for a whole series of sensory nerve impulses, followed by motor directives from the brain.

Learning

During the period in which a skilled operation is being learned, we can distinguish between two processes:
(a) *learning the movements;*
(b) *adaptation of the organs involved.*

From a physiological point of view, *learning is essentially a matter of imprinting a "template" of the necessary movements upon the medulla of the brain.* To begin with, all the movements must be performed consciously, but as training progresses the conscious element is gradually reduced. New pathways and junctions are built up in the brain, and control of movements is gradually taken over completely by cerebral nerve centres. In simpler terms: *acquiring a skill consists mainly of creating reflex arcs which replace conscious control.*

Learning to write is a good example. First the child learns one letter after another by consciously copying what the teacher draws. Writing is very laborious and difficult at this stage. Gradually, after months or years perhaps, the different letters, and later on complete words, become "impressed on the brain as templates for the necessary hand movements". Writing has become "automated", at least as far as finger movements are concerned, and conscious awareness is diverted more and more to finding words and constructing sentences.

Another phenomenon appears during the learning phase: the gradual elimination of all muscular activity that is not essential to the skilled work in hand. The skilled man is relaxed and economical in his movements, whereas the novice's work is cramped and tiring. Hence the level of energy consumption for a given task decreases during the training period, as non-essential movements are gradually eliminated.

Recommendations for training

The following recommendations will make training easier:

(a) *short training sessions*. To acquire a skill calls for the highest concentration. A tired person easily develops bad habits, which are difficult to unlearn afterwards. It may be said that the higher the level of skill to be acquired, the shorter should be the training sessions. For very delicate skilled work four sessions per day, each of 15–30 minutes length, are enough to begin with, gradually becoming longer later on;

(b) *splitting the job into separate operations*. A work study should be made, to enable the job to be divided into a number of distinct operations or processes. It is then possible to make tests to determine which parts of the job are the most difficult and to set the expected performance accordingly. It is also possible, with advantage, to practise the most important operations by themselves, only later practising several operations in sequence, and especially switching from one to the other;

(c) *strict control and good examples*. As we have seen, the learner must avoid acquiring bad habits, so it is important that he should be strictly supervised throughout his training. Young persons learn a good deal by direct and largely subconscious imitation, so they should be given highly skilled instructors to learn from. Older learners depend less on imitation and more on visual aids such as diagrams.

Adaptation of organs

The second process – adaptation of the bodily organs involved – is a matter of making gradual changes in the muscles and occasionally other organs, such as the heart or the skeleton. Muscular adaptation involves thickening the muscle fibres, and thereby increasing the total power of the muscle. Training for very rapid movement means not only increasing muscle power, but concurrently reducing internal friction by getting rid of some of the non-contractile material, such as connective tissue and fat.

Maximum control of skilled movements

With the object of increasing the precision and speed of skilled work, many authors since World War II have made detailed studies of several typical movements of the hand and forearm. The most important of their results include the following:

*Position of
the arms*

Ellis [79] and *Tichauer* [307] studied how the height of the working surface affected manual performance. *Ellis* was able to confirm an old empirical rule: *the maximum speed of operation for manual jobs held in front of the body is achieved by holding the elbows down to the sides, and the arms bent at right angles.* In a similar situation in practice, *Tichauer* [307] studied 12 female packers of foodstuffs, to find out how the position of the upper arms affected performance and working metabolism. The results, reproduced in *Figure 78*, showed that the best performance was achieved with the forearms held out sideways to make an angle of 8—23° with the vertical.

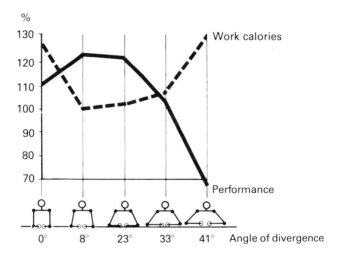

Fig. 78 **Performance and metabolism for various attitudes of the upper arms (angles of abduction) when packing groceries.** After *Tichauer* [307].

In another study *Tichauer* [308] found that, if the arms were held out as much as 45° to the sides, the shoulders took up a balancing posture, causing fatigue in the shoulder muscles. The need to do this was often a result of too low a seat.

Criteria often used to discover the optimum sequence of movements were the time consumed and the degree of precision; modern ergonomists would say that not enough attention was given to signs of fatigue and to monotony. With this reservation, the most important findings are summarised below.

Time and motion   Time studies show first of all that reaction time (interval between the signal and the beginning of the motor reaction) has a fairly constant value of 0.25 s, which is hardly affected by either the nature of the movement or its distance from the body. Hence the grasping time and grasping distance are not quite linearly related, as *Barnes* [14] had already shown in 1936; thus:

Grasping distance (cm)	Grasping time (%)
13	100
26	115
39	125

Other studies by *Brown* and his colleagues [41] [42] showed that turning movements with the right hand were quicker from right to left than from left to right. If the distance was increased from 10–40 cm, the time for the movement increased from 0.7–0.95 s.

Precision of movement   *McCormick* [220] reported some unpublished work by *Briggs*, showing that the precision of hand movements is increased if the objective of the movement is 60° diagonally forwards (measured from a horizontal line in front of the body). Furthermore, outward movements were more precise than inward movements, and precision fell as the distance of the movement increased. In similar fashion *Schmidtke* and *Stier* [281] were able to show that the speed of a horizontal movement with the right hand is greatest in a direction 45° to the right, measured from a horizontal line in front of the body, and slowest at 45° to the left. These findings of *Briggs* [220] and *Schmidtke* [281] show that a hand operation can be performed faster and more precisely if most of the movement comes from the forearm. The movement is both slower and less precise if the upper arm participates in it.

Optimal working field   Experimental studies of the speed and precision of manual operations should be useful when deciding the layout of controls in vehicles of all kinds, including aircraft, but they are less significant when it comes to most industrial work places, since skilled operations involved are much more complex. The work of *Bouisset* and his co-workers [31] [33] is more relevant to these. These authors made their research subjects shift weights up to 1 kg, and studied the best conditions so far as physical stress was concerned. Oxygen consumption and electrical activity were measured in several muscles of the arm and trunk, both for one-handed and two-handed action. The weight had to be moved over a distance of 30 cm, in different directions, starting from a point immediately in front of the centre of the body, and repeated 24 times/min. The oxygen consumption for various directions is shown in *Figure 79*.

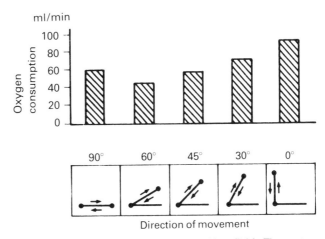

Fig. 79 Energy expenditure and working field. The test persons must move weights of 1 kg repeatedly in the given directions. The direction of 60° is the most efficient, provided that the movement does not have to be controlled by eye. After *Bouisset* and *Monod* [33].

The lowest oxygen consumption corresponds to the direction of 60° from the front, which agrees with the findings of *Briggs* [220] and *Schmidtke* [281]. This was also the best angle for two-handed action, for all weights tested, and for all the different frequencies.

Electromyography explained these findings. Electrical activity in various muscles was distinctly less when the arms were held out to the side, than when they were held forwards, the latter setting up heavy static loads in the arms, shoulders and trunk.

These optimal angles cannot, however, be applied in practice without reserve. A control placed at an angle of 60° is ideal only when there is no need to look directly forwards at the same time. When forward vision is involved, or when both arms are used, angles of only a few degrees right or left are to be preferred.

Typing

A form of work that is certainly skilled is to use a typewriter, or any other equipment fitted with a keyboard. The design of keyboards, and the planning of typing work must take into account the maximum speed at which the fingers of the two hands can operate the keys. According to *McCormick* [220] it seems that most people can manage up to five strokes per second.

Special importance attaches to the position of the forearms and hands. Typewriters usually require the hands to be turned slightly outwards so that the fingers are parallel to the lines of keys. This requires a tiring static effort in the muscles of the forearm that are involved, and must be the main cause of the inflammation of the tendon sheaths that results from prolonged periods of typing.

Lundervold [205] carried out detailed electromyographic studies, by which he showed that the working height is of critical importance. If the hands are higher than the elbows, muscular activity in the forearms and shoulders is greatly increased, whereas with the hands slightly below elbow level, only the hands and fingers are active. This finding is in general agreement with the observation of *Grandjean* and *Burandt* [107], that pains in the neck and shoulders occurred with more than average frequency among typists whose table was too high (see *Chapter 5*). These facts have prompted several ergonomists to redesign typewriters or punched-card equipment [344], but unfortunately these designs have not yet been adopted.

Handgrips

The design of hand grips has a high priority in skilled work. Hand grips which are not shaped to fit the hand properly, or which pay too little attention to the biomechanics of manual work, may lead to a poor performance, and may even be injurious to the health of the operator. The hand and fingers are capable of a wide range of movements and grasping actions, which depend partly on being able to rotate the wrist and forearm. The most important of these many movements are illustrated in *Figure 80*.

  Finger-tips   7–14 kp

  Thumb and sides of fingers   7–14 kp

  Clasping in thumb and fingers 30–54 kp

Fig. 80 Three ways of grasping by the hand. The figures indicate the range of finger-pressure, according to *Taylor* [301].

The maximum grasping force can be quadrupled by changing over from holding with the fingertips to clasping with the whole hand. The power of the fingers is at its greatest when the hand is slightly bent upwards (dorsal flexion). In contrast, grasping power, and consequently level of skilled operation, is reduced if the hand is bent downwards, or turned to either side.

Tichauer [308] says that inclining the hand either outwards (ulnar direction) or inwards (radial direction) reduces rotational ability by 50%, and if the hands are held in such positions every day and frequently, this may cause inflammation of the tendon sheaths. This possibility was mentioned in *Chapter 1*, where *Figure 9* showed an example of using a pair of wire-cutters at work. The conclusion from this is that, on biomechanical grounds, *the hands should always be kept in line with the forearms as much as possible.*

Design of handgrips   The work of *Barnes* [15] can be taken as a basis for the design of hand grips for the whole hand, with a cylindrical shape as the best. *This should be at least 100 mm long, and its effectiveness increases with thickness up to 30–40 mm.*

It is not possible to enumerate all the range of different tools that should be designed to suit the shape of the hand, but *Figure 81* shows a few examples of good design.

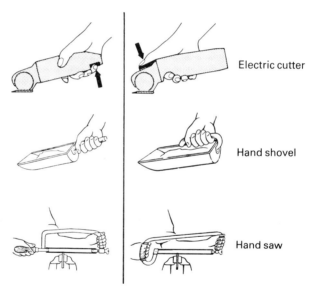

Electric cutter

Hand shovel

Hand saw

Fig. 81   Handgrips sensibly shaped to the anatomy and functioning of the hand. Left bad; right good.

Facilitating skilled work

A threefold process   Every skilled operation involves three distinct processes:
(a) *perception of signals;*
(b) *alertness;*
(c) *consequential movements.*
Maintaining one's alertness over long working periods is the most arduous of these three, as well as being the key to the other two: not missing any signals, nor making any wrong movements.

We know from everyday life that it is a natural human reaction to automate one's movements as much as possible. Think how difficult walking would be if we had to take each step by making each muscular contraction deliberately!

Automation of some actions means that the whole of one's alertness can be reserved for the highly skilled movements.

Guidelines

Table 20 summarises briefly the most important ways in which skilled work can be made easier.

Table 20 Facilitation of skilled work.

Process	Requirements
Perception	Good visual control of work. Optimal display of information, and output. Lighting and colour adequate.
Alertness	Protection against distractions, including noise. Work place in full view, and sensibly arranged.
Movements	Ergonomic arrangement of working field, with working movements as economic as possible. Make one movement at a time, as far as possible, and in rhythmical sequence.

Some of these requirements are discussed in detail elsewhere, in the relevant chapters, and the following comments are limited to the requirements of skilled work only.

Improving perception

Skilled operations depend primarily on visual control, and the following recommendations are specially important:

(a) *every movement should be in full view of the operator.* Control levers, switches, hand-wheels and recording instruments should be set out in such a way that they can be seen and operated without moving the body from its normal position;

(b) *colour, shape and illumination of controls should be such that they are easily seen and interpreted, using optical aids if appropriate;*

(c) *small items should contrast strongly in colour with their immediate surroundings.* Big areas of brightness, such as windows should not come into the working field of view, because they are dazzling. Sometimes it is necessary for the operator to notice acoustic signals or some particular sound; this is easier if background noise is 10 dB less loud than the sound to be detected.

Improving alertness

The biggest distraction is other people's conversation and noise in general. Nemecek and *Grandjean* [243] carried out a survey in

open-plan offices, and found that of the 519 persons questioned:
(a) 35% were badly disturbed by noise;
(b) 60% had difficulty in concentrating.

The cause of distraction in 65% of the cases was noise, including talking. The importance of noise and conversation as distractions is emphasised by the fact that the above example relates even to unskilled work. Hence the recommendation that *only one kind of skilled work should be carried out in one room*, because different kinds of work are distracting to one another.
Visual distraction may be caused by passers-by, and to a lesser extent by moving parts of machinery, by reflections from these, or just by garish colours in the surroundings. So a working place needs to be protected from distraction optically as well as acoustically.

Rules for maximum skill

Some of the important conditions for the maximum control of skilled movements have already been discussed earlier in this Chapter. The following ten points summarise ways of making skilled work easier to perform:
(a) the working field must be so arranged *that manual operations can be performed with the elbows lowered, and the forearms at an angle of 85–110°* (see also *Figure 39*);
(b) *for very delicate work the working field must be raised up to suit the visual distance*, the elbows lowered, head and neck lightly bowed, and the forearm supported (in this context see the seven guidelines given in *Chapter 3*, and also *Figure 22*);
(c) *skilled operations should not call for much force to be exerted*, since heavily loaded muscles are more difficult to control and to coordinate with others. Above all, avoid imposing a static stress at the same time. It is equally bad to have to exercise skill immediately after physical effort. These recommendations are illustrated in *Figures 82* and *83*;

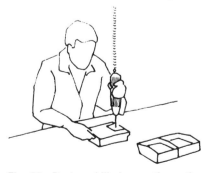

Fig. 82 **During skilled operations, the muscles should be kept free from having to exercise force, and especially from static effort.** The electric screwdriver is suspended from a spring support which reduces manual effort to a minimum.

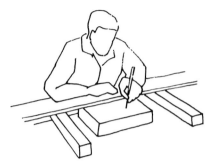

Fig. 83 A good example of relieving muscular tension during skilled work. Suitable supports for the forearms make the precision work of the lithographer much easier.

(d) *concentration on a manual operation is improved if it is not necessary to do other things at the same time*. Hence it is helpful to support the work, and if possible to have foot pedals with which to clamp and unclamp it, and to switch the machine on and off.

Chutes are useful for taking finished work away and delivering fresh items;

(e) the arrangement of working materials, parts and controls should be such as to *allow the operations to follow rhythmically in a sensible sequence*;

(f) *a free rhythm is better than any kind of imposed tempo, whether it is time-controlled, or cyclical, or on a conveyor belt.* A free rhythm consumes less energy (because there are fewer secondary movements), motor control is easier, fatigue is reduced, and monotony and boredom less frequent. It should be mentioned here that too slow a rhythm is bad because the work has to be supported, and too fast a rhythm is even worse, because of the nervous stress it imposes, and the fatigue it engenders. The operator usually finds his own rhythm instinctively, to suit his liking;

(g) when both hands are used in the work, the working field should extend as little as possible to each side, to give the best visual control. The muscular effort should be symmetrical, i.e. as nearly as possible equal for the two hands, which should begin and end each movement together;

(h) *movements of the forearms and hands are at their most skilful, both in speed and precision, if they take place within an arc of 45–50° to each side*. Both for grasping things and for working, their optimum arc is two-thirds of their maximum reach, i.e. over a radius of 35–45 cm from the tip of the lowered elbow;

(i) horizontal movements are easier to control than vertical ones, and circular movements than zig-zag ones. Each operation should end in a good position for starting the next one;

(j) *handles of controls and tools should be shaped to fit the hand*, and should operate when the hand is held in line with the forearm.

9
"Man–machine systems"

A "man–machine system" means that the man and his machine have a reciprocal relationship with each other. Figure 84 shows a simple model of such a system.

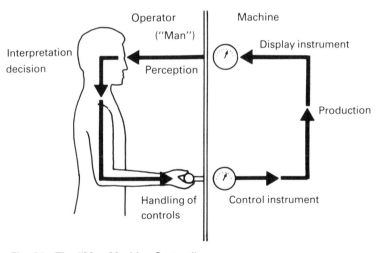

Fig. 84 The "Man-Machine System".

Obviously such a system is a closed cycle in which the man holds the key position because the decisions rest with him. The pathways of information and direction are, in principle, the following: *the recording display* gives information about the progress of production; the operator absorbs this information visually (*perception*), and must understand and assess it correctly (*interpretation*). On the strength of his interpretation, and in the light of his previous knowledge, he takes a *decision*. The next step is to communicate this decision to the machine by *using the controls. A control display* tells the operator the result of his action (e.g. how much water has been mixed in with the reagents). The machine then carries out the *production process* as programmed. The cycle is completed when various significant parts of the process, such as temperature or quantities, are displayed for the operator to see.

In a report of the World Health Organisation, *Singleton* [291] points out that the machine is capable of high speed and precision, as well as being very powerful, whereas the man is sluggish, releasing only small amounts of energy; on the other hand, he is much more flexible and adaptable. Man and machine can combine to form a very productive system, provided that their respective qualities are sensibly used.

The control of machines was no great problem until recently, when the development of electronics, more elaborate controls and higher output, with the consequent need for accurate interpretation of the information displayed, made the operator's task both more delicate and more demanding. As a result, the "human factor" in such systems became increasingly important. In an aircraft the speed of the pilot's, or the engineer's reaction can be vital; in chemical processes alertness and correct decision-taking may alone avert catastrophe. So modern "man-machine systems" need to be ergonomically sound.

The "points of interchange" *from man to machine* and *from machine to man* – "interfaces", in the jargon of electronic data-processing — are of paramount importance. The ergonomic "interfaces" of "man–machine systems" are therefore:

(a) *perception* of all the information on display;
(b) *manually operating the controls.*

In the next few subsections we shall consider the following problems from an ergonomic point of view:

(a) the physiology and psychology of visual and acoustic perception;
(b) display of information;
(c) design of controls.

Visual perception

Perception

The word "perception" will be used a good deal in this and the next subsection. It means the senses of seeing and hearing, the associated sense organs, and their interpretation in the cerebral cortex as a personal experience.

This personal experience may be understood either as an *interpretation of the sense data* themselves, or as *the formation of pictures of the outside world with their aid.*

The eyes and ears, acting as receptor organs, pick up energy from the outside world in the form of light waves and sound waves, and convert this into a form of energy that is meaningful to a living organism, i.e. into bioelectric nerve-impulses. It is only through integration by the brain of sensory impulses from outside the body that we are "aware of the existence of the outside world". If the afferent sensory nerves are cut, we become blind and deaf.

Perception in itself does not give us an exact photocopy of the world outside. *Our impressions are a subjective modification of what we have perceived.* Thus:

(a) a particular colour seems darker when we see it against a lighter background than against a darker one;

(b) a straight line appears distorted against a background of curved or radiating lines;

(c) a steady sensation over a long period becomes gradually weaker, until it ceases to register (the mechanism of adaptation and habituation involved here will be discussed later);

(d) individual variations in interpretation of sense data may be critical in certain situations, for example in reading from a display. People differ in experience, in their attitude to their work, and in their preconceived ideas;

(e) people also differ greatly in the intensity with which they react to sense data. There is an element of artistic creativity in everyone.

A discussion of the multiplicity of factors that can modify sensory perception is outside the scope of this book. The ergonomically important factors are those concerned with handling the controls and display equipment.

In spite of all these effects, we can say that perception gives us an accurate enough picture of the outside world for everyday purposes, and supplies us with enough information to react suitably to it.

Vision

The successive stages of seeing are as follows: light rays from the object pass through the iris of the eye and fall on the retina. Here the light energy is converted into the bioelectric energy of a nerve stimulus which then passes as a nervous impulse along the nerve fibre to the brain. At the first series of intermediate neurones, new impulses are generated which branch off to the centres which control the eyes, varying the width of the pupil, the curvature of the lens, and the movements of the eyeball. There is a feedback of accumulated "information" between eye and brain, which keeps the eyes continuously directed at the object, and this is automatic, not under conscious control. At the same time the original sensory impulses travel further into the brain, and after various filtering processes end up in the cerebral cortex, the seat of consciousness.

Here the signals from outside are integrated into a picture of the external world. Here, too, arise new impulses which are responsible for coherent thought, decisions, feelings and reactions. These processes are shown diagrammatically in *Figure 85*.

The visual apparatus

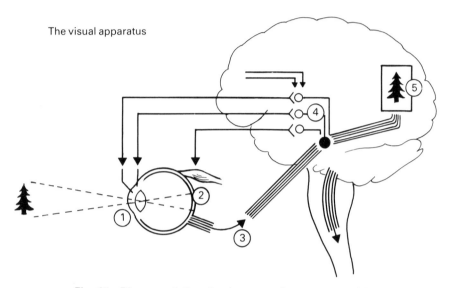

Fig. 85 Diagram of the visual system. 1 = cornea and lens. 2 = light received on the retina. 3 = transmission of optical information along the optic nerve to the brain. 4 = synapses and feedback to the eye. 5 = visual perception of the external world in the conscious sphere of the brain.

In reality, the essential processes of vision are nervous functions of the brain; the eye is merely a receptor organ for light rays. The complete assembly of organs and nervous structure which contribute to vision is called the optical apparatus, which controls 90 % of all our activities in everyday life. It is even more important in a great many precise jobs in industry and commerce. If we consider the very many nervous functions that are under stress during seeing, then it is not surprising that the eyes are a common and important source of fatigue.

The eye

The principal parts of an eye are shown in *Figure 86*. The eye has many parts in common with a photographic camera: the retina corresponds to the light-sensitive film, while the transparent cornea, the lens and the pupil with its variable iris represent the optics of the camera. Cornea and lens together refract the incoming rays of light and bring them to a focus on the retina.

The actual receptor organs are the visual cells embedded in the retina, consisting of "cones" for daylight vision, and specially sensitive "rods" for vision in dim light. The visual cells convert

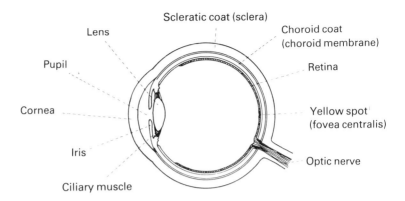

Fig. 86 Diagram of the eye in longitudinal section.

light energy by photochemical reaction into nervous impulses, which are then transmitted along the optic nerve. The human eye contains about 130 million rods and 7 million cones, each of which is approximately 0.01 mm long and 0.001 mm thick. On the posterior surface of the eye, a few degrees away from the optical axis, is the retinal pit, or *fovea centralis*, recognisable by having a thinner covering than the surrounding area. Because of this thin covering, the light rays pass directly to the visual cells, which at this point consist entirely of cones, here at their maximum density of about 10 000 per mm². Each foveal cone has its own fibre connecting it to the optic nerve. For these reasons the vicinity of the fovea has the highest resolving power of any part of the retina, up to about 12 seconds of arc. Since vision is most acute in the region of the fovea, it is instinctive to look at an object closely by turning the eye until the image falls upon this area of the retina, which is called the area of central vision.

Any object that is to be clearly seen must be brought to this part of the retina, which covers a visual angle of only 1°.

Outside the foveal area the cones are considerably fewer, and one nerve fibre serves several cones and rods. Rods are distinctly more abundant than cones, and they become even more numerous as the angle from the fovea increases, while the number of cones declines. Although rods are more sensitive to light than are cones, they do not detect such fine differences of either shape or colour. They become the more important light-detecting organs in poor visibility and at night.

So only objects focused upon the fovea are seen clearly, and other images become progressively less distinct and blurred as distance from the fovea increases. Normally the eyes move rapidly about, so that each part of the visual field falls on the fovea in turn, allowing the brain to build up a sharp picture of the whole surroundings.

The visual field

The visual field is that part of one's surroundings that is taken in by the eyes when both they and the head are held still. Only objects within a small cone of 1° apex are focused sharply. Outside this cone objects become progressively more blurred and indistinct. If we keep our eyes still when reading we can focus only a few letters; in practice we move our eyes in jerks, taking in about 12 letters at a time.

As shown in *Figure 87*, the visual field can be divided up as follows:

(a) area of distinct vision: vertical angle 1°
(b) middle field: vertical angle 40°
(c) outer field: vertical angle 40°–70°.

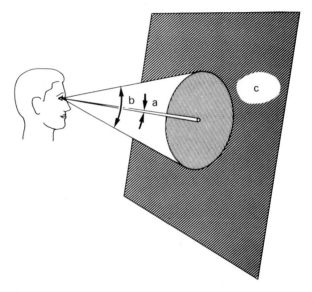

Fig. 87 Diagram of the visual field. a = zone of sharp vision: angle of view of 1°. b = middle field: vision unsharp; angle of view from 1° to 40°. c = outer field: movements perceptible: angle of view from 41° to approximately 70°.

Objects in the middle field are not seen clearly, but movement and strong contrasts are noticed: alertness is maintained by quickly shifting the gaze from one object to another. The outer field is bounded by forehead, nose and cheeks, and objects in this area are not noticed unless they move.

Accommodation

Accommodation means the ability of the eye to bring into "sharp focus" objects at varying distances from infinity down to the nearest point of distinct vision (the near point). If we hold up a finger in front of the eye, we can focus sharply on the finger, leaving the background blurred, or focus the background sharply, leaving the finger indistinct. This is a demonstration of the phenomenon of accommodation.

An object is clearly seen only when refraction through the cornea and lens produces a tiny but sharp image on the retina, the three components forming an optical system. Focusing on near objects is achieved by changing the curvature of the lens, by contraction of the muscles of accommodation (the ciliary body, or circular ciliary muscle).

When the ciliary muscle is relaxed, and the mechanism of accommodation is in its resting position, the refraction of the cornea and lens is such that parallel rays from distant objects are focused on to the retina. In the absence of accommodation, the image of an object nearer to the eye would fall behind the retina, which would receive a blurred impression. The ciliary muscle compresses the lens so that it bulges just the right amount to bring the sharp image back into the plane of the retina.

When attention is allowed to wander over distant objects, the eyes are focused on "infinity", and the ciliary muscles are relaxed. The eyes are being rested. It is quite possible that blue and green colours are restful because they give an impression of distance.

The size of the pupil is also a factor in accommodation. Narrowing the pupil reduces the optical aberrations, which are greatest towards the periphery of the lens, and thereby sharpens the image. The pupil is narrowed when the eye is focused on near objects, and relaxed for distant vision.

So the nearer the object focused upon, the greater is the load on the ciliary muscle. The shortest distance at which an object can be brought into sharp focus is called *the near point*, and the furthest away is *the far point*. The near point depends on the power of the ciliary muscle to compress the lens, and moves further away as the muscle becomes fatigued after a long spell of close work.

Age and accommodation

Age has a profound effect on powers of accommodation, because the lens gradually loses its elasticity. As a result, the near point gradually recedes, while the far point usually remains unchanged.

The average distance of the near point at various ages is as follows:

(a) at 16 years: 8 cm
(b) at 32 years: 12.5 cm
(c) at 44 years: 25 cm
(d) at 50 years: 50 cm
(e) at 60 years: 100 cm.

When the near point has receded beyond 25 cm the condition is called presbyopia, the long-sightedness of age. Spectacles can be prescribed which will bring the near point back to a more convenient distance and so restore normal vision. It is particularly important to do this before undertaking close and delicate work. Another effect of age is to slow down the speed with which the eye accommodates itself.

Lighting and accommodation

The level of illumination is a critical factor in accommodation. When the lighting is poor the far point moves nearer, and the near point recedes, while both speed and precision of accommodation are reduced. Contrast is important, too; the more the object stands out against its background, the quicker and more precise the accommodation.

Aperture of the pupil

Just as the iris diaphragm of a camera is used to avoid under- or over-exposure, so in the human eye the aperture of the pupil (iris) is under reflex control to protect the retina. As more light falls on the retina, nervous impulses pass along the optic nerve to one of the switching centres (synapses) in the brain that were mentioned earlier, and from there to the centre which controls the motor nerve which causes the iris to contract. This is a reflex arc, not under conscious control. The adjustment of the aperture of the pupil takes a measurable time, which may vary from a few tenths of a second to one second. If the level of lighting changes frequently and suddenly, there is a danger of over-exposure, since the reaction time of the pupil is comparatively slow.

This reflex mechanism also adjusts the aperture of the pupil to suit a steady level of lighting: during daylight the aperture may have a diameter of 3–5 mm, increasing at night to 7–8 mm.

The aperture of the pupil is also affected by two other factors:

(a) the pupil contracts when near objects are focused, and opens when the eye is relaxed;
(b) the pupil reacts to emotional states, dilating under strong emotions such as alarm, joy or pain, intense concentration or deep thought; narrowing under fatigue and sleepiness.

Under normal conditions, however, the general level of lighting is the dominant factor, the others being of secondary and comparatively minor importance.

Adaptation of the retina

If we look at the headlights of a car at night we are dazzled, but the same headlights do not dazzle us so much in daylight. If we go from daylight into a darkened cinema, where the film has already started, we can see very little at first, but after a few minutes objects gradually become visible. These are everyday examples of how the sensitivity of the retina is continuously adapted to the prevailing light conditions. This sensitivity is many times greater in darkness than in daylight.

The process is called *adaptation*, and comes about through photochemical and nervous regulation of the retina. Thanks to this facility we can see equally well in moonlight and in the brightest sunlight, even though the level of illumination has increased more than 100 000 times.

Dark adaptation

Adaptation to darkness (dark adaptation) or to brightness (light adaptation) takes up a comparatively long time, and the greater the difference in light intensity, the longer the time taken. Going from daylight into a very dark room the rate of adaptation is very quick in the first five minutes, afterwards becoming progressively slower; 80 % adaptation takes about 25 minutes, and full adaptation as much as one hour. Hence sufficient time must always be allowed for dark adaptation, at least 25–30 minutes for good night vision.

Light adaptation

Light adaptation is quicker than dark adaptation. The sensitivity of the retina can be reduced by several powers of ten in a few tenths of a second. Yet light adaptation, too, continues over a measurable time, of the order of several minutes. Two phases can be distinguished. The first phase (α-adaptation) lasts only 0.05 s, during which retinal sensitivity is abruptly reduced to one-fifth of its original value. This first phase is obviously a nervous reflex. The second phase (β-adaptation) progresses more slowly and, like dark adaptation, it clearly involves a change in the balance between breakdown and reconstitution of the photosensitive materials of the retina.

The abrupt reduction of sensitivity during light adaptation affects the entire retina. Whenever the image of a brightly lighted surface (a window, a light source, or a bright reflection) falls on to any part of the retina, sensitivity is reduced all over, including (and this is most important for precision work) the *fovea centralis*. This phenomenon is illustrated in *Figure 88*.

Partial adaptation

If the visual field contains a dark or bright area, adaptation will occur in the corresponding part of the retina, a local or partial adaptation. There is, in addition, an effect on the whole retina, including the fovea.

Furthermore, adaptation of one eye has some effect on the other, a fact that may be significant at working places where only one eye is employed. It can be turned to advantage by closing one eye when both are dazzled by on-coming headlights. The closed eye retains some of its dark adaptation and recovers quickly afterwards.

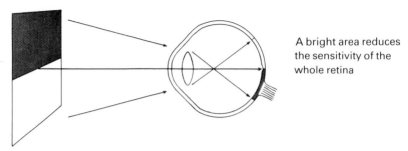

A bright area reduces
the sensitivity of the
whole retina

Fig. 88 Effects of bright and dark surfaces on the retina. A bright patch reduces the sensitivity of the entire retina, and thereby reduces the visual acuity in the fovea. This form of disturbance is called relative glare.

Two general principles are based on these arguments:
(a) *all surfaces within the visual field should be of the same order of brightness, to avoid dazzle effects;*
(b) *the general level of illumination should not fluctuate rapidly because adaptation is a relatively slow process.*

Distance perception

Binocular vision requires the optical axes of the two eyes to meet in the object being looked at, so that the image falls on the corresponding part of the retina in each eye. If the gaze is shifted to a second object, further away, the angles of the two eyes must be opened out until the optical axes again cross through the object. This movement is brought about by activity of the outer eye muscles, and is a very delicate adjustment, upon which distance perception depends. The necessary sensitivity is gradually developed in infancy, until we have finally learned by experience to estimate distance from the angular convergence of our two eyes.

With only one eye (monocular vision) distances must be guessed from the apparent size of objects, foreshortening by perspective, cast shadows, and apparent speed of movement, if appropriate.

Glare

Physiologically speaking, *glare is a gross overloading of the adaptation mechanism of the eye, brought about by "overexposure" of the retina to light.* Three types of glare (dazzle) may be distinguished:

(a) *relative glare*, caused by excessive contrast between different parts of the visual field;
(b) *absolute glare*, when a source of light is so bright (e.g. the sun) that the eye cannot possibly adapt to it;
(c) *adaptive glare*, a temporary effect during the period of adaptation: e.g. on coming out of a dark room into daylight outside.

Practical hints

In this context the following hints are important in the layout of work places:

(a) the shorter the period of dazzle, the quicker the recovery. Isolated changes in lighting lasting less than one second have little effect (e.g. a short flash on the headlights during night driving). On the other hand, a succession of quick changes is very disturbing;

(b) the effect of relative glare is greater the nearer the source of dazzle is to the line of sight, and the bigger its area;

(c) a bright light above the line of sight is less dazzling than one below, or to either side;

(d) the risk of dazzle is appreciably greater if the general level of illumination throughout the visual field is low, since the retina is then at its most sensitive: a window is less dazzling when the room is brightly lighted; a headlight is not dazzling in daytime.

Visual capacity

The various functions of the eye are not usually pushed to the limits of their performance in everyday life, but this may sometimes occur in industry or under modern traffic conditions.
The most important functions are:

(a) acuteness of vision;

(b) sensitivity to contrast;

(c) speed of perception.

Acuteness of vision

Acuteness of vision is the ability to see the finest details of objects and surfaces, the visual separation of points lying close to each other, and the appreciation of form and shape. By and large, this is a function of the resolving power of an optical system. Since the retinal light-receptors quickly tire, the eyes must make continual small movements of about one minute of arc (fixational restlessness) to preserve normal acuteness of vision.

To measure acuteness of vision, opticians use a series of symbols (letters, numbers or circles of various sizes), the thickness of the lines being so designed that at the appropriate distance they subtend an angle of less than one minute of arc at the eye.

Then the minimum distance between two points in the image on the retina is 5μ,[1] which covers about four photoreceptors.

Acuteness of vision is related to illumination and to the nature of the objects being seen, as follows:

(a) acuteness increases with the level of illumination, reaching a maximum at 5000 asb.[2] The increase is 150% between 1 and 5000 asb;

(b) acuteness increases with the contrast between the test symbol and its immediate background. The effect is greatest at the low end of the contrast scale;

[1] $1\mu = 10^{-6}$ m.
[2] For a definition of an apostilb (asb) see *Chapter 15*.

(c) acuteness is greater for dark figures on a light background than for the reverse.

Sensitivity to contrast

Sensitivity to contrast means the ability to perceive the smallest differences in luminance, and thus to appreciate niceties of shading, small variations in tone, and the smallest nuances of brightness, all of which may be decisive in deciding about shape and form, i.e. for vision in depth. Sensitivity to contrast may be even more important than acuteness of vision in many jobs of inspection and control.

Sensitivity to contrast is subject to the following rules:

(a) it is greater where big areas are involved than for small ones;
(b) it is greater over sharp boundaries than when the change is gradual or indefinite;
(c) it increases with the luminance of the surroundings, and is greatest within the range 200–10 000 asb;
(d) it obeys the Weber–Fechner law, and within the range given in c) a contrast equal to about 2 % of the surrounding luminance can be detected;
(e) it is greater when the outer parts of the visual field are darker than the centre, and less in the reverse contrast. Maximum values are 1200–1500 asb in the centre and 100–300 asb in the periphery.

Figure 89 repeats findings from studies by *Luckiesh* and *Moss* [203], from which it appears that raising the luminance from

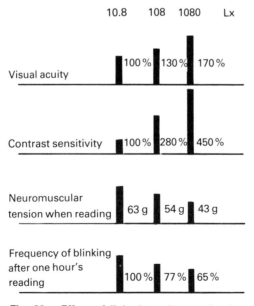

Fig. 89   Effect of light intensity on visual acuity, contrast sensitivity, nervous strain, and frequency of blinking the eyelids. After *Luckiesh* and *Moss* [203].

approximately 10 lx to 1000 lx increases acuteness of vision from 100–170 % and sensitivity to contrast by 450 %. At the same time the investigators recorded a decrease of muscular tension (measured from the continuous pressure of a finger on a key) and of the rate of blinking of the eyelids. This was interpreted as a reduction of nervous tension as a result of better lighting.

Speed of perception

The speed of perception is the time interval that elapses between the appearance of an object in the visual field and its perception by the brain. This speed increases with improved lighting and increased contrast between the object and its surroundings, as do acuteness of vision and sensitivity to contrast. This means in practice that improving acuteness of vision by better lighting and better contrast also gives greater sensitivity to differences in contrast, and more rapid perception, and vice versa.

Speed of perception is commonly measured by the technique of tachistoscopy. This apparatus presents the test subjects with a series of words or pictures for a fraction of a second, and measures either the speed of perception or, conversely, the minimum display time required for the object to be correctly identified. Equipment such as this is used in most research to find the best ways of presenting information, either a printed symbol or visual display.

Speed of perception can be vital in transport. We need only think of an airliner flying at the speed of sound, and how much can happen during a perception time of 0.2 seconds, a common figure. But perception speed is also important during the reading of printed matter. During reading, the eye makes a series of little "jumps", taking in and fixing several letters, or even one or two words, at a time. According to *Dubois–Poulsen* [71] the following times are normal:

(a) fixed gaze between jumps: 0.07–0.3 s
(b) jumping time: 0.03 s
(c) transition to next line: 0.12 s
(d) number of jumps per line: about 6.

Reading speed is greatly affected by level of lighting and, by contrast, between letters and background, but speed can be increased only up to the maximum speed of perception.

Perception of sound

Sound perception

The physiological processes of perception of sound are essentially the same as those already discussed for visual perception. In this case the inner ear provides the "interface" at which sound-waves are converted into nervous impulses along the auditory nerve. *The actual perception of sound is the integration and interpretation of these sensory impulses in the brain, or more precisely in the auditory cortex.*

Perception of sound is not a faithful reproduction of the whole band of frequencies, which is simply "played" in the brain. This fact is especially important in people's reaction to noise, which varies greatly from person to person. What is noise to one may be music to another. Another example of varying perception is the assessment of loudness in relation to pitch. In practice low-pitched sounds seem less loud than shrill ones, even though the energy content may be the same.

Sounds

Any sudden mechanical movement sets up fluctuations in the air pressure which spread out as waves, just like waves when water is stirred. As long as these variations of pressure occur with a regular frequency and intensity, the human ear reacts to them as sounds. *The extent of pressure variation is the sound pressure, and this determines the intensity of the sensation.*

The frequency of a sound is the number of fluctuations or vibrations per second, expressed in herz (Hz), subjectively perceived as pitch. Most noises contain a mixture of sounds of different frequencies; if high frequencies predominate, we regard it as a high-pitched noise, and vice versa.

The decibel

The physical unit of *sound pressure is the microbar* $(1\mu b = 10^{-6}\text{ b})$. The human ear registers sound pressures over a very wide range between $2 \times 10^{-4}\mu b$ and approximately $200\mu b$, which encompasses every sound from the gentle murmur of a stream to the scream of a jet engine.

To accommodate such a wide range with a practical scale, *a logarithmic unit, the decibel (dB)*, was devised.

A difference of 20 dB means that two sound pressures are in the ratio 10:1; in other words, raising the sound pressure by a factor of ten is recorded as an increase of 20 dB.

Sound pressures are recorded logarithmically by using the sound level L (sound intensity level), according to the formula:

$$L_{dB} = 20 \log \frac{p_x}{p_o} \text{ (p measured in microbars)}$$

where L_{dB} = sound level in dB

p_x = sound pressure in μb

p_o = basic sound pressure, internationally fixed at $2 \times 10^{-4}\mu b$ (10^{-16} watts/cm²)

Pitch and loudness

As we have seen, the apparent loudness of a sound depends a good deal on its pitch or frequency. The human ear is sensitive to sounds in the frequency range from 16 to 20 000 Hz, a span of nearly nine octaves. Sounds pitched below 16 Hz (infrasonic) are perceived as vibrations: above 20 000 Hz they are ultrasonic and are used therapeutically in medicine. *Low-pitched sounds seem much less loud than high-pitched ones.* This is shown very clearly by the curve of auditory threshold for different frequencies, the lowest curve in *Figure 90*. This curve shows that the greatest sensitivity lies in the range 2000–5000 Hz, although the pitch of the human voice is much lower than this, mainly between 300–700 Hz.[1]

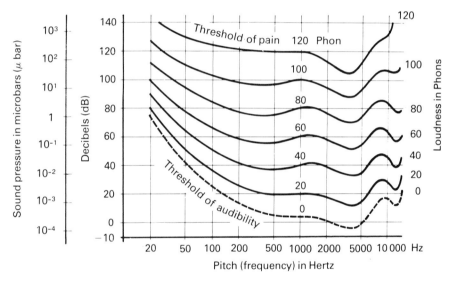

Fig. 90 **Sound pressure (in microbars), sound level (in decibels) and the curves of equal subjective loudness (in phons).** The lowest curve indicates the threshold of audibility, the least sound level that can be perceived. After *ISO-Norm (1957).*

Curves of equal loudness

As long ago as 1933, *Fletcher* and *Munson* [82] had plotted curves of equal loudness in relation to sound pressure and frequency. For this purpose they worked from an arbitrary base of 1000 Hz and tones of higher and lower frequencies, determining the sound pressures necessary to give the research subjects the impression of equal loudness. In this way they obtained curves of equal subjective loudness, which were afterwards converted to a scale of phons. *Robinson* and *Dadson* [262] carried these studies further, and their results have been incorporated in the ISO-Norm since 1957. They are displayed in *Figure 90*.

---

[1] Most vowels are pitched below 1000 Hz, but consonants may be pitched up to 10 000 Hz, especially if they are sibilant.

These curves of equal loudness, on the scale of phon values, are valid only for pure tones; they no longer agree with subjective impressions of loudness if the noise includes many different frequencies. Since nearly all noises, and many signal sounds, are a mixture of frequencies, the use of phon values as a measure of loudness has become obsolete. Nevertheless these curves of equal loudness still retain their value for assessing the effects of different frequency ranges on the human ear.

Weighted noise level

Nowadays the so-called weighted sound levels have come into use as measures of loudness. Weighted sound level is essentially a process of filtering out the sound energy in the lowest and highest frequencies, at which, according to the curves of equal loudness, sensitivity is less. Hence sound pressure has less significance in these frequency ranges. The term weighted noise level has a similar derivation. *Figure 91* sets out three weighted curves in dB, (A), (B) and (C), that are in current use.

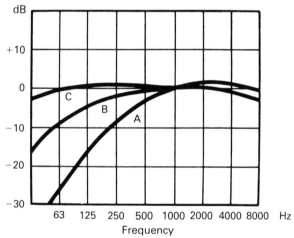

Fig. 91   Curves showing the relationship between dB(A), dB(B) and dB(C). They show how much of the sound energy is filtered out in each frequency range.

The curve of weighted sound level in dB(A) is most often used, because many psychological studies have shown that *noise levels measured in dB(A) give a reliable assessment of subjective disturbance from noise.*
Other units of measurement, which are useful in particular circumstances, are discussed later, in *Chapter 16.*

Hearing organs

A sensation of hearing is produced when sound waves pass through the external auditory passages, into the inner ear, where the energy from sound pressure is converted into nervous impulses. These travel along the auditory nerve to the brain, where the sound is "heard".

The most important components of the hearing organs are shown in *Figure 92*. Sound waves cause the eardrum or tympanic membrane to vibrate, and these vibrations are transmitted by the auditory ossicles to the inner ear. Here fluids (the perilymph and endolymph) transmit the vibrations along the cochlea and back to the round window. The so-called basilar membrane divides the cochlea longitudinally into two chambers, and contains the organ of Corti with its sound-sensitive cells; it is these which convert the pressure waves into nervous impulses. Each cell is sensitive to a particular range of frequencies, that is to sounds of a certain range of pitch, and passes its stimulus to a single nerve fibre, which conveys it to the brain.

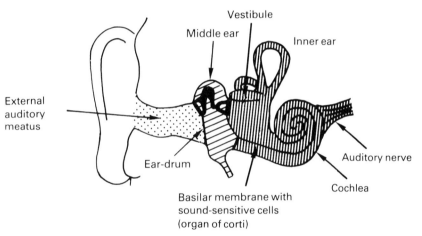

Fig. 92 **The anatomy of the ear.** The shaded parts belong to the inner ear, in which the cochlea detects the sound, whereas the curved tubes overlying it (the vestibule) perceives accelerations.

The basilar membrane has a length of about 30 mm. The sense cells at the apex of the cochlea are stimulated by low-frequency sounds, whereas those at the large entrance (adjoining the basal plate of the stapes) respond to the high frequencies.

At this point, movements of the stapes (the last of the auditory ossicles) create a series of waves, and the distance from the stapes to the crest of each wave (i.e. wavelength) is a function of its frequency. High notes set up short waves, with their crests near to the beginning of the cochlea; low notes create longer waves, with their crests nearer the apex of the cochlea. Hence the point at which the basilar membrane receives maximum pressure depends on the frequency (i.e. the pitch) of the sound.

Müller–Limmroth [235] compares this phenomenon with the breaking of waves in surf; short waves (high frequencies) "break" early at the beginning of the cochlea, whereas longer waves (low frequencies) travel progressively further along before "breaking". In this way the inner ear carries out a sort of "frequency analysis" of the incoming sound waves, and transmits information to the brain of all the frequency components making up the sound. The cerebral cortex integrates these components again so that we are conscious only of the sound as a whole, and not of its constituent parts.

Auditory pathways

The nervous impulses generated by the sound waves travel along the auditory nerve to the brain, entering into the brain stem or medulla, and passing through two synapses or nerve connections to the auditory sphere of the cerebral cortex. Here the separate nerve impulses are localised, as if the cochlea of the ear were spread out over the cerebral cortex. This is where the brain finally integrates these impulses into an impression of the sound. *Figure 93* shows these auditory pathways diagrammatically.

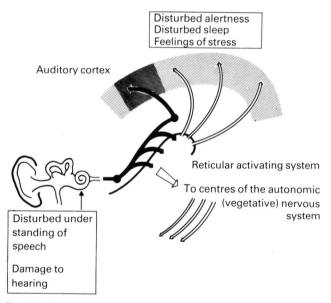

Fig. 93 The auditory pathways (black arrows), and the accessory tracks, which set up secondary effects of noise.

It should be emphasised once again that conscious hearing is a phenomenon of the brain, more precisely of the cerebral cortex. The inner ear and the auditory pathways are no more than the transmitting mechanism, the "interface" between the atmospheric sound and the conscious perception of the brain.

Side-effects

Figure 93 also shows that between the two synapses, mentioned above, nerve fibres lead out to the reticular activating system, of which more will be said in *Chapter 11*. From this system fibres run to the entire conscious sphere of the cerebral cortex, so that excitation of the system by incoming acoustic signals may excite alarm throughout the consciousness of the individual, disturbing sleep, reducing concentration when awake, and producing other distressing symptoms. The effects shown in *Figure 93* are more fully discussed in *Chapter 16* under "Noise and vibration". This kind of "alarm call" has a biologically important role in alerting a person, giving him an opportunity to interpret the noise and react suitably to it. Here is an example: a pedestrian is strolling dreamily along a beautiful country road, and the noise of an approaching motor car grows steadily louder. At first the pedestrian remains unconscious of this, but when the noise has reached a certain level the reticular activating system is excited and sends a signal to the brain. The pedestrian is immediately alerted, so that he can:

(a) take conscious notice of the sound;

(b) interpret it;

(c) get out of the way of the car.

Hearing is an alarm system

It is clear that hearing has two principal functions:

(a) to convey specific information, as *a basis for communication between individuals*: this function is highly developed in man;

(b) as an alarm system: by activating secondary pathways leading to the brain *it plays an essential part in waking, increased alertness, and finally alarm*.

The alarm function of the sense of hearing may be used to advantage in planning transport and in industry. It is essential to recognise dangerous situations quickly, whether supervising a control panel or a switchboard, driving a locomotive, or flying an aeroplane. To this end a suitable combination of acoustic signals and visual aids is needed. The acoustic signals serve to "alert" the brain, and the visual aids convey the necessary information.

Display equipment

A display conveys information to the human sense organs by some appropriate means. In "man-machine systems" it is usually a matter of visual presentation of dynamic processes, such, for example, as fluctuations of temperature or pressure throughout a chemical production.

Display equipment used commercially falls into three categories:

(a) *a window in which figures can be read directly;*

(b) *a circular scale with moving pointer;*

(c) *a fixed marker over a moving scale.*

Each of these has its advantages in certain circumstances, some of which are listed below.

Reading off values

If it is simply a matter of reading off the value of some quantity, the window display is best, provided that only one figure is visible at a time.

If it is necessary to see how a process is going on, or to note the amplitude of some change, then the pointer moving over a fixed scale gives more information. The same applies to a slow change, which needs to be checked from time to time. A moving scale with a fixed indicator mark may also be used for these purposes, but suffers from the disadvantage that it is not easy to memorise the previous reading, nor to assess the extent of movement.

Presetting

If the instrument is to be preset to some particular reading (e.g. steam pressure or electrical voltage) it is easier to do this with a moving pointer, because a second, hand-operated pointer can be preset and the process accurately controlled as the two pointers come together. If the preset process goes slowly, then either of the other two types of instrument will serve equally well. *Figure 94* illustrates the three types of dial and processes for which they are appropriate.

Type of display	 Moving pointer	 Fixed marker moving scale	798 Counter
Ease of reading	Acceptable	Acceptable	Very good
Detection of change	Very good	Acceptable	Poor
Setting to a reading: controlling a process	Very good	Acceptable	Acceptable

Fig. 94   Different ways of displaying information.

Fixed or moving scale

McCormick [220] considers that in general a fixed scale with a moving pointer is better than the converse arrangement. The moving pointer catches the eye and is quick to read off. On the other hand fixed scales have their limitations. If they have to cover a wide range of values, the scale becomes too big and it is then better to see only a part of the scale moving against a fixed indicator.

Reading errors

In the years after World War II it was realised how important is the layout of instruments in aircraft and transport vehicles, in the interests of speed and accuracy of reading, and many studies were carried out to find the best design and arrangement. This usually involved a use of the tachistoscope, for which the criterion was the number of reading errors when the information was displayed for a fraction of a second. (see *"Speed of perception"*, earlier in this chapter.)

Figure 95 shows results from a study made by *Sleight* [292]. Each of the scales was shown 17 times to a total of 60 research subjects, for a period of 0.12 seconds each. The results show significant differences in the frequency of reading errors, the "open window", with 0.5 % errors, being undoubtedly the best. It appears that with the open window no time is lost in locating the pointer.

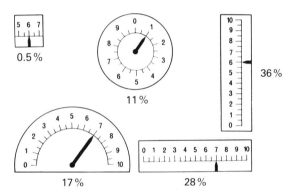

Fig. 95 How the type of dial affects the precision of reading. The figures give percentages of reading errors during a display time of 0.12 seconds. After *Sleight* [292].

Sleight's results [292] have not gone unchallenged. Later work has shown that reading errors depend on the lapse of time the figure is shown [305]. Furthermore *Murrell* [238] and other authors have cast doubt upon the suitability of the tachistoscope, because under practical conditions the operator looks at the indicator for as long as is necessary to read off the figure. In spite of these objections we can agree with *McCormick* [220] that *in general the "open window" type of instrument is less prone to reading errors than the others, and that a vertical scale is less satisfactory than a horizontal one.* *McCormick* [220] does admit, however, that vertical scales are better when several need to be compared side by side.

Altimeters

An instructive example of the need for sound instrumentation is the design of altimeters, which have not infrequently been the cause of aircraft accidents. The traditional altimeter had three pointers (hands), one reading hundreds of feet, the second thousands and the third tens of thousands. Many studies, detailed by *McCormick* [220], have shown that both in the laboratory and in flight, reading errors could be reduced by designing the altimeter differently. *Hill* and *Chernikoff* [146] were, in 1965, the first to arrive at the instrument which is currently the most troublefree and the most pleasant to use. It consists of a round dial with a single moving pointer for the 100-foot scale, and two windows, one of which shows the higher ranges and the other the setting for the atmospheric pressure.

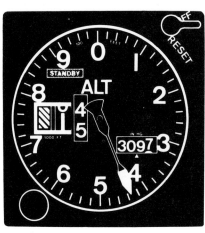

Fig. 96 An aircraft altimeter designed to reduce reading errors. According to *Hill* and *Chernikoff* [146], however, a weakness is that the pointer sometimes covers up one or other of the counters.

Matching instrument and information requirement

It is important that the instrument should give the operator only the information he requires, for instance by displaying the smallest unit that he is likely to read off. Thus if he needs to read pressures to the nearest 100 mm Hg, then the smallest division should be 100 mm Hg.

Sometimes the instrument is required to show, not a precise reading, but a range, say between a lower, safe limit and an upper danger line, or "cold", "warm" and "too hot". Here a moving pointer is best, and the various ranges should be picked out in different colours, or patterns. *Figure 97* shows an example of an instrument on which information is kept to a minimum.

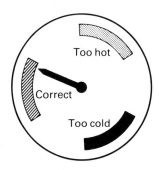

Fig. 97 A display instrument should convey the required information as simply and unmistakeably as possible.

Scale graduation

Even more important than the shape of the dial itself is the size of the scale graduations. Earlier American studies fixed minimum heights for the strokes and their distance apart [119] [171].

When the lighting and contrast were good, these authors recommended quite small dimensions for the graduations. For example, viewing the dial from a distance of 72 cm, the strokes could be as little as 1.3 mm apart, while *Murrell* [239] arrived at even smaller lettering, with separations of 0.5 mm for the same viewing distance. Since, however, lighting and contrast are not always ideal, and since other adverse factors are present at a real work place, we recommend somewhat bigger graduations, as follows: if a is the greatest viewing distance to be expected, in mm, then the minimum dimensions of graduations ought to be as follows:

Height of biggest graduations: $\dfrac{a}{90}$

Height of middle graduations: $\dfrac{a}{125}$

Height of smallest graduations: $\dfrac{a}{200}$

Thickness of graduations: $\dfrac{a}{5000}$

Distance between two small graduations: $\dfrac{a}{600}$

Distance between two big graduations: $\dfrac{a}{50}$

Recommendations for design of scale-graduations

From the above discussion, plus one or two other obvious considerations, the ergonomic design of scale graduations can be summarised as follows:

(a) height, thickness and distance apart of scale graduations must be such that they can be read off with minimum likelihood of error, even if lighting conditions are not ideal;

(b) the information presented should be what is actually wanted: scale divisions should not be smaller than the accuracy required; qualitative information should be simple and unmistakeable;

(c) scale graduations should give information that is easy to interpret and to make use of. It is laborious to have to multiply the reading of the instrument by some factor, and if this is unavoidable, then the factor should be as simple as possible, say $\times$ 10 or $\times$ 100;

(d) subdivisions should be ½ or 1/5: anything else is difficult to read off;

(e) numbers should be confined to major scale graduations and, once again, subdivisions should be ½ or 1/5. Good and bad examples are shown in *Figure 98*;

Subdivisions of a scale

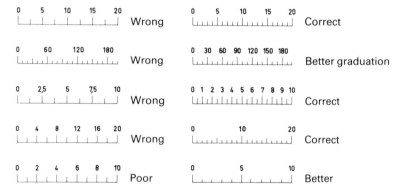

Fig. 98 Good and bad subdivisions of a scale. These should be easily related to the numbered divisions (5 or 2 or 1).

(f) the tip of the pointer should not obscure either the numbers or the graduations, and if possible should not be broader than a scale line. It is best if the tip of the pointer comes as close as possible to the scale, without actually touching it. *Figure 99* shows good and bad pointers;

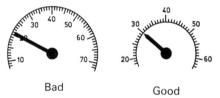

Bad   Good

Fig. 99 Bad and good arrangements of numbers and pointer on a dial. The tip of the pointer should be only as broad as one of the scale lines, and it should not cover the figure.

(g) the pointer should be as nearly as possible in the same plane as the graduated scale, to avoid errors of parallax, and the eye must be positioned so that the line of sight is at right angles to the dial and pointer.

Letters and figures

The Anglo-Saxon literature contains many studies of the effects of size and shape of letters and figures on visual perception, but these have mostly concerned aircraft. A good review of this literature will be found in the papers of *McCormick* [220] and *Murrell* [238]. The current view can be summarised as follows:

(a) the question whether white letters on a black ground are or are not easier to read than the opposite is still not fully decided; (b) black letters on a white ground are to be preferred, in principle, because white letters tend to blur by halation [23], and a black background may set up relative glare against its lighter surroundings. On the other hand white symbols show up better in poor lighting, especially if they and the pointer are luminous.

Size of symbols

Sizes of letters and figures, thickness of lines, and their distance apart must all be related to the viewing distance between the eye and the display. The following formula may be used:

$$\text{height of letters or figures in mm} = \frac{\text{viewing distance in mm}}{200}$$

Table 21 gives examples:

Table 21   Recommended heights of lettering.

Distance from eye in cm	Height of small letters or figures in cm
Up to 50	0.25
50– 90	0.5
90–180	0.9
180–360	1.8
360–600	3

Capitals and lower case are easier to read than letters all of the same size. Most letters and symbols should have the following proportions:

(a) breadth: 2/3 of height
(b) thickness of line: 1/6 of height
(c) distance apart of letters: 1/5 of height
(d) distance between words and figures: 2/3 of height.

These proportions are illustrated in *Figure 100*.

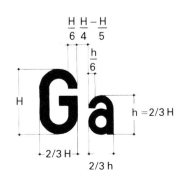

Fig. 100   Recommended proportions for letters or figures. H = height of capitals: h = height of lower case letters. The absolute sizes recommended are listed in *Table 20*.

If appropriate, the information may be given in symbols or graphically rather than in words. Research has shown that information can sometimes be conveyed more quickly by a simple diagram than in the form of words. Symbols come between the two. Symbols and diagrams have the further advantage of avoiding the language barrier.

Controls

Adequate controls

Controls constitute a second "interface" between man and machine. We may distinguish between:

(a) *controls which require little manual effort*: push-buttons; tumbler switches; small hand-levers; rotating and bar knobs, all of which can easily be operated with the fingers;

(b) *controls which require muscular effort*: hand-wheels; cranks; heavy levers and pedals, which involve the major muscle groups of arms or legs.

The right choice and arrangement of controls is essential if machines and equipment are to be operated correctly.

The following guidelines should be followed:

(a) controls should take account of the anatomy and functioning of the limbs. *Fingers and hands should be used for quick, precise movements; arms and feet for operations requiring force*;

(b) hand-operated controls should be easily reached and grasped, between elbow and shoulder height, and in full view;

(c) distance between controls must take account of human anatomy. *Two knobs or switches operated by the fingers should not be less than 15 mm apart; controls operated by the whole hand need to be 50 mm apart*;

(d) push-buttons, tumbler switches and rotating knobs are suitable for operations needing little movement or muscular effort, small travel and high precision, and for either continuous or stepped operation (click-stops);

(e) long-armed levers, cranks, hand-wheels and pedals are suitable for operations requiring muscular effort over a long travel, and comparatively little precision.

There is an abundant literature dealing with the ergonomic design and layout of controls. Good reviews are given by *McCormick* [220], *Kroemer* [186], *Morgan* and others [228], *Schmidtke* [278], *Woodson* and *Conover* [335] and in *DIN 33 401* [70]. The practical recommendations from these sources are summarised as follows.

Coding

Big machines in industry, agriculture and transport often have control panels with many similar controls and it is important to be able to pull the right lever or turn the right knob, even without looking at it. According to *McFarland* [221] the American airforce in World War II suffered 400 crashes in 22 months because the pilots mistook some other lever for that controlling the undercarriage.

For this reason *any controls that might be mistaken for each other should be so designed that they can be identified without difficulty.* Identification can be assured by:

(a) *arrangement.* For example sequence of operation, or the difference between vertical and horizontal movement. Only a small number of controls can be identified in this way;

(b) *structure and material. Figure 101* shows knobs of 11 different shapes, developed from experiments by *Jenkins* [162]. These were the shapes that were least often confused by blindfolded operators.

Fig. 101 Types of hand grips (knobs) that are easy to distinguish. After *Jenkins* [162], modified from *Kroemer* [186].

Besides their shape and size, knobs can be made still more distinctive by their surface texture (smooth, ridged, etc.). These characteristics are most helpful if the control must be handled unseen, either in darkness, or while the attention is being directed elsewhere;

(c) *colour and labelling* can be useful, but only in good light and under visual control.

Distance apart If controls are to be operated freely and correctly, without unintentionally moving adjoining controls, they must be a certain minimum distance apart. *Table 22* shows minimum and optimum separations.

Table 22 Distance apart of adjoining controls.

Control	Method of operation	Distance apart in mm Minimum	Optimum
Push button	With one finger	20	50
Toggle switch	With one finger	25	50
Main switch	With one hand	50	100
	With both hands	75	125
Hand wheel	With both hands	75	125
Rotating knob (round or bar-shaped)	With one hand	25	50
Pedal	Two pedals with same foot	50	100

Resistance	Controls should offer a certain amount of resistance to operation, so that their action is more positive and they are less likely to be triggered off by a slight tremor of the finger. *Schmidtke* [278] recommends the following resistances in mkp:

    (a) single handed rotation: 0.2–0.25
    (b) pressing with one hand: 1.0–1.5
    (c) pedal pressure: 4–8.

Higher resistances than these are sometimes desirable to isolate particular controls from the others.

Among *controls suitable for precision operation with little force*, are the following: tumbler switches, hand levers, and knobs.

Push-buttons for finger or hand operation	Push-buttons to be operated by the pressure of either finger or hand take up little space and can be made distinctive by colour or other marking. The surface area of the knob must be big enough for the finger or hand to be able to press the knob easily and apply the necessary pressure without slipping off. Recommended dimensions:

    (a) diameter: 12– 15 mm
    (b) (for an isolated emergency stop: 30– 40 mm)
    (c) travel: 3– 10 mm
    (d) resistance to operation: 250–500 g.

Figure 102 shows two push-buttons, one of the correct size and the other too small.

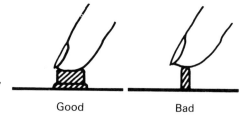

Fig. 102 Push-buttons. The button must be big enough (12–15 mm) to ensure that it can be pressed with confidence, and the finger will not slip off.

         Good   Bad

Finger-operated push-buttons should be slightly concave, whereas buttons to be pushed by hand should be mushroom-shaped.

Dimensions of the latter should be:

    (a) diameter: 60 mm
    (b) travel: 10 mm
    (c) resistance to operation: 1 kp.

Toggle switches	Toggle switches are easily seen and reliable in operation. They should preferably have only two positions ("off" and "on"). Several toggle switches can be placed side by side, provided that each is clearly identified.

The direction of travel should be vertical, and the "off" and "on" positions marked above and below (convention whether the "on" position is up or down varies in different countries). If a toggle switch with three positions is used, there should be at least 40° of travel between two adjoining positions, which should all be clearly identified. Recommended dimensions for toggle switches can be seen from *Figure 103*.

$\alpha = 45°$
$d = 3{-}25$ mm
$l = 12{-}50$ mm
Resistance to operation
$0.25{-}1.5$ kp

Fig. 103   Dimensions of toggle switches.

Hand levers

When the toggle is longer than 5 cm, it is spoken of as a *hand lever* and allows greater force to be exerted than a mere toggle. Here, too, the direction of movement should always be either up–down or forwards and backwards. When a hand lever has several positions, not just "on" and "off", each position should have a definite notch. If the lever is capable of fine adjustment, then some suitable form of support should be given to the elbow, forearm or wrist. Hand levers may have different knobs, according to their function:

(a) a finger grip, diameter 20 mm;

(b) a grip for the palm, diameter 30–40 mm;

(c) a mushroom-shaped grip, diameter 50 mm.

Figure 104 shows a hand lever suitable for delicate control movements with the fingers.

Fig. 104   Control knob for precise operation with the fingers. Lower arm and wrist should be supported by a smooth piece of wood, on which they can easily move.

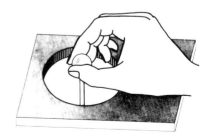

Hand levers which need considerable force to operate them are called *switch levers*, and come into the category of heavy controls. Suitable dimensions for a switch lever are indicated in *Figure 105*.

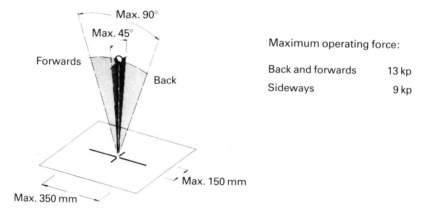

Maximum operating force:

Back and forwards   13 kp

Sideways   9 kp

Fig. 105   **Large switch lever requiring moderate forces to operate it, with ranges of movement.**

Rotating knobs

Rotating knobs may take a variety of shapes: round, arrow-shaped, combinations of knob and handle, or even several knobs on the same spindle.

An important requirement for all types is that *they must fit the hand comfortably and turn easily, and they should be in full view throughout the operation.* Good and bad designs are shown in *Figure 106.*

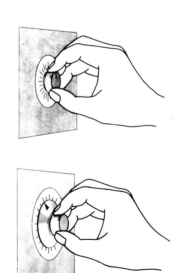

Fig. 106   **Knobs should be easily grasped in the fingers, which should not cover the dial.**

Rotating switches

Figure 107 summarises data about rotating switches with click stops. These should have a somewhat higher resistance than knobs for continuous rotation, so that the operator receives a clear "tactile signal" at each position. A resistance level of 1.5 kp cm is recommended. Furthermore, the successive positions should be not less than 15° apart if the switch is kept in view, and at least 30° apart if it is operated solely by feel.

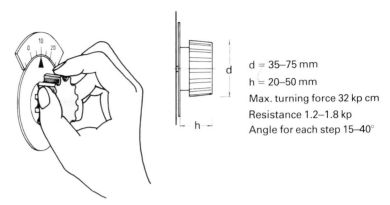

d = 35–75 mm
h = 20–50 mm
Max. turning force 32 kp cm
Resistance 1.2–1.8 kp
Angle for each step 15–40°

Fig. 107 Knob for step-by-step adjustment (click stops). The notched margin makes it easier to control.

Knobs for continuous rotation

Knobs without click stops are suitable for fine and precise regulation over a wide range. An arc of 120° can be turned without shifting the grip, and under precise control; to turn further, the grip of the hand can be changed without difficulty. Such a knob can be turned either with the fingers or with the whole hand, and it helps if the surface of the knob is slightly grooved or roughened.

The following dimensions can be recommended:
(a) diameter for use with two or three fingers: 10–30 mm
(b) diameter for use with the whole hand: 35–75 mm
(c) depth for finger control: 15–25 mm
(d) depth for hand control: 30–50 mm
(e) maximum turning moment for small knobs: 8 kp cm
(f) maximum turning moment for large knobs: 32 kp cm
(g) resistance to operation: 0.35–5 kp cm.

Figure 108 shows a suitable knob for continuous rotation by either two or three fingers.

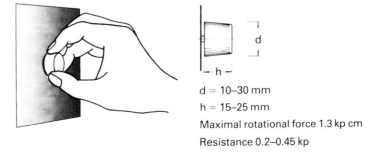

d = 10–30 mm
h = 15–25 mm
Maximal rotational force 1.3 kp cm
Resistance 0.2–0.45 kp

Fig. 108 Knob for continuous rotation.

Pointed bar knobs

Arrow-shaped or pointed bar knobs have the advantage of locating the reading quickly and easily, and should be 25–30 mm, measured along the bar. *Figure 109* shows such a bar for use with click-stops.

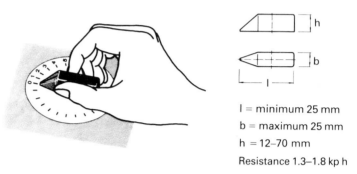

l = minimum 25 mm
b = maximum 25 mm
h = 12–70 mm
Resistance 1.3–1.8 kp h

Fig. 109   Pointed bar knob for click stops.

Controls which call for muscular strength and long travel, but not a high degree of precision, can be operated by a crank, a hand-wheel, or a pedal.

Cranks

Geared levers, or cranks, are appropriate for either setting a control, or providing continuous adjustment, when a wide range of movement (long travel) is necessary. The gearing may be coarse or fine, according to the degree of precision required. The crank can be operated more quickly if the handle can rotate on its own axis, but a fixed handle is more precise.

The following dimensions are recommended:
(a) length of lever arm for low torque
        (up to 200 rpm):   60–120 mm
(b) length of lever arm for high torque
        (up to 160 rpm):   150–220 mm
(c) length of lever arm for quick setting: up to 120 mm
(d) torque: 0.0–2.5 kg.

According to *Morgan* and others [229] the following are the rotation speeds (rpm) in relation to the crank radius:

mm	*rpm*
20	270
50	255
120	185
240	140

Hence it appears that the slower the speed of rotation of the crank, the longer it should be, and vice versa.

The relation between crank length and resistance to operation should be:

crank 240 mm long, resistance 5–25 kp cm
crank 120 mm long, resistance 30–40 kp cm
This range, 120–240 mm is the best crank length for fine and
precise control (see "Hand grips" in Chapter 8).
Dimensions of hand grips should be as follows:

diameter:	25– 30 mm
length for one-handed operation:	80–120 mm
length for two-handed operation:	190–250 mm.

Handwheels

Handwheels are recommended when large forces must be
applied, because they allow the use of both hands and a relatively
long leverage where the turning speed is low.
Notches inside the rim of the wheel give a surer grip and allow
force to be applied more efficiently.

Pedals

Pedal controls do not often have pedal pressures of more than 10
kp, though some such occur in machinery used for agriculture,
building and occasionally in industry. The brake pedals of motor
vehicles also fall into this category. Pedals are very suitable for this
purpose – as already mentioned in *Chapter 3*, *"best use of bodily
strength"* – because the human foot is capable of very high
pressures, up to 200 kg.
In order to exert heavy pedal pressures, it is necessary to have:
(a) a high back rest;
(b) angle at knee beteen 140–160°;
(c) angle at ankle 90°;
(d) slope of lower leg 20–30°.
For even higher pedal pressures the whole leg must be brought
into use ("leg pedal" rather than foot pedal), and the initial
resistance must be enough to support the weight of the leg resting
on the pedal.
Pressure should be applied with the ball of the foot, and the axis of
tread should be a line from the ankle to the point of pressure on
the backrest.
The following recommendations are made for *pedals with a heavy
tread*:

(a) travel of pedal:	5–15 cm
(the smaller the knee angle, the longer the travel)	
(b) minimum resistance to operation:	6 kp.

*Pedals such as the accelerator of a car, which require only light
pressures*, are operated solely with the foot, and their initial
resistance is low. It is a help if this type of pedal can be operated
with the heel on the floor and the foot resting on the lower edge of
the pedal [16]. The following are recommended dimensions:

(a) travel of pedal:	at most 6 cm

(b) maximum pedal angle:[1] 30°
(c) optimum pedal angle.[1] 15°
(d) resistance to operation: 3–5 kp.

Pedals of all types should have a non-slip surface.

As already mentioned in *Table 22*, the distance apart (clearance) of adjacent pedals should lie between 5–10 cm. Under special circumstances, e.g. for use with heavy shoes or gumboots, the pedals must be even wider apart.

Pedals for standing use

When a machine has a multiplicity of controls, foot pedals are sometimes used to relieve the hands, but this is undesirable if the operator stands at his work. In any case they should be restricted to operating an on-and-off switch. If a pedal cannot be avoided, then it should be of the type recommended in *Figure 110*.

Fig. 110 Pedals are undesirable for standing work, since they set up heavy static loads in the legs. Left: bad arrangement, heavy loads on one leg. Right: a good arrangement, a treadle that can be operated with either foot, at will.

Summary

Figures 111 and *112* illustrate the good and bad points of various types of control.

Fig. 111 Evaluation of various control systems. Part I.

Component	Speed	Accuracy	Effort required	Working range	Load
Lever horizontal	Good	Poor	Poor	Poor	Up to ca. 9 kp
Lever vertical	Good	Moderate	Good	Poor	Up to ca. 13 kp
Control lever small	Good	Moderate	Poor	Poor	Ca. 1 kp
Control lever large	Good	Poor	Good	Poor	Ca. 2–9 kp

[1] Pedal angle = operating angle between the two extreme positions.

Gear lever 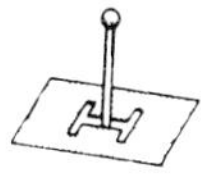	Good	Good	Poor	Very Poor	Up to ca. 9 kp
Crank small 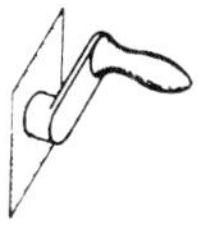	Good	Poor	Very poor	Good	0.9–2.5 kp Lever arm up to 120 mm
Crank large	Poor	Very poor	Good	Good	Over 3.5 kp Lever arm 150–220 mm

Fig. 112 Evaluation of various control systems. Part II.

Component	Speed	Accuracy	Effort required	Working range	Load
Handwheel 	Poor	Good	Moderate	Moderate	2–25 kp Diameter 180–500 mm
Knob Small (continuous) 	Poor	Good	Very poor	Moderate	0.2–0.4 kp Diameter 10–30 mm
Large	Very poor	Moderate	Poor	Moderate	Up to 2.5 kp Diameter 35–75 mm
Knob (clock stops) 	Good	Good	Very poor	Very poor	1.3–1.8 kp Diameter 25–100 mm
Push-button 	Good	Very poor	Very poor	Very poor	0.25–0.5 kp
Pedal 	Good	Poor	Good	Very Poor	1.5–90 kp

Relationship between controls and display instruments

Relative speeds of movement

To set up a particular reading on a display instrument, the operator starts by moving the control quickly, until the reading is approximately correct, then he moves the control slowly to make the precise adjustment. The relative distances of travel of the control lever or knob and of the pointer of the instrument are of importance during this second, precision stage.

A coarse adjustment is easier when the pointer moves more quickly than the control, but *for precise adjustment the control should travel faster than the pointer.* The best ratio varies so much in different situations that it is not possible to formulate quantitative rules that would apply generally. Here we shall merely quote a recommendation of *Shackel* [287] for precise adjustment using a rotating knob and a pointer moving over a scale. One complete rotation of the knob should. make the pointer move through 50–100 mm, allowing a reading tolerance of 0.2–0.4 mm. For a higher reading tolerance of 0.4–25 mm the pointer movement should be 100–150 mm per round.

Controls should move in the obvious direction

When steering an unfamiliar motor-car, we have a right to expect that turning the steering wheel clockwise will turn the wheels to the right: no-one would expect to turn the wheel to the left in order to steer to the right.

Similarly it is reasonable to expect that when a control knob is turned to the right, the pointer of the corresponding instrument will also move to the right. Such expectations depend on so-called *stereotyped reactions*, or fixations: experience has "engraved the corresponding template" on the brain, as we have already indicated during the discussion on learning to speak (*Chapter 8*). *Stereotypes are thus conditioned reflexes which have become subconscious and "automatic".*

Obviously, therefore, controls and instruments that are functionally linked ought also to make corresponding movements that seem sensible to our own stereotypes. *Figure 113* shows examples of sensible coordination between the direction of movement of knobs and levers and instruments with vertical or horizontal scales.

Recommendations

Summary of recommendations:

(a) when a control is moved or turned to the right, the pointer must also move right over a round or a horizontal scale: on a vertical scale the pointer must move upwards;

(b) when a control is moved upwards or forwards, the pointer must move either upwards or to the right;

(c) a right-handed or clockwise rotation instinctively suggests an increase, so the display instrument should also record an increase;

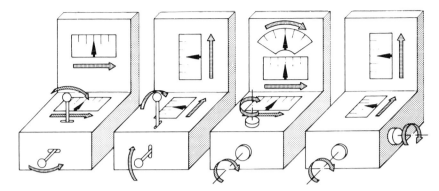

Fig. 113 Examples of logical relationship between movements of controls and movements of indicators.

(d) *Hoyos* and others [150] recommend that a moving scale with a fixed indicator should move to the right when the control is moved to the right, *but* the scale values should increase from *right to left*, so that a rotation of scale to the right gives increased readings;

(e) when a hand lever is moved upwards, or forwards, or to the right, the display readings should increase, or a switch should move to the "on" position. To reduce the reading, or to switch to "off", it is instinctive to pull the lever towards the body or move it to the left, or downwards. These stereotypes affecting a hand lever are illustrated in *Figure 114*.

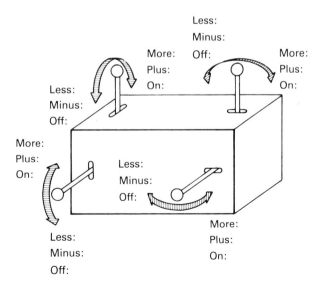

Fig. 114 Expected movements in relation to the movements of a hand lever. After *Neumann* and *Timpe* [245].

Exceptions	Not all stereotypes are equally firmly established, and occasionally there are considerable individual aberrations. According to *Chapanis* and *Gropper* [49] layouts designed for right-handed operation are not equally suitable for left-handed people. Not enough research has yet been made into this question.
	There is one important exception to the rule that clockwise rotation means an increase in whatever is being controlled. Both water and gas are controlled by stopcocks which turn off to the right. This can lead to problems when one person has to control water/gas and electricity at the same time. In such a case some solution must be sought that involves the least risk of accident, and some degree of conscious control is necessary.
Sensible control panels	A sensible layout of both controls and display instruments will make supervision easier and reduce the risk of confusion caused by false readings.
Five principles	Five principles should be taken into account when designing control panels:

(a) as far as possible, display instruments should be located close to the controls which affect them. The controls should be placed either below the display, or, if need be, to the right of it;

(b) if it is necessary for controls to be in one panel and display instruments in another, then the two sets should be set out in the same order and arrangement;

(c) identification labels should be placed *above* the control, and identical labels *above* the corresponding display;

(d) if a number of controls are normally operated in sequence, then they and the corresponding displays *should be arranged on the panel in that order, from left to right*;

(e) if, on the other hand, the controls on a particular panel are not operated in a regular sequence, then both *they and the corresponding displays should be arranged in functional groups*, in order to give the panel as a whole some degree of orderliness. The grouping can be emphasised by choice of colours, by labelling, by using knobs of different sizes and shapes, or simply by arranging members of a group in a row. Those controls and displays that are used most often should be directly in front of the operator, the less important ones to each side.

Figure 115 shows an example of a logical grouping of controls and displays on a control panel.

These recommendations may seem trivial and pedantic to the reader, but we must remember that long and often monotonous work can cause fatigue, and lead to reduced alertness and an

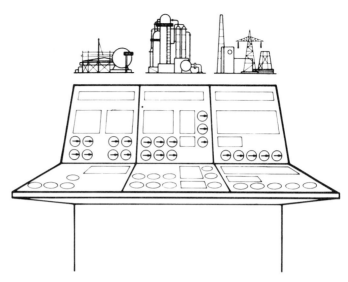

Fig. 115 **Logical layout of a control panel, with controls and display instruments in associated groups.**

increase in errors. Under these conditions a logical layout of controls and displays, which takes advantage of stereotyped, "automated" behaviour, is beneficial, since a weary operator tends to fall back upon his conditioned reflexes.

More everyday examples

It is not only in industry that sensibly planned control panels are important. Every model of motor car would be improved if more attention were given to the layout of switches and the instrument panel.

The study of stereotypes has its uses even in the home. Electrical switches ought always to be "on" in the same direction, though convention has this "up" in some countries and "down" in others. Nor is it easy to ensure that oven controls are foolproof. *Chapanis* and *Lindenbaum* [50] seized upon and studied a very practical, everyday problem, that of finding a sensible arrangement for the electrical switches for the hot plates. For this purpose they tested the arrangements shown in *Figure 116* on 15 research subjects, with 80 tests each. The test subjects were instructed to switch on a particular hotplate, and their reaction time and the number of mistakes were noted. The four arrangements and the number of mistakes are shown in *Figure 116*.

The result was decisive. The arrangement in *Model I*, with the sequence of switches matching the sequence of hotplates, produced no errors, and the average operating time was also the shortest.

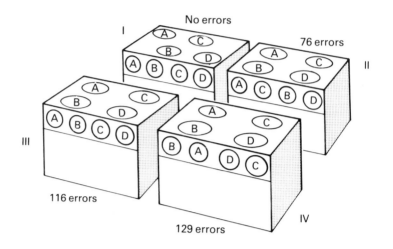

Fig. 116 Tests carried out on a simulated oven to associate controls correctly with the burners they operate. The numbers of errors relate to 1200 tests on 15 research subjects. After *Chapanis* and *Lindenbaum* [50].

Concluding remarks   Finally we must emphasise yet again the importance of taking into account stereotyped behaviour when designing and arranging controls and display instruments, which are important "interfaces" in the "man–machine system". Display instrument–man–control constitutes a minor circuit which directs part of the larger production process.

Man and equipment must be compatible and work smoothly together, to ensure speed and precision, to minimise the risk of errors, and to avoid incidents which might end in catastrophe.

10
Mental activity

What constitutes
mental activity?
Until a few years ago there was a simple line of demarcation between manual work, performed by operatives, and brain work which was the domain of white-collar workers, but nowadays this distinction is less clear on two grounds. Some jobs call for a good deal of mental activity without really coming into the category of brain work, e.g. information processing, supervisory work, taking important decisions on one's own responsibility. Moreover, this sort of work is by no means restricted to white-collar workers but is often delegated to manual operatives.

Hence the expression "mental activity" as a general term for any job where the incoming information needs to be processed in some way by the brain. Such activity can be divided into two categories:

(a) brain work in the narrow sense;

(b) information processing as part of the "man–machine system".

Brain work
Brain work in the narrow sense is essentially a thought process which calls for creativity to a greater or lesser extent. As a rule the information received must be combined with knowledge already stored in the brain and committed to memory in a new form. Decisive factors include knowledge, experience, mental agility and the ability to think up and formulate new ideas. Examples include: constructing machines, planning production, studying the files and extracting the essential facts from them, making a precis of these, giving instruction and writing reports.

Information
processing
Information processing as part of "man–machine systems" has been discussed in the previous chapter, but nevertheless the essentials will be summarised again. They are:

(a) perception
(b) interpretation } of the information transmitted by the sense organs.
(c) mental processing

This "processing" consists of combining the new information with what is already known, so providing a basis for decision-taking.

The mental load at work places such as we have discussed is conditioned by the following:

(a) *the obligation to maintain a high level of alertness over long periods*;

(b) *the need to take decisions* which involve heavy responsibility for the quality of the product and for the safety of work people and plant;

(c) occasional lowering of concentration *by monotony*;

(d) *lack of human contacts*, since one work place is often isolated from any others.

The barrier that confronts us

Neurophysiology, psychology and other branches of science try very hard to get some degree of insight into the basic processes of mental effort. The well-known neurophysiologist *Penfield* makes the following comparison: "Anyone who studies mental processes is like a person who stands at the foot of a mountain range. He has cut himself a clearing on the lowest foothill, and from there he looks towards the mountain-top, since that is his objective, but the summit is obscured in dense cloud." It is this "clearing" that we are about to examine.

Mental performances that are important in ergonomics include:

(a) uptake of information;

(b) memory;

(c) sustained alertness.

Uptake of information

Information theory

The information theory of *Shannon* and *Weaver* [288] has made an important contribution to the understanding of how information is taken up. These authors made a mathematical model that would represent quantitatively the transfer of information, and they devised the term *bit* (binary unit) for an item of information. The simplest definition of a bit is that it is the quantity of information conveyed by one of two alternative statements. For example: in olden times one flash of light from a watchtower might mean "enemy approaching from the sea", whereas two flashes would mean "enemy approaching from the land". These alternative pieces of information were a bit.

As soon as there are more than two choices, of varying probability, the situation becomes much more complicated. So this theory has its limitations when applied to human beings, since the full significance of a stimulus conveying information cannot be interpreted by information theory. This theory is valid only for comparatively simple situations which can be split into units of information and coded signals. It is already useless when applied, for example, to the information being received by the driver of a car.

We need not go further into information theory at this point, since the interested reader needs only to consult items [2] and [12] in the Bibliography.

Channel capacity theory

Another theory is based upon comparison of information uptake with the capacity of a "channel" (channel capacity theory). According to this theory the sense organs deliver a certain quantity of information to the input end of the channel, and what comes out at the other end depends upon the capacity of the channel.

If the input is small, very little of it is absorbed by capacity of the channel, but if the input rises, it soon reaches a threshold value, beyond which the output from the channel is no longer a linear function of the input. This threshold is called the "channel capacity" and can be determined experimentally for a variety of different sorts of visual and acoustic information.

Human beings have a large channel capacity for information communicated to them by the spoken word. Thus it has been calculated that a vocabulary of 2500 words requires a channel capacity of 34–42 bits per second [254]. This capacity is very modest, when compared with the channel capacity of a telephone cable, which can handle up to 50 000 bits per second.

In everyday life the incoming information is much greater than the "channel capacity" of the central nervous system, so that a considerable "reduction process" must be carried out. It is estimated that this results in the following number of bits being characteristic of different parts of the system:

Process	Information stream in bits/s
Registration in sense organ	1 000 000 000
At nerve junctions	3 000 000
Conscious awareness	16
Lasting impression	0.7

Although these are only rough figures [295], yet it is clear that only a minute fraction of the information available is consciously absorbed and processed by the brain. The brain selects this minute fraction by reduction, i.e. by some kind of filtration process, about which we know virtually nothing.

Memory

<table>
<tr>
<td>Information
store</td>
<td>Memory is the process of storing incoming information in the brain, often of only a selected portion of it, after it has been processed. How this selection is carried out, we do not know.</td>
</tr>
</table>

Information store

Memory is the process of storing incoming information in the brain, often of only a selected portion of it, after it has been processed. How this selection is carried out, we do not know.

We do know that the process is subject to the emotions of the moment, which release their own bits of information, and we must further assume that information to be stored must have some relevance to what is already there. Each person determines to a great extent what is relevant to himself, and what is not.

Two kinds of memory can be distinguished:

(a) *short-term or recent memory*;

(b) *long-term memory*.

Short-term memory comprises immediate recollection of instantaneous happenings, up to the remembrance of events which occurred a few minutes, or an hour or two ago.

Recall of events months or years after they occurred becomes long-term memory.

Items of information which become part of the memory leave "traces", or *engrams* in certain areas of the brain. Information thus stored can be recalled at will, though, regrettably, not always as fully as one could wish!

Short-term memory

There is a model for short-term memory. The information received leaves a "trace", which continues to circulate as a stimulus through the nervous tracts, and which, by a kind of feedback, can be recalled into the conscious sphere at any time within a few hours.

This reservoir of short-term memory is expendable. This is what happens during "retrograde amnesia", following some mechanical or emotional disturbance to the brain, whereby the memory of events immediately before the shock is obliterated. Periods of hours or even weeks can be lost from memory in this way. We must assume, therefore, that there is a period during which memories are being consolidated, or "engraved" on the brain, and that during this period they are vulnerable to destruction. Afterwards they become more stable and surprisingly resistant "engrams", which constitute long-term memory.

The role of the limbic system in memory

Experiments on animals, as well as clinical observations on humans, have shown that parts of the *limbic system* of the brain, especially that in the hippocampus, play an important role in the engraving process of short-term memory.

Before we go into that, let us take a general look at the anatomy of the brain, which may be considered in three sections:

(a) *fore brain* including the cerebral cortex, where, among other functions, memory, learning, consciousness, perception and all the thinking processes are located;

(b) *mid brain* lies beneath the fore brain and links it with the medulla. This is where autonomic functions are localised, such basic sensations and reactions as hunger, thirst, anger, defensiveness, flight, as well as vegetative control of the internal organs. Among its important constituents are the thalamus and hypothalamus;

(c) *hind brain* forms the link with the spinal cord and also includes the cerebellum. Here are located such vital functions as respiration, control of heart and blood-circulation, as well as hiccuping, coughing, swallowing and vomiting. Nearly all the nerves from the brain connect with the medulla, which has a dense and complex network of synapses, the *formatio reticularis*, or reticular formation.

Figure 117 shows a few parts of the brain in a simple diagram.
The limbic system is an elongate structure, lying beneath the

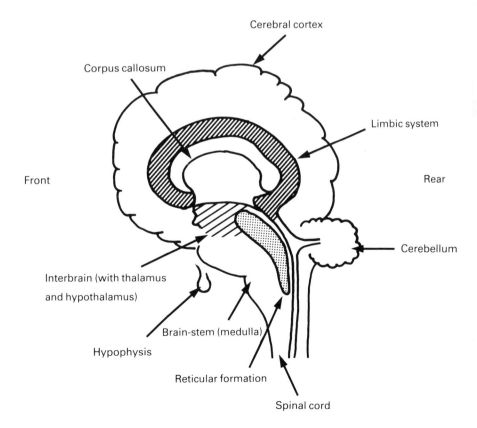

Fig. 117 General plan of the layout of the brain. A longitudinal section to one side of centre line: the hypophysis requires a median section.

cerebral hemispheres (the two parts of the cerebral cortex) and forming a link with the deep parts of the brain (mid brain). It consists in part of cortical tissue and in part of nervous structures belonging to the underlying nerve-centres. The limbic system is believed to play an important part in generating emotions and the physical excitement that goes with them. In addition, this system is involved in the direction of day/night rhythms, appetite, sexual behaviour, motivation and excitements such as rage and fear.

If the lateral part of the limbic system (the hippocampus) is destroyed it results in serious deficiencies in short-term memory, though long-term memory remains intact, but we still do not know the exact location of short-term memory.

The *formatio reticularis* is an activation system essential to the processes of learning and memory,[1] since it is only during waking hours that memories can be stored away. The more active the brain, the more learning and memorising go on, and doubtless the higher the level of motivation.

Long-term memory is very stable and resistant to both brain disturbance and electric shock. This fact has led to the conclusion that long-term memory must depend upon some form of intra-molecular storage of stimuli, that is, to changes in the chemical substratum of the nerve cells. Many experiments suggest that this may be a durable engraving on the ribonucleic acid (RNA). According to *Ganong* [71], fascinating experiments with planarians have shown that acquired behaviour is not lost when these flatworms are cut up. If a worm is trained to react in a particular way to a certain stimulus, and is then cut into two, both head and tail regenerate complete worms, and both of these retain the acquired behaviour! This leads to the hypothesis that acquired behaviour is based upon changes in the RNA. It is well-known, of course, that the RNA provides the template for the synthesis of proteins during cell-division; hence it is assumed that the cells of each cut-half are replicated into the regenerated half complete with their "marked RNA". This hypothesis is confirmed by a further experiment: if either half is treated with ribonuclease, an enzyme that destroys RNA, then the acquired behaviour is lost, and the regenerated worm is no longer conditioned.

Many other experiments suggest that long-term memory is en-graved upon the RNA, including several studies of learning in mammals, which were accompanied by an increase in the RNA in the brain. In one study, cited from *Müller–Limmroth* [235], rats were trained to use only the left forepaw to reach for food, and the corresponding centre in the brain showed an increase in RNA content. Information theory also gives interesting insights into memory processes. The number of nerve cells in the brain that are

[1] An activation system is discussed in detail in *Chapter 11: "Fatigue"*.

concerned in memory storage may be estimated at *10 milliards,*[1] and assuming that all these cells are actively involved, *the storage capacity of the human memory may be as great as 10^8–10^{15} bits* [93]. This is an inconceivably great storage capacity, exceeding that of present-day computers.

A problem for human beings is to be able to "recall" the stored information. Everyone knows the difficulties that older people have, and how sadly we say "I can't call it to mind any more." So far science can do nothing about this problem.

Sustained alertness (vigilance)

Jobs in industry and transport that call for sustained alertness are specially demanding mentally. Before we consider the actual problems of remaining alert, it is sensible first to discuss some of the activities that are often used as indicators of the level of mental efficiency.

Reaction time

Psychologists and ergonomists have particularly concerned themselves with *speed of reaction*: psychologists because a study of reaction times gives them an insight into mental problems, and ergonomists because reaction time can often be used as a way of assessing the ability to perform mental tasks. Reaction time means the interval between the receipt of a signal and the required response. According to *Wargo* [320] this time is made up of the following parts:

		Milliseconds
(a)	conversion into a nerve impulse in the sense organ:	1–38
(b)	transmission along a nerve to the cerebral cortex:	2–100
(c)	central processing of signal:	70–300
(d)	transmission along a nerve to musculature:	10–20
(e)	latent time of response of muscle:	30–70
	Total time, say:	100–500

It will be seen from this table that a substantial part of the reaction time is taken up with mental processing of the signal in the brain. The considerable range in each of these times must be attributed to the wide variety of sense organs studied, varying lengths of sensory and motor pathways, and differences in the type of signal to be processed.

Simple reaction time

A simple reaction time is one that involves a simple signal, and one that is expected, which is answered by a simple motor

---

[1] 1 milliard = 10^9.

reaction. The average time for such a reaction is 0.15–0.20 seconds. According to *Swink* [300] *simple reaction times* under various conditions of study, produced the following average figures:

		Seconds
(a)	light signal:	0.24
(b)	siren:	0.22
(c)	electrical shock on skin:	0.21
(d)	light signal + siren:	0.20
(e)	all three at once:	0.18

Selective reaction time

If a variety of signals must each be answered with a different reaction, we speak of *the selective reaction time*. For example, a signal may be green, red or yellow, and each must be answered by pressing a different key. These reaction times are substantially longer than simple reactions, because the decision requires the incoming signal to be processed by the brain, and the number of possible answers is an essential factor.

A summary by *Damon* and his colleagues [60] suggests the following relationship between the number of possible answers and the consequent reaction time:

Number of answers:	1	2	3	4	5	6	7	8	9	10
Approximate reaction time in <u>seconds</u>: 100	20	35	40	45	50	55	60	60	65	65

Thus it can be assumed, to a first approximation, that selective reaction time increases linearly with the number of bits of information, up to a threshold of about 10 bits [145].

Anticipation

As we have said, both types of reaction time are shortened if the signals are anticipated. This is always the case in experimental work, whereas in everyday life most reactions are to unexpected stimuli.

In one investigation [321] shorthand typists were given a knob to be pressed whenever they heard a buzzing tone. This signal was sent out no more than once or twice a week, over a period of six months, in the middle of the typists' ordinary work. Reaction times under these conditions averaged 0.6 s, in place of the normal 0.2 s.

Movement times

So far no allowance has been made for the time taken up by the movement of the control, which can be especially important when controlling a vehicle, since it may be at least 0.3–0.5 s. The total elapsed time may then amount to 0.5–1 s. The importance of this factor to the driver of a vehicle reaches its extreme in the pilot of a supersonic aircraft. If the aircraft is flying at 1800 km/h it will travel a distance of 300 m during the total reaction time of 0.6 s.

Limits of mental load

It has long been known that thinking and other mental processes become less effective as time goes on. It is common experience that the longer one reads, the harder it becomes to take in the information, and the more often it is necessary to read a passage again before the words and sentences make sense. Who has not suffered the wandering of attention that occurs during a long and boring speech?

Bills' blocking theory

These simple, everyday observations have been amplified by the work of *Bills* [24], who demonstrated by psychological experiments that people cannot concentrate on a mental activity without respite. In practice there occurred at comparably frequent intervals interruptions in the processing of information, which *Bills* called "blockings" or "blocks". The duration of a block extends to at least twice the length of the average processing time. *Bills* sees these blockings as *a kind of autonomic regulatory mechanism, which has the effect of allowing the level of mental effort to be as high as possible for as long as possible.*
Long-continued mental effort brings about more frequent and longer blocks, which can be seen as a fatigue symptom. *Broadbent* [36] studied fatigue symptoms of this kind in an experiment during which the test subjects had to remain continuously alert for the appearance of a weak optical signal of short duration, which was given only infrequently (15 times per hour). *Broadbent* found that various distractions (noise, heat, being deprived of sleep) made blocks appear more quickly, and more signals were missed. The author compared blocking with the closing reflex of the eyelids, which is an interruption of continuous perception of light, and which also becomes more frequent and lasts longer as fatigue increases.

Variability of heart rate

In recent years *the variability of the heart rate has been used as an indicator of mental stress*. In practice heart rate is not regular from one beat to the next, but constantly varies both up and down. The physiological term for this variation is *sinus arrhythmia*, which is mainly linked to the act of breathing. At each inspiration the heart rate rises, to slow down again during the following expiration. This arrhythmia seems to be directed, in the first instance, by the autonomic nervous system, with the vagus nerve playing an important role as "pacemaker". Several authors, among whom *Kalsbeck* [169] and *O'Hanlon* [250] may be mentioned, agree in finding that the variability of the heart rate is less during either physical or mental stress. In practice both arithmetical problems and carrying out exacting observation work led to a reduction in the variability. *Kalsbeck* [169] is of the opinion that the reduction in variability is a function of the mental stress caused by the processing of information. *O'Hanlon* [250] found that when his research subjects relaxed their concentration, their heart rate became more variable, and he suggested that this rise in variability might be used as an indicator of the level of concentration. In practice he found a significant individual correlation ($r = -0.46$) between variability and concentration. So it may be said, in simple terms, that:

(a) *a fall in variability of heart rate is a sign of increasing concentration*;

(b) *a rise in variability accompanies any fall in concentration*.

Sustained concentration or vigilance

Sustained concentration is the ability to maintain a given level of alertness over a long period of time. Vigilance is another term for this ability which was of topical importance during World War II; it was noticed that the frequency with which look-outs noticed U-boats on the radar screen diminished with the length of the period on watch. 50% of all sightings were reported during the first 30 minutes on watch; during subsequent 30-minute periods these fell to 23%, then to 16%, and finally to 10%. Obviously alertness became less, the longer the time on watch. This wartime experience led to many studies of alertness, which came to be called "vigilance research".

Missed signals

In an early investigation, which has become almost a classic, *Mackworth* [209] set up a number of research subjects in a quiet situation and made them watch an electric clock. Each rotation of the hand occupied one minute and was divided into 100 steps of 1/100 minute each. Occasionally the hand jumped two steps at once, and the observers were required to notice this and respond to it. The experiment lasted two hours, and during each 30-minute period there were 12 critical signals (i.e. double jumps of the hand) at irregular intervals. The results are shown in *Figure 118*.

These results confirm the experience of the radar observer and the submarines, referred to earlier: the number of missed signals increased as time went on. Following this pioneer work of *Mackworth*, there have been hundreds of other experiments to find out how the level of alertness was affected by different kinds of signals, critical and non-critical, by their number and arrangement, and by the duration of the experiment. *Leplat* [197] rightly criticised vigilance research, which dissipated itself in the study of too many variables, using questions that had too little relevance to practical conditions. In spite of these limitations it is possible, with

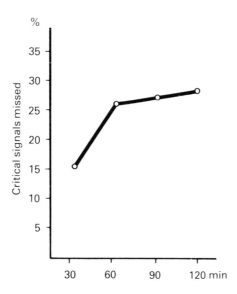

Fig. 118 Frequency of missed signals during a test of vigilance lasting 2 hours by 25 sailors, as a percentage of that of the preceding 30-minute period. Critical signal: a double jump of a pointer. After *Mackworth* [209].

hindsight, to pick out a number of results that have a practical value, and to formulate certain laws. Good reviews of the literature are given by *Broadbent* [36], *Leplat* [197], *Schmidtke* [279], *Jane F. Mackworth* [208], as well as *Davis* and *Tune* [61].

Signal frequency and performance

Among all these results there is one that is significant for our later preoccupation with the problem of boredom. Several investigations have shown unmistakeably that the frequency with which signals are noticed rises when there are more such signals per unit time. *Schmidtke* [279] has shown that this rise continues up to an evidently optimal frequency of 100–300 critical signals per hour. If this limit is appreciably exceeded, then the observational performance falls off again. These results are set out in *Figure 119*.

Schmidtke [279] comes to the conclusion that the relation between the frequency per unit time of the critical signals presented and the observational performance has the shape of an inverted "U". Hence we may assume that too few signals leave the research subjects understressed, and conversely that excessively frequent signals make too heavy demands on them.
Jerison and *Pickett* [164] have observed in addition that the frequency of irrelevant signals also affects performance; the more there are, the poorer is alertness. The obvious conclusion is that habituation is the cause of this loss of performance (see *"Boredom"* in *Chapter 12*).

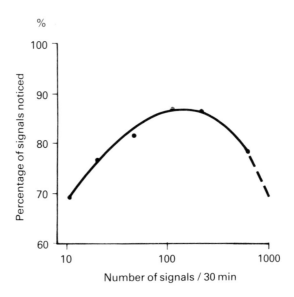

Fig. 119 Relation between frequency of signals requiring reaction and the observed performance. After *Schmidtke* [279].

Results from vigilance research

The most important results to date from vigilance research may be summarised as follows:

(a) sustained alertness (measured by the number of missed signals) decreases the longer the period of supervisory duty. The decline begins to be evident, as a rule, after the first 30 minutes;

(b) within certain limits the observational performance is *relatively improved*, if:
   (i) the signals are more frequent;
   (ii) they are stronger;
   (iii) the subject is informed about his own performance;
   (iv) the signals are more distinct in shape or contrast;

(c) performance is *worse*, if:
   (i) intervals between signals vary a great deal;
   (ii) the research subject has previously been under physical stress, or aroused from sleep;
   (iii) the research subject has performed under unfavourable conditions of noise, temperature, humidity, etc;

(d) many results of vigilance tests are strikingly similar to analogous research into reaction times. Both show a close relationship with the strength of stimulus, with the interval between signals, and with the current state of information available to the subject.

Theories

The psychologists, too, have not missed their opportunity to construct theories about vigilance! Thus we have between six and ten different theories to explain the varied observations made during concentration experiments. The three theories most worth considering may be mentioned briefly, as follows:

(a) *Mackworth's* *internal inhibition theory* [209] invokes our knowledge of conditioned reflexes and explains the decline in performance as a consequence of inhibition because of the absence of "rewards" and "incentives";

(b) the *"arousal" or activation theory* explains the fluctuations in concentration in terms of the level of activity of the cerebral cortex, as we know it from neurophysiology (see *Chapter 11*), and its relation to the activity of the *formatio reticularis* (*Figure 117*). According to this theory, a lack of external stimuli reduces the activity of the *formatio reticularis*, which in turn induces changes in the cerebral cortex, leading to increased drowsiness, and a lower level of concentration.

Numerous experiments which have clarified different aspects of vigilance can be readily explained by the effect of the activation system on the functional state of the brain, and hence on observational efficiency;

(c) *Broadbent's filter theory* [36] is based upon "channel capacity", whereby a filtering mechanism allows only signals with certain characteristics to pass through in the stream of information. During the course of a task requiring alertness the filter becomes less and less discriminating, and allows irrelevant signals to pass. These overload the channel, exceeding its capacity and crowding out some of the relevant signals.

All these theories are still disputed, and none of the numerous experiments has been satisfactorily explained. Nowadays it seems doubtful whether sustained concentration can be considered as an isolated process, complete in itself. Perhaps that is why no one theory can fully explain it.

From a physiological point of view there is support for the belief that *sustained concentration depends on the functional state, i.e. the level of activity, of the cerebral cortex*. This is a dynamic equilibrium, controlled by a variety of influences, by stimuli which excite, or which damp down activity. No doubt the activation system of the *formatio reticularis* also plays a decisive role, but this is certainly not the only motivation. There are good reasons for assuming that other neurophysiological processes affect the level of activity and hence the degree of sustained concentration. Among these may be mentioned *habituation* (becoming accustomed to unimportant stimuli), *adaptation* (decline in intensity of stimuli transmitted by the sense organs), and finally *the part played by the limbic system of the brain in motivation and emotional reactions.* The processes of habituation and adaptation are discussed in more detail in *Chapter 12.*

The obvious conclusion is that the arousal (activation) theory best fits the observed facts of sustained concentration. If we consider all the many processes such as habituation, adaptation and the influence of the limbic system, there are few experimental results in vigilance research that cannot be accounted for by neurophysiological processes in the brain. Still, to say it is all a neurophysiological phenomenon is hardly to formulate a theory.

11
Fatigue

Terminology

"Fatigue" is a state that is familiar to all of us in everyday life. The term usually denotes a loss of efficiency and a disinclination for any kind of effort, but it is not a single, definite state. Nor does it become clearer if we define it more closely as physical fatigue, mental fatigue, and so on.

The term "fatigue" has been used in so many different senses that its applications have become almost chaotic. A reasonable distinction is the common division into *muscular fatigue* and *general fatigue*.

The former is an acutely painful phenomenon, which arises in the overstressed muscles, and is localised there. General fatigue, in contrast, is a diffused sensation, which is accompanied by feelings of indolence and disinclination for any kind of activity. These two forms of fatigue arise from completely different physiological processes, and must be discussed in separate sections.

Muscular fatigue

External symptoms

Figure 120 illustrates the external signs of muscular fatigue as they appear in an experiment with an isolated muscle from a frog. The muscle is stimulated electrically, causing it to contract and perform physical work by lifting a weight. After several seconds it is seen that:

(a) the height of lift decreases;

(b) both contraction and relaxation become slower;

(c) the latency (interval between stimulation and beginning of the contraction) becomes longer.

Essentially the same result can be obtained, using mammalian muscle. The performance of the muscle falls off with increasing

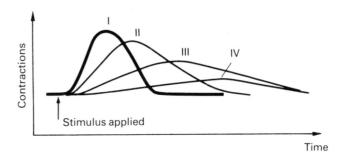

Fig. 120 Physical manifestations of fatigue in an isolated muscle from a frog's leg. I: contraction and relaxation of a fresh muscle. II: the same, after moderate stress. III: the same, after heavy stress. IV: the same after most severe stress.

strain until the stimulus no longer produces a response.

Human beings show this process, whether the nerve or muscle is stimulated electrically, or whether the research subject makes voluntary and rhythmical contractions of a muscle over a period.

This phenomenon of reduced performance of a muscle after stress is called "muscular fatigue" in physiology, and is characterised not only by reduced power, but also by slower movement. Herein lies the explanation of the impaired coordination, and increased liability to errors and accidents that accompany muscular fatigue.

Biochemical changes

We know that during muscular contraction chemical processes occur which, among other things, provide the energy necessary for mechanical effort. After contraction, while the muscle is relaxed and resting, the energy reserves are replenished. Thus both energy-releasing breakdown, and energy-restoring synthesis are going on in a working muscle. If the demand for energy exceeds the powers of regeneration, the metabolic balance is upset, resulting in a loss of muscular performance.

After a muscle has been heavily stressed, its energy reserves (sugar and phosphorus compounds) are depleted, while waste products multiply, the most important of these being lactic acid and carbon dioxide. The muscular tissue becomes more acidic.

Electrophysiological phenomena

There are many references in the literature to the fact that even after a muscle has been exhausted by repeated voluntary contractions, it will still respond to an electrical stimulus applied to the skin, suggesting that this form of fatigue is a phenomenon of the central nervous system, i.e. of the brain, not of the muscle itself.

This interpretation cannot, however, be confirmed in every case. Many physiologists have made the same observation, that when the muscle itself is in an exhausted state it does not contract any more, even though further motor nerve impulses are visible on the electromyogram.

It seems that fatigue has now become a peripheral phenomenon, affecting the muscle fibres, as *Scherrer* [277] suggests. We must presume that one lot of people were looking for the first sign of fatigue, whereas the others were studying muscles that were already in a state of exhaustion.

Electromyograms of fatigued muscle

Comparing all the experimental results, we are led to the assumption that the central nervous system acts as a compensatory mechanism during the early stages of fatigue. Thus many studies with the electromyograph have shown that when a muscle is repeatedly stimulated its electrical activity increases, even though its contractions remain at the same level or decline. This must mean that more and more of its individual fibres are being stimulated into action (recruiting of motor units). The electromyogram in *Figure 121* demonstrates this increasing electrical activity as fatigue increases.

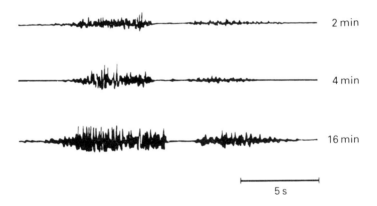

Fig. 121 Three sections from the electromyogram of the extensor muscle of the upper arm, after it has been fatigued by a long series of contractions of equal strength. From above downwards, the electromyogram after 2, 4 and 16 minutes' work. Redrawn after *Scherrer* [277].

These electromyographical phenomena have also been observed under practical conditions. For example, after 60–80 minute periods of perforating punched cards, increased electrical activity was recorded in the muscles of forearms and shoulders [344]. It is reasonable to conclude from this that the muscular fatigue that arises in industry is still at the stage where it can be compensated for by increased activity of the CNS.

If we now turn to a state of exhaustion, which is certainly located in the muscles themselves, we find that this is accompanied by a reduction of muscular strength. At first this can be partly compensated for by increasing the load on the motor neurones, but later on, as muscular fatigue increases, the reduced performance needs to be offset by an effort of will, i.e. by bringing more neuro-muscular elements into play.

Theories of
muscle fatigue

There are essentially only two current theories of muscular fatigue: chemical, and CNS. The former sees the loss of performance in fatigued muscle as a result of chemical processes (consumption of energy-producing substances, and accumulation of waste products) with the electrical phenomena in the muscles and nerves playing only a secondary role.

In contrast, the CNS theory sees the chemical processes as being merely the releasing stimulus for sensory impulses which travel along the nerves to the brain and cerebral cortex. It is here that muscular fatigue makes itself evident as a sensation of weariness. In addition, the afferent impulses inhibit that centre of the brain that is responsible for motor control of movements, and so bring about a reduction in the number and frequency of impulses along the motor neurones. This in turn brings about the external signs of muscular fatigue, namely the reduction in muscle power, and, where rhythmic movements are concerned, shortening and slowing down of muscle travel.

Neither of these two theories explains everything. All that can be said with certainty is that *both CNS phenomena (recruiting of motor neurones) and chemical processes in the muscles (using up of energy reserves, and accumulation of waste products) contribute something to muscular fatigue.* Just how the two processes are linked together is not fully understood, and must be left for future experiments and theorising.

General fatigue

A sensation
of weariness

A major fatigue symptom is a general sensation of weariness. We feel ourselves inhibited, and our activities are impaired, if not actually crippled. We have no desire for either physical or mental effort; we feel heavy and drowsy.

A feeling of weariness is not unpleasant if we are able to rest, but it is distressing if we cannot allow ourselves to relax. It has long been realised, as a matter of simple observation, that weariness, like thirst, hunger and similar sensations, is one of nature's protective devices. Weariness discourages us from overstraining ourselves, and allows time for recuperative processes to take place.

Different kinds
of fatigue

Leaving aside purely muscular fatigue, the following other kinds can be distinguished:

(a) visual fatigue, arising from overtiring the eyes;

(b) general bodily fatigue: physical overloading of the entire organism;

(c) mental fatigue, induced by mental or intellectual work;

(d) nervous fatigue, caused by overstressing one part of the psychomotor system, as in skilled work;

(e) monotony of either occupation or surroundings;

(f) chronic fatigue, an accumulation of long-term effects;

(g) circadian or nyctemeral fatigue, part of the day–night rhythm, and initiating a period of sleep.

This classification of types of fatigue is based partly on the cause, and partly on the way in which the fatigue manifests itself, with the obvious corollary that the two should be linked as cause and effect. This should be particularly true for the different sensations of fatigue, which vary according to source. We believe, however, that there are certain regulatory processes in the brain that are common to fatigue of all kinds, and we shall now consider these in more detail.

Functional states

At any moment the human organism is in one particular functional state, somewhere between the extremes of sleep on the one hand, and a state of alarm on the other. Within this range there are a number of stages, as shown in the following summary:

Deep sleep	Light sleep, drowsy	Weary, hardly awake	Relaxed, resting	Fresh, alert	Very alert, stimulated	In state of alarm

Seen in this context, fatigue is a functional state which in one direction grades into sleep, and in the opposite direction into a relaxed, restful condition.

Before we go into the neurophysiological basis of this functional state it would be sensible to take a look at the most important method by which fatigue can be studied: *the electroencephalograph*, which records the electrical activity of the brain. When studying human subjects the electrodes are usually applied to the skin of the head, where they detect and register the waves of electrical potential in the cerebral cortex. The resulting encephalogram makes it possible to study the varying amplitudes and frequencies of these waves.

The electro-encephalogram

In greatly simplified form, the most important features recorded in an electroencephalogram are as follows;

(a) the *alpha rhythms* comprise waves in the frequency band 8–12 Hz. Alpha waves are present during waking hours and are blocked by sensory impulses, so that a high alpha wave component indicates a relaxed condition and a reduced readiness to react to stimuli. A lower alpha component, coupled with a higher beta component indicates a more alert state;

(b) the *theta rhythms* (4–7 Hz) are slow, long-period waves, which replace the alpha waves when we go to sleep. Sleep is characterised by a whole series of other phenomena, which cannot be discussed here;

(c) the *delta rhythms* are also slow waves, of less than 4 Hz, which are present only during sleep;

(d) *desynchronisation*, producing beta rhythms of 14–30 Hz, is an irregular electrical activity of very low amplitude. It occurs after the receipt of a sensory stimulus, and is the expression of an interruption of the synchronised activities of neurones which make up the alpha rhythm. (Synchronisation is the simultaneous excitation of a large group of nerve cells.) *Desynchronisation is the sign of a state of increased alertness, and it is also known as an arousal reaction.* Awaking and taking alarm are examples of this. It may be mentioned here that adrenalin (a stimulant derived from the adrenal medulla), which is released into the blood as a reaction to stress, acts by inducing desynchronisation;

(e) *evoked potentials* are single fluctuations of potential which are induced by isolated sensory stimuli. They are located in that area of the cerebral cortex where the ascending nerve tract from that particular sense organ terminates (this has made it possible to plot a "map" of the cortex, showing which area serves which sense organ). The first of "primary" evoked potentials is followed by other "diffuse secondary potentials", about which no more can be said here.

Figure 122 shows five extracts from electroencephalograms, each characteristic of a different functional state of the body.

What has been said up to now gives the impression that for every functional state that we can describe by terms such as "weary", "lively", etc. there should be a distinctive pattern on the electro-encephalogram. Unfortunately this is not the case. Those patterns shown in *Figure 122* indicate no more than a very rough correspondence between the encephalogram trace and the underlying physiological state. Nevertheless it has been possible in the last 20 years to make some interesting discoveries about how various functional states are controlled, using electroencephalography in association with electrical stimuli and behaviour studies.

The reticular activation system

Several neurophysiologists, particularly *Magoun* and his school [211], have revealed nervous structures deep in the brain, in the

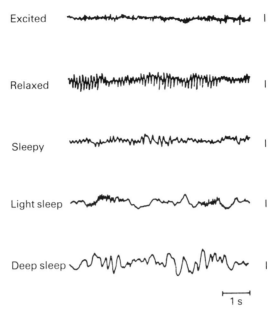

Excited

Relaxed

Sleepy

Light sleep

Deep sleep

1 s

Fig. 122 Five sections from electroencephalograms, characteristic of various functional states. The vertical lines indicate the scale for 1 μV. Redrawn after *Jasper* [160].

so-called reticular formation of the medulla, which can increase the sensitivity of the cerebral cortex to some extent (see *Figure 117*). It is in the cerebral cortex that conscious awareness is localised, including the powers of perception, subjective feeling, reflection and will-power. Increasing the sensitivity of the cortex will therefore bring all the conscious functions to a higher level of alertness.

Thus it appears that the reticular formation controls the degree of alertness, including attention, and readiness for action. Its level of activity is very low during deep sleep, increases when sleep becomes shallow, and rises steeply on awakening. The higher the level of reticular activity, the higher the level of alertness, culminating in a state of alarm.

These reticular structures, which control the sensitivity of the conscious functions, are called *the ascending (arousal) reticular activating system* (ARAS), whereas those structures that increase the readiness of the skeletal musculature are the descending activating system. In what follows we shall be concerned solely with the arousal activating system.

The activating structures of the reticular formation do not, however, act on their own initiative. They themselves must be activated by afferent nerve stimuli, which come in the main from *two sources, the conscious sphere of the cerebral cortex and the sense organs.* What significance do these two have?

Feed-back control   *Nervous tracts coming from the cerebral cortex* carry impulses from the conscious sphere into the reticular activating system. Such impulses arise for example, when an idea, or something noticed outside, seems ominous and calls for increased alertness. A closed circuit is then set up; the reticular activating system arouses the cerebral cortex and alerts conscious perception. If this results in any significant signals being received, then these stimuli send impulses back along the nerve tracts from the cortex to the reticular activating system. The result is a feed-back system, analogous to that in many electronic devices.

Afferent   *The other sources of stimuli to the reticular activating system are*
sensory system   *the stream of afferent stimuli from the sense-organs;* nerve fibres branch off from all the afferent nerve tracts and pass to the reticular activating system, and this sensory inflow has its effect upon the level of reticular activity. Every strong impulse that comes from the ear, the eye, or from nerves conveying pain, can raise the level of reticular activity in a flash. The importance of this is obvious. Signals coming into the body from the outside world are passed over to the reticular activating system, and make this more active; this alerts the cerebral cortex and so ensures that the brain is ready to notice and act on what is happening outside the body. *This close linkage between the reticular activating system and the afferent sensory system is an essential prerequisite for conscious reaction to the external world.*

For example, suppose that a loud noise is heard. Any sudden noise stimulates the reticular activating system by way of the afferent sensory stream, and thereby increases the alertness of the cerebral cortex. There the noise is perceived and interpreted, perhaps resulting in precautionary action against possible danger. In that case the reticular activating system is operating as a distributor and amplifier of signals from the outside world, and makes sure that the organism is alerted as much as is necessary for the preservation of its life.

Figure 123 shows diagrammatically the flow of stimuli from the cortex and the sense organs to the reticular activating system, as well as the presumed pathways in the cortex itself.

Limbic system   The reticular activating system is not the only nervous organ that
and level of   affects the state of readiness of the cerebral cortex, and through
activity   this, of the entire body. Yet another of the numerous functions of the limbic system must be mentioned: the part it plays in excitement, emotion and motivation. *Figure 117 (Chapter 10)* showed the anatomy of the limbic system and briefly enumerated its functions. The general level of alertness is greatly dependent upon the limbic centres for circadian rhythm (day–night periodicity), fear, rage and calmness, as well as for motivation.

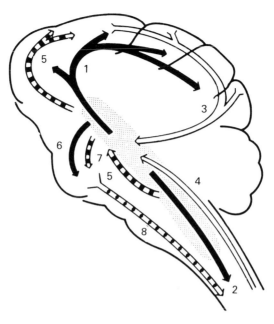

Fig. 123 The activating and inhibitory systems in the brain. Dotted area = reticular formation. 1 = ascending reticular activating system (ARAS). 2 = descending activating system. 3 = pathways from the cerebral cortex. 4 = incoming sensory pathways. 5 = inhibitory (damping) system. 6 and 7 = links with vegetative (autonomic) centres. 8 = vegetative tracks leading to the autonomic nervous system of the internal organs.

The limbic system has many links with other parts of the brain structure. The so-called cross-switching is interesting for our purposes. Connections from the limbic system run to particular structures (the mamillary body, and the thalamus), and extend diffusely over the cerebral cortex. Return tracts lead from the cortex, across the hippocampus, back to the limbic system. Thus the limbic system and the cerebral cortex are united by a feed-back system similar to that of the reticular formation.

Since the limbic system (together with the hypothalamus) is involved in the generation of emotional states and the build-up of motivation, it seems obvious that *the activity level of the cerebral cortex, and through it of the entire organism, will be influenced by what goes on in the limbic system.* Links between the flow of afferent sensory impulses and the reticular activating system lead to the further assumption that the two systems have a reciprocal effect on each other.

Inhibiting system   These two activating systems are not alone, however, but have other structures which clearly work in opposition to them. In fact many physiologists have conducted experiments showing the existence of nervous elements both in the interbrain and in the medulla which have an *inhibitory or damping effect on the cerebral cortex*. These studies trace back to *W. R. Hess* [140], who was the first to show experimental inhibition, by applying electrical stimuli to particular areas of the deep structure of the brain (the somnogenous zone of the thalamus) in cats. First the pupils contracted, then the eyelids closed frequently; the cats looked for a suitable place to lie down, and finally fell into a natural sleep. We must conclude from this behaviour that stimulation of the brain first of all induced the feelings and symptoms of weariness, and these in turn progressed into actual sleep. This discovery by *W. R. Hess* was subsequently confirmed by various authors (e.g. *Akert, Caspers, Monnier,* etc.). In particular, these authors were able to demonstrate that an inhibition can spread from these structures in the thalamus and extend to the cerebral cortex, and that its effects are evident not only in the animals' behaviour, but also in the electroencephalogram.

Jasper [161] and his school have made more detailed studies of structures in the thalamus (which lies immediately in front of and above the reticular formation in the interbrain), and came to the conclusion that a diffuse "projection system" spreads out from this somnogenous (sleep inducing) region and extends over the cerebral cortex.

Later on other research workers, notably *Moruzzi* [228] and *Magoun* [212], found other nerve centres in the posterior part of the reticular formation (in the bulbopontine structures), which inhibited the conscious functions in the cortex. The authors think that these centres can exercise an inhibiting function directly on the reticular activating system, reducing activity until sleep supervenes.

Many authors treat all these inhibiting structures as being independent, but others refuse to do so. Certainly we do not know enough about the relationship between the somnogenous zone in the thalamus, with its extensions over the cerebral cortex, and those structures which inhibit the reticular formation. On the other hand almost all modern neurophysiologists are of the opinion that *there are structures in the interbrain and in the medulla which actively inhibit the alertness of the organism and which produce effects on the electroencephalogram resembling those seen during sleep.* Without committing ourselves as to whether there are several centres, or one unified system, we can safely say that these inhibiting and sleep-inducing centres collectively constitute a damping system. The working of this damping system is shown diagrammatically by the broken arrows in *Figure 123*.

Interrelations with the autonomic nervous system

In this connection *W. R. Hess* [140] has already shown that the nervous structures in the interbrain (mostly located in the thalamus and hypothalamus), which induce weariness and sleep, also have their effect on other vital functions. Thus, for example, intake of food and water, temperature control, and even emotionally controlled defensive and flight reactions are directed by this part of the brain. Some of the feelings and impulses that may have a vital effect on the behaviour of either animals or humans obviously arise in the interbrain. Feelings of weariness are one of the symptoms that accord well with the functions of the interbrain.

We now know that there are close links between the vegetative (autonomic) nervous system, which controls the activities of the internal organs, and these activating and inhibitory systems. In fact any increase in stimulation of the reticular activating system is accompanied by a whole series of changes in the internal organs, of which we may mention the following:

(a) increase in heart rate;
(b) rise in blood pressure;
(c) more sugar released by the liver;
(d) increased metabolism.

Ergotropic
tuning-up

This increased sensitivity spreads from the activating system to all parts of the body, brain, limbs, internal organs, until the entire organism is braced up for a period of high energy consumption, whether this is for work, for fighting, for flight, or whatever. *W. R. Hess* coined the term *"ergotrope Einstellung"*, or "ergotropic tuning-up" for this process.

Trophotropic
tuning-up

By an analogous process, increased activity of the inhibitory system lowers the heart rate and blood pressure, cuts back respiration and metabolism, and relaxes the muscles, while the digestive system works more vigorously to assimilate more energy. *W. R. Hess'* term for this is *"trophotrope Einstellung"* or "trophotropic tuning-up" by which he means that this process accelerates the recuperative functions by assimilating more food and replacing lost energy.

Thus we see that the nervous structures in the interbrain and brain stem (medulla) play a key role in the regulation and coordination of functional states of the body. *The cerebellum as well as the internal organs are directed in a logical manner by these nerve centres*, as two examples will show:

(a) a fire alarm sounds in a factory. The acoustic signal raises the level of activity of the reticular formation in all the operatives within the danger zone. On the one hand this activates the cerebral cortex to a greater awareness of its surroundings, while alerting the internal organs ready for a greater consumption of energy;

(b) the same person goes home after his work. The stream of afferent sensory impulses dies down, while the inhibitory centre shows more activity. The level of activity of the cerebral cortex is diminished, and the internal organs go into "low gear", i.e. they concentrate on assimilative and recuperative processes. The nervous centres in the interbrain and in the brain stem (medulla) damp down both the cerebral cortex and the internal organs.

Humoral

Humoral control means regulation by means of chemical substances which circulate in the body fluids. We have seen that the reticular activating system is not autonomous, but is dependent upon stimulation from the cerebral cortex as well as from afferent sensory signals. In addition to these purely nervous mechanisms there are also the humoral effects which have a part to play in tuning-up the activating system as well as regulating the sensitivity of the limbic system.

Adrenalin

We have already mentioned desynchronisation, the effect produced by the performance hormone *adrenalin*, and which must be interpreted as a raising of the degree of alertness and of the activation level of the body. *Bonvallet* and his colleagues [29] deduced from what we now know that every time the activation level is raised by stimuli from outside the body there is a release of adrenalin, which in turn stimulates activity of the reticular formation. The increased activity from outside stimuli would soon fade away again if this hormone did not intervene to maintain it. On this theory, the reticular activating system is responsible for short-term reactions, while effects of longer duration depend upon the action of adrenalin.

We can say, therefore, that *the level of activity of the reticular activating system depends upon:*
(a) *the inflow of sensory stimuli;*
(b) *stimulation of the cerebral cortex;*
(c) *level of adrenalin.*

Effect of hormones on the inhibitory system

Much less is known about how the inhibitory system comes under nervous and hormonal control. Nervous mechanisms are scarcely known, but the existence of hormonal factors has been theoretically assumed for many years. The protracted inhibition that occurs during sleep or during a state of exhaustion can hardly be represented as a purely nervous mechanism. Moreover, the fact that feelings of weariness disappear after a period of sleep is another reason for supposing both the onset and the continuance of inhibition to have a chemical basis.

A few pieces of research [193] [252] indicate that if animals are prevented from sleeping, some kind of "fatigue substance" appears in their cerebrospinal fluid. Both the research and its interpretation are still disputed, however. The experiments of *Monnier* and *Schoenenberg* [225] may be mentioned in this context. They were able to extract a substance from the blood of sleeping rabbits which, when injected into waking rabbits, would put them to sleep.

Is serotonin an inhibitory substance?

Serotonin is a substance naturally produced in the body and present in high concentrations in the blood-platelets, in the intestinal tract and in the brain. It is assumed to have a function in transmitting stimuli. *Koella* [181] and *Jouvet* [166] carried out interesting experiments, in which they both observed a sleep-inducing effect after the application of serotonin to the brain. The serotonin seems to have accumulated in the brain stem (raphe nuclei of the medulla), whence there are linkages to the hypothalamus, the limbic system and the cerebral cortex. Either lesion of a raphe, or reduction of the serotonin content brings about a permanently waking state. In contrast, any substance that increases the serotonin content of the brain promotes sleepiness, along with more slow waves in the electroencephalogram. So there are good grounds for assuming that *serotonin-sensitive nerve fibres in the brain stem influence the inhibiting system by some means that are still unknown.*

In everyday life we are aware that weariness and sleep can result from changes in the content of the blood (e.g. a fall in the concentration of sugar), from many forms of illness, and finally from drugs and medicines. We still know almost nothing about how all these factors actually operate on the inhibitory centres in the interbrain and brain stem. Is there, perhaps, some structure in the brain that holds a key position in controlling internal changes in the body, analogous to the key position that the reticular activating system holds in relation to external processes? The odds are in favour of such an analogy, but it is by no means certain as far as our present knowledge goes. We can, however, say very generally that activation is promoted by outside influences, and inhibition by changes within the body.

A neuro-physiological model of fatigue

The processes illustrated in *Figure 123* may be considered to be scientifically verified neurophysiological mechanisms. On the basis of these data we shall try to construct a mental picture that will explain how fatigue and its opposite, readiness for action, are regulated.

The six essentials for this purpose are as follows:

(a) fatigue is a functional state, one of several intermediate conditions between the two extremes of alarm and sleep;

(b) functional states are determined by the level of activity of the cerebral cortex, which, in turn, is recognisable by the degree of synchronisation of the discharges seen in the electroencephalogram (frequency distribution of various frequency ranges among the waves), along with the correlated subjective impressions such as "weary", "fresh", etc.;

(c) the functional state in which a person happens to be at any particular moment is regulated by the level of activity of various structures in the interbrain and the medulla (brain stem);

(d) there are two types of control structure: those systems that raise the level of activation of the cerebrum and, conversely, those that reduce it;

(e) the reticular activating system, and also to some extent certain centres in the limbic system, increase the activation level of the cerebral cortex. If the one declines, so does the other;

(f) inhibitory centres in the interbrain and the medulla constitute a damping system, which acts by lowering the activation level of the cerebral cortex.

It follows from points (e) and (f) that *the level of cerebral activation, and hence the functional state of the entire organism, is in a state of balance, an equilibrium between the opposing systems of activation and inhibition.*

If the activating influences predominate, then the organism is placed in a state of increased readiness to react to stimuli, indicated by the physiological signs of freshness, excitement and eager alertness.

On the other hand, if inhibitory influences predominate or the activating processes are greatly reduced, then the organism is less ready to react to stimuli, indicated by the physiological signs of weariness and somnolence, accompanied by a lessening of motivation and awareness.

Figure 124 compares this neurophysiological model with a balance.

Our mental concept can be summarised in different terms, as follows: the level of readiness to act lies somewhere between the two extremes of sleep and the highest state of alarm. Control mechanisms in the medulla and the interbrain regulate this and adjust it to meet the momentary demands of the organism. If external influences are dominant, then the activating systems prevail; the person feels keyed up, even in a state of alarm, and he is ready for action both physically and mentally. If, however, inhibiting influences from inside the body predominate, then the damping system prevails; the person feels sluggish, drowsy and lethargic.

Fatigue in industrial practice

Causes of general fatigue

We know from everyday experience that fatigue has many different causes, the most important of which can be seen in

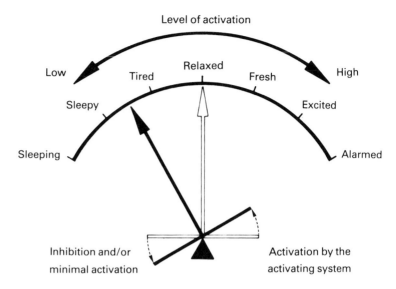

Fig. 124 A theoretical model to illustrate the neurophysiological mechanism which regulates the functional state of the organism. The level of activation of the cerebral cortex, the degree of readiness for action, and the level of alertness all increase from left to right.

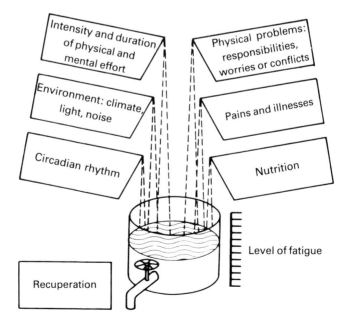

Fig. 125 Theoretical diagram of the combined effect of everyday causes of fatigue, and the recuperation necessary to offset them. The total stresses must be balanced by the total recuperation, within the 24-hour cycle.

Figure 125. This is a block diagram illustrating that the degree of fatigue is an aggregate of all the different stresses of the day. Visualise this diagram as a barrel partly filled with water. The recuperative rest periods would be the outflow from the barrel. To make sure that the barrel does not overflow we must ensure that inflow and outflow are of the same order of magnitude. In other words, to maintain health and efficiency the recuperative processes must cancel out the stresses. Recuperation takes place mainly during nighttime sleep, but free periods during the day and all kinds of pauses during the work also make their contribution.

It must be emphasised that stress and recuperation must balance out over the 24-hour cycle, and that neither of these should be carried over to the next day. If rest is unavoidably postponed until the following evening, this can be done only at the expense of well-being and efficiency.

Symptoms
of fatigue

Fatigue symptoms are both subjective and objective, the most important being:
(a) subjective feelings of weariness, somnolence, faintness and distaste for work;
(b) sluggish thinking;
(c) reduced alertness;
(d) perception poor and slow;
(e) unwillingness to work;
(f) decline in both bodily and mental performance;
Some of these symptoms result in a measurable drop in bodily and mental efficiency.

Clinical or
chronic fatigue

Some of the fatigue states that arise in industrial practice are of a chronic nature. These are conditions that are brought about, not by *a single* instance of overstrain, but by stresses which recur everyday over long periods. Since conditions such as these are usually also accompanied by signs of ill-health, this may correctly be called *clinical or chronic fatigue.*

Under these conditions the symptoms not only occur during the period of stress or immediately afterwards, but are more or less latent all the time. Tired feelings are often present on waking up in the morning, before work has even begun. This form of fatigue is often accompanied by feelings of distaste, which have an emotional origin. Such persons often show the following symptoms:
(a) increased psychic instability (quarrelsomeness and associated behaviour);
(b) fits of depression (baseless worries);
(c) general weakening of drive and unwillingness to work;
(d) increased liability to illness

The aforesaid ailments are mostly vague and come under the heading of psychosomatic complaints. This term is applied to functional disturbances of the internal organs or the circulation which are judged to be external manifestations of psychological conflicts and difficulties. Some of the commoner of these symptoms are:

(a) headaches;
(b) giddiness;
(c) loss of sleep;
(d) irregular heart-beat;
(e) sudden sweating fits;
(f) loss of appetite;
(g) digestive troubles (stomach pains; diarrhoea; constipation).

More ailments mean more absences from work, especially short absences, indicating that the cause of the absenteeism is the need for more rest.

People who have psychological problems and difficulties easily fall into a state of chronic fatigue, and it is often difficult to disentangle their psychic from the physical problems. In practice cause and effect are hard to distinguish in cases of clinical fatigue. This may be caused by dislike of the occupation, of the immediate task, or of the work place, or conversely may itself be the cause of maladjustment to work or surroundings.

Measuring fatigue

Why measure fatigue?

The science of ergonomics is just as interested in the quantitative measurement of fatigue as is industry itself. Science studies the relationships between the fatigue and the level of stress, or more accurately fatigue and the output of work; otherwise, the reaction of the human body to different stresses can be measured in order to develop ways of improving work and making it less laborious. The question finally asked by industry is often simply whether the working conditions make excessive demands on the operatives, or whether the stresses involved are physiologically acceptable.

We are measuring only "indicators" of fatigue

Discussion of measuring methods is subject to one serious limitation: *to date there is no way of directly measuring the extent of the fatigue itself.* There is no absolute measure of fatigue, comparable to that of energy consumption, expressed in kilocalories. All the experimental work carried out so far *has merely measured certain manifestations or "indicators" of fatigue.*

Methods of measurement

Currently used methods fall into six groups:
(a) *quality and quantity of work performed;*

(b) *recording of subjective impressions of fatigue;*
(c) *electroencephalography (EEG);*
(d) *measuring subjective frequency of flicker-fusion of eyes;*
(e) *psychomotor tests;*
(f) *mental tests.*

Measurements such as these are frequently taken *before, during* and *after* the task is performed, and the extent of fatigue is deduced from these. As a rule the result has only relative significance, since it gives a value to be compared with that of a fresh subject, or at least with that of a "control" who is not under stress. Even today we have no way of measuring fatigue in absolute terms.

Correlation with subjective feelings

More recently it has been the practice to study a combination of several indicators so as to make interpretation of the results more reliable. It is particularly important that subjective feelings of fatigue should also be taken into account. *A measurement of physical factors needs to be backed up by subjective feelings before it can be correctly assessed as indicating a state of fatigue.* The five indicators listed above will now be briefly discussed.

Quality and quantity of output

The quality and quantity of output is sometimes used as an indirect way of measuring industrial fatigue. Quantity of output can be expressed as number of items processed, as time taken per item, or conversely as the number of operations performed per unit time. Sometimes it is found that production can be improved by making the work easier. In addition output may vary periodically, being dependent on different times of day or on different days of the week. Fatigue and rate of production are certainly interrelated to some extent, but the latter cannot be used as a direct measure of the former because there are many other factors to be taken into account: production targets, social factors and psychological attitudes to the work.
Sometimes fatigue needs to be considered in relation to the *quality* of the output (bad workmanship, faulty products, outright rejects), or to the frequency of accidents, once again with the reservation that fatigue is not the only causal factor.

Subjective feelings

Special questionnaires have been used *to assess subjective feelings*, among which may be mentioned the so-called bipolar questionnaires, very simple to handle and easy to interpret. Two opposing, mutually exclusive states are placed at opposite ends of a line 7 cm long, and the research subjects are required to place a mark somewhere along the line to indicate how they assess these characteristics in themselves at that particular moment. This gives a quantitative value to the subjective feeling, though as a rule what is used is the difference between marks made at the beginning and end of the task.

Bipolar questionnaires

The contrasting pairs of characteristics that give particularly sharp differences after the performance of hard work are the following:

Fresh	weary
Sleepy	wideawake
Vigorous	exhausted
Weak	strong
Energetic	apathetic
Dull, indifferent	ready for action
Interested	bored
Attentive	absent-minded.

Barmack [13], *Haider* [129] and *Edith Groll* and *Haider* [122] tried out questionnaires of this type under industrial conditions and used them successfull. *Yoshitake* [345] studied 160 employees of a Radio-TV firm and of a bank, using a very simple scale of numbers 1–9, the extreme states being "I feel fresh and rested" at the one end, and "I feel extremely weary and exhausted" at the other. The author found a very good correlation between this simple process of self-assessment and the frequency of occurrence of 30 different fatigue symptoms. Other authors have suggested more complicated questionnaires, among which may be mentioned the comparison process of *Pearson* [253] and the very comprehensive questionnaire used by *Nitsch* [248].

The electro-encephalograph

The electroencephalograph is particularly suitable for standardised research in the laboratory, where variations in the trace in the sense of increasing synchronisation (increase of alpha and theta rhythms, reduction of beta waves) are interpreted as indicating states of weariness and sleepiness (see *Figure 122*).
The techniques of detecting and recording have been improved recently, so that the electroencephalograph can now be used successfully to monitor sedentary activities, such as driving a vehicle [251] [346].

Flicker-fusion frequency of the eye

During the last 25 years the *flicker-fusion frequency of the eye* has been increasingly used as an indicator of the degree of fatigue. This is measured as follows. The research subject is exposed to a flickering lamp, and its frequency is increased until the flickers appear to fuse into a continuous light. The frequency at which this occurs is called the subjective flicker-fusion frequency. The source of light should have an area that subtends an angle of 1–2° at the eye, and should be so placed that it does not call for any optical accommodation.

Improved techniques

Much better methods have recently come into use. Instead of the single lamp previously used, it is an improvement to have two lighted surfaces, each being half of a circular disc. If one of these flickers while the other remains steady, it is easier for the observer to decide the precise moment at which flicker disappears. A further improvement was achieved through the adoption of *Paule Rey's* [259] "straddling" technique. A light source with a frequency above the fusion frequency and another with a frequency below this flash alternately, and their frequencies are gradually brought together, thus "straddling" the target and fixing it more precisely. Finally it should be mentioned that it is considered better to produce the flashes by mechanically interrupting the current than by electrical discharges.

With these improvements in technique it is possible, on the one hand, to reduce the spread of the measurements, and on the other to make fairly sure that the research subjects are not being influenced by the investigator. *Baschera, Martin* and *Gierer* [19] give a short account of these innovations in method.

Lowering of the flicker-fusion frequency

It has been observed that reductions in the flicker-fusion frequency of 0.5–6 Hz take place after mental stress, as well as under various industrial stresses. Yet a survey of the literature shows that not every kind of stress brings about such a reduction. Experience to date can be summarised thus, to a first approximation:

(a) *a distinct lowering* of the flicker-fusion frequency can be expected during unbroken mental stress of a high level. Examples are calculating in one's head; working as a telephonist; piloting an aircraft; carrying out demanding visual work.

We shall see later (in *Chapter 12: "Boredom"*) that dull, repetitive and monotonous situations produce a distinct lowering of the flicker-fusion frequency;

(b) *either little lowering or none at all* result from work that requires only moderate mental effort, and which allows the operative comparative freedom of action, or which involves physical effort. Examples are office work; sorting jobs; repetitive work at a moderate level.

Figure 126 summarises the results of a piece of research by *Schmidtke* [280]. After calculation, either in the head or on paper, as well as a microscope test, he observed an average lowering of flicker-fusion frequencies, which after 3 hours reached values of from five to almost seven Hz. While it can be assumed that microscope work involves an element of optical stress which contributes to this lowering, that occurring after calculation in the head must be the result of some central process in the brain stem or the cerebral cortex. Later studies have several times confirmed the lowering of flicker-fusion frequency that accompanies sustained

calculation or mental stresses of a similar nature [18] [44] [103]·
even though the extent of the lowering was usually distinctly less
than that found in *Schmidtke*'s studies [280], already mentioned.

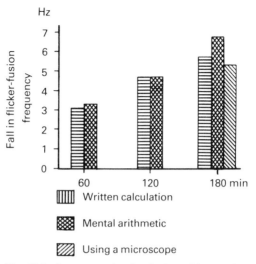

**Fig. 126 Average reduction in the subjective flicker-fusion frequency after
mental stress caused by calculating and using a microscope.** The columns
indicate the amount by which this frequency has declined from its original
value. After *Schmidtke* [280].

During several of these studies the reduction of flicker-fusion
frequency was accompanied by parallel changes in other signs of
fatigue, notably by increased feelings of weariness and sleepiness.
As an example, here is an experiment by *Annette Weber* and
others [322], in which eight research subjects in each of three tests
were given a dose of 5 mg of diazepam (Valium) to produce
"pharmacological fatigue". In a control experiment the same
persons were given a placebo (a similar tablet without the drug).
Measurements were taken of both the flicker-fusion frequency and
the subjective feelings of fatigue, the latter by using a bipolar
questionnaire slightly modified from that of *Edith Groll* and
Haider [122].

The results set out in *Figure 127* show that after the administration
of Valium there was a distinct lowering of the flicker-fusion
frequency, amounting to about 2 Hz on the average. Simultaneous
records of subjective feelings by the bipolar questionnaires
showed that in 10 out of 15 of the contrasting pairs of feelings
there was a shift in the direction indicating increased fatigue.
Correlations calculated by *Spearman* are summarised in *Table 23*.

This result shows a good correlation between flicker-fusion
frequency and subjective feelings of fatigue. The statistics show

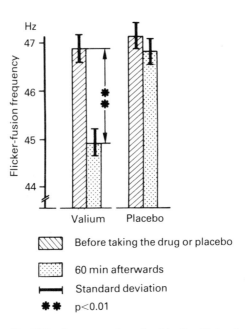

Fig. 127 **Average value of subjective flicker-fusion frequency before and after taking 5 mg diazepam (Valium), or a placebo.** 24 experiments with each of 8 test subjects. After *Annette Weber* and colleagues [322].

Table 23 **Individual correlations between the lowering of flicker-fusion frequency and the increase in subjective fatigue symptoms after 5 mg of Valium.**

r = correlation coefficient according to *Spearman*
p = probability
FF = flicker-fusion frequency.
After *Annette Weber* and colleagues [322].

Correlation between	r	p
FF and "fresh–weary"	0.81	< 0.01
FF and "strong–weakened"	0.83	< 0.01
FF and "wideawake–sleepy"	0.98	< 0.01
FF and "vigorous–exhausted"	0.83	< 0.01

that the same persons who showed a marked lowering of flicker-fusion frequency also exhibited considerable subjective fatigue.

In spite of these and other similar observations we can still say nothing of general application about the relationship between flicker-fusion frequency and subjective feelings of fatigue, since the individual correlations recorded here have not been confirmed by other experiments.

These experiments collectively have encouraged most authors to interpret a *lowering of the flicker-fusion frequency as a sign of fatigue*. Hence many authors think that this lowering is associated with a deactivation of the cerebral cortex [134], but this must certainly be thought of as no more than a hypothesis which needs to be tested by further experiment.

Psychomotor tests

Psychomotor tests measure functions that involve perception, interpretation and motor reactions. Tests very often used are the following:

(a) simple and selective reaction times;
(b) tests involving touching or pricking squares in a grid;
(c) tests of skill;
(d) driving tests under artificial conditions;
(e) typing;
(f) tachistoscopic tests (see *Chapter 9*) to measure performance involving perception.

Several examples of such test methods are described by *Welford* [324].

Reservations

In tests like these it is also assumed that a falling off of performance can be taken as a sign of a state of fatigue. Since, however, ability to perform a psychomotor test is dependent on other factors such as, for example, motivation, it is sometimes doubtful whether a general state of fatigue is really the main cause.

A further disadvantage of psychomotor tests arises from the fact that often the test itself makes heavy demands on the subject, thereby raising the level of excitability. In view of what we have said previously, it is very likely that such tests will cause some kind of cerebral activity, which may at least temporarily mask any possible signs of fatigue.

Performance of mental tests

Performance of mental tests often involves:

(a) arithmetical problems;
(b) tests of concentration (e.g. crossing-out tests);
(c) estimation tests (e.g. estimation of time intervals);
(d) memory tests.

The same reservation must be made as for psychomotor tests: the test itself may excite the interest of the person being examined and so cancel out any signs of fatigue. Other disturbing factors are the effects of training and experience, and, if the test is protracted, fatigue brought on by the test itself.

Field studies

In recent decades more and more use has been made of fatigue studies carried out under industrial conditions, in traffic, in schools and in various other everyday situations. As a rule their significance is limited to a particular problem in a particular setting, and almost nothing can be deduced from them that is of wider application or which would lead to generalisations about the relationship between stress and fatigue.

Traffic fatigue

A particular importance attaches to studies of fatigue in traffic, because it is reasonable to suppose that fatigue is an important contributory factor in mistakes and accidents. As long ago as 1936 *Ryan* and *Warner* [275] tried to record driving fatigue by studying six truck drivers, comparing the efficiency of various nervous functions on days when they had done no driving, and on days when they had driven more than 450 km. The results showed that on days with long journeys a form of fatigue appeared which expressed itself in the following functional disorders: slowing down of the toe-curling reflex; increase in the number of balancing movements when standing up; less precise coordination of eye and hand; poorer optical accommodation; loss of speed and accuracy in mental arithmetic. From these results the authors concluded that long periods of driving led to a reduced ability to discriminate between certain sensory impressions and a loss of efficiency in the associated processes such as motor functions, which placed a particularly heavy stress on the driver. These observations lead to the conclusion that prolonged driving brings a danger of accidents caused by fatigue.

Since these studies were carried out there have been many other surveys of drivers of motor vehicles, which are described in detail in the survey of literature given by *Lecret* [191].

The driver's activation level

Several authors have shown unmistakeably that about four hours of continuous driving is enough to bring on a distinct lowering of the level of alertness and thereby increase the risk of accidents. Thus *Pin* [255] as well as *Lecret* and others [192] observed a progressive increase of the alpha waves in the electroencephalo-grams of drivers of motor vehicles, with a simultaneous increase in the rate of blinking the eyes and a fall in heart rate.

Several authors, too, have studied these symptoms as indicators of driving fatigue. *Harris, Mackie* and others [133] report that after a few hours at the wheel the performance of drivers of buses and of heavy lorries became distinctly poorer, especially their judgment of the edge of the road and the number of corrective movements of the steering wheel they needed to make. Simultaneously there was an increased irregularity of heart beat and a growing feeling of fatigue which the authors interpreted as signs of a fall in the level of cerebral activation. Those drivers who made the most errors also showed the greatest fall in heart rate.

This confirms the research results of *O'Hanlon* [250], who found a distinctly more variable heart rate among drivers of motor vehicles after long test-drives. *O'Hanlon* interpreted the observed rise in variability as a sign of diminished alertness (see *Chapter 10*). Furthermore, the variability fell back to the normal value after a stop for a rest or after a street incident. These studies of driving fatigue may be summarised as follows:

(a) *the first signs of fatigue appear after only a few hours' driving;*

(b) *they manifest themselves in a diminished accuracy of driving, accompanied by a fall in the activation levels of the CNS (fall in heart rate, with more variable rhythm; rise in alpha waves in EEG; more frequent blinking);*

(c) *the fall in activation level is accompanied by reduced alertness, sufficient to increase significantly the risk of accidents;*

(d) *this fall in activation levels is especially marked in motorway driving and at night;*

(e) *the degree of driving fatigue is determined by the combined effect of several factors, specially important of these being the extent and duration of mental stress, boredom, and circadian (day–night) rhythm.*

Fatigue among telephonists and other occupational groups

In a survey of 15 female telephonists in the long-distance exchange and in the information service, as well as 14 office workers, spread over a year and a half, we studied seven nervous functions as indicators of fatigue [99]. Each function was measured twice daily for 2–4 weeks, once before the start of the afternoon's work and again shortly before the afternoon's work finished. On average, three functions of the telephonists showed the following distinct and statistically significant changes:

(a) simple optical reaction times were increased by 10%;

(b) an exercise in manual skill showed a loss of efficiency; subjective flicker-fusion frequency fell by about 2 Hz when the lighting of the room was unsuitable, and by about 0.6 Hz when the work was carried on in daylight.

Figure 128 shows the results of measuring the subjective flicker-fusion frequency of 13 female telephonists who were exposed to unsuitable room lighting (not bright enough, using fluorescent tubes not dephased).

Parallel studies were carried out on 14 female office workers. This control group showed little or no changes between the beginning and the end of their work period. These results lead to the conclusion that *the impairments of nervous function observed in the telephonists are symptoms of a state of fatigue, which is caused by the almost uninterrupted mental stress in the long-distance exchange and at the information desk, together with unhealthy surroundings while at work.*

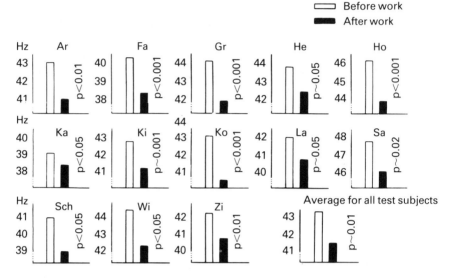

Fig. 128 **Subjective flicker-fusion frequency of 13 female telephone operators, during the afternoon, before and after work.** Each column represents the average value for a total of 20 days spread over 4 weeks. Initials (Ar, etc.) indicate which operator is concerned. After *Grandjean* [99].

Shunters, signalmen, punched card operatives

For comparison we may mention here the results of measuring the flicker-fusion frequencies of other groups of public service employees in the postal and railway services [114]. The average drop in frequency found during the course of three hours of work are set out in *Figure 129*.

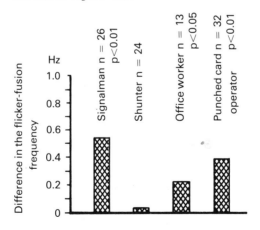

Fig. 129 **Average reduction in subjective flicker-fusion frequency during three hours work by four different occupational groups in the public service (Posts and Railways).** Each column represents the average difference found between each pair of measurements taken on at least three successive days. After *Grandjean* and others [114].

It can be seen from this diagram that the signal box operatives (first column) and those of the punched card section (fourth column) showed a small, but well marked, drop in flicker-fusion frequency; whereas the groups of shunters amd office workers showed hardly any difference. Although the four groups are not strictly comparable in a scientific sense, yet the results lead one to suppose that occupations which make moderate, but sustained demands on the operative's alertness cause a small, but significant drop in flicker-fusion frequency, whereas the purely manual and not excessively demanding job of a shunter has no such effect.

Fatigue among air-traffic controllers

We had the opportunity to carry out a large-scale study [100] [115] [116] of the air-traffic controllers at Zurich airport. The surveys concentrated on questions relating to fatigue, because this is important in relation to mistakes and accidents to aircraft. The studies comprised an analysis of the work carried on, measurements of indicators of fatigue, an interrogation, and (for a small group) a study of the catecholamine excretion in the urine (as a measure of stress reactions).

One group of air-traffic controllers spend their time watching the movements of aircraft on a radar screen, while at the same time they exchange information and instructions on the RT with the pilots and with other members of the control organisation.

Work analysis shows that a controller concentrates his attention on the radar screen for a total of 3½ hours a day, and that he passes on about 800 items of coded information every day. The following were used as indicators of fatigue:

(a) subjective flicker-fusion frequency;
(b) a tapping test (maximum number of tapping movements in 10 s);
(c) a grid-pricking test (maximum number of pricks in 10 s);
(d) a bipolar questionnaire designed by *Barmack* [13] for surveys of subjective impressions.

The tests took place nine times every 24 hours at an average interval of 2½ hours, extending over three weeks and involving 68 controllers. Part of the result is shown in *Figures 130 and 131* as average values,[1] in relation to the duration of the work.

During the first six working hours the flicker-fusion frequency showed a drop of 0.5 Hz and afterwards fell more quickly until 10 hours of work, when the drop totalled 2.3 Hz. The psychomotor performance in tapping and pricking tests followed a similar

---

[1] Since air-traffic controllers work very irregular shifts, averages must be worked out for the people actually present, even though there were never fewer than 50 of them there.

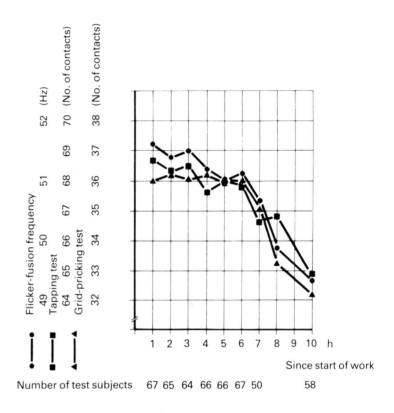

Fig. 130 **Average values for the flicker-fusion frequency, and for two psychomotor tests of air-traffic controllers, in relation to the time elapsed since the start of their working shift (but unrelated to clock time).** After *Grandjean* and colleagues [100] [115] [116].

course. All three psychomotor tests showed a moderate fall in the first six working hours, but after the seventh hour the drop became much steeper.

The sequence of subjective impressions shows first a gradual shift towards disinclination for action. The middle line (line 4) of the ordinate of *Figure 131* represents a neutral level between the two extremes of feeling, and all five of the emotional states are in this vicinity between the fourth and the seventh working hours (abscissa of graph). From then onwards the decline is particularly striking in the direction of "sleepy" (5) and "weary" (3).

Comparison of the results in the two diagrams shows, therefore, a similar result: *during the first 4–7 hours there is only a moderate falling-off of readiness for action and of psychophysiological efficiency; after the seventh hour fatigue increases much more quickly, as shown by both objective and subjective indicators.*

The individual correlations as calculated by *Spearman* are as follows:

1. Strong
2. Relaxed
3. Fresh
4. Vigorous
5. Wide-awake

1. Weakened
2. Tense
3. Weary
4. Exhausted
5. Sleepy

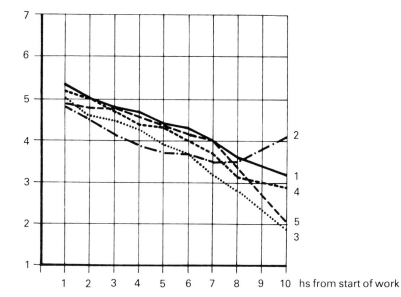

Number of test subjects   67 65 64 66 66 67 60 50   58

Fig. 131 Average values for self assessed, subjective feelings of fatigue in relation to time elapsed since start of work (but unrelated to clock time) among air-traffic controllers. The ordinate represents a line 7 cm long, upon which the test subjects have to indicate their position between the two extremes with the same number in the upper and lower lists (e.g. 5 awake/5 sleepy). Ordinate No. 4 is the approximate neutral level between these extremes. After *Grandjean* and colleagues [100] [115] [116].

Correlation between:	r	p
Flicker-fusion frequency – "weary–fresh":	0.26	< 0.05
Flicker-fusion frequency – grid–pricking test:	0.31	< 0.02
Tapping test – "vigorous–exhausted":	0.29	< 0.05
Grid-pricking test – "weary–fresh":	0.32	< 0.02

Even though the correlation coefficient (r) is very low, it can nevertheless be concluded from this calculation that those air-traffic controllers who showed a marked diminution in flicker-fusion frequency as well as in both the tactile tests, also showed a marked decline in their subjective condition, with increasing fatigue. *This confirms our assumption that the subjective flicker-fusion frequency of the eyes can be used as an indicator of the level of fatigue.*

The above experimental results show that the occupation of air-traffic controller makes heavy demands on sustained alertness and carries a heavy responsibility. Both surveys by questionnaire and direct measurement of the excretion of catecholamine have shown that this job gives rise to stress reactions.[1]

Comparison with bus driver

A comparison between these results for air-traffic controllers and those for drivers of buses and heavy vehicles, obtained by *Harris* and *Mackie* [133], *O'Hanlon* [250] and *Lecret* [191], shows interesting parallels. Both occupations call for sustained vigilance; in both cases the first signs of reduced efficiency appear after about four hours, and this becomes very marked after seven or eight hours. *This decline is a symptom of a fatigue state, which shows itself in both groups as:*
(a) *subjective fatigue;*
(b) *fall in flicker-fusion frequency;*
(c) *decline in psychomotor efficiency;*
(d) *decline in driving precision;*
(e) *more irregular heart beat;*
(f) *fall in heart rate;*
(g) *rise in alpha waves in EEG.*

It is difficult to deny the obvious assumption that all these symptoms are expressions of a decline in the level of activation (arousal) of the central nervous system.

One final conclusion is inevitable. Occupations that demand sustained vigilance must be so planned, with working periods and rest periods, that the risk of accidents is not increased through fatigue of the operators. The research work detailed above shows that these conditions are not being fulfilled at the present time.

---

[1] By stress reaction we understand the body's response to adverse circumstances, which manifests itself biochemically in an increased excretion of the so-called performance hormone, catecholamine.

12
Boredom

*A monotonous environment is one that is lacking in stimuli, and,
at least in Anglo-Saxon countries, the reaction of an individual to
monotony is called boredom. Boredom as conceived by Anglo-
Saxon scientists is a complex mental state characterised by both
symptoms of decreased activation of higher nervous centres, with
concomitant feelings of weariness, lethargy and diminished alertness; and sometimes increased activation of presumably lower
brain structures responsible for the experience of unpleasant
emotions and the motivation to escape from the monotonous
environment.*

Boring situations are common in industry and commerce. They can
be found, for example, at a control desk if there are too few
moments when the operator needs to respond to a signal by taking
action of some kind. An engine driver can be in a similar situation
if signals are too far apart. An example of a monotonous job is
being in charge of a stamping press and having to carry out exactly
the same operation 10–30 times per minute, for hours, days and
years on end. Occupations such as this are *repetitive* as well as
monotonous and boring.

During the last 20 years many psychologists, as well as a few
physiologists, have concerned themselves with the problem of
boredom. The psychologists have mainly described the external
causes of boredom and the behaviour of persons suffering from it.
The physiologists have concerned themselves more with the
nervous mechanisms of boredom and related these to the
measureable indicators of this condition. Although boredom is a
single condition, it will be considered from these two opposite
points of view.

Boredom from the standpoint of psychology

Review of the psychological aspects of boredom can be found in the work of *Bartenwerfer* [17], *Haider* [129], *Leplat* [197], *Gubser* [124] and *Martin* [215].

External causes

Experience shows that the following circumstances give rise to feelings of boredom:

(a) *prolonged repetitive work that is not very difficult, yet which does not allow the operator to think about other things entirely;*

(b) *prolonged, monotonous supervisory work, yet which calls for continuous vigilance.*

The decisive factor in these situations is obviously that there are not enough matters that call for action. *Bartenwerfer* [17] established this in 1957, when he defined scope of attention (*Beachtungsumfang*) by the frequency with which the attention had to be directed to a fresh point. Any restriction of the scope of attention leads to a lowering of psychic tension, which is at the root of boredom; in fact *Bartenwerfer sees boredom as being a state of diminished psychic activity.*

Observations in industry have shown that certain conditions make boredom more likely. Examples are a very brief cycle of operations, short periods of training and few opportunities for bodily movements. Others are dimly lighted or warm workrooms, and solitary working without contacts with fellow-workers.

Personal factors

Yet working conditions are not the only and often not even the decisive factors in boredom. We know, in fact, that personal factors have a considerable effect on the incidence of boredom, or to put it differently on the ability to withstand boredom. To enumerate a few of these personal factors, *liability to boredom is greater for*:

(a) people in a state of fatigue;

(b) night workers, until they are adapted to night work;

(c) people with a low motivation;

(d) people with a high level of education, knowledge and ability;

(e) keen people, who are eager for a demanding job.

Conversely the following are very resistant to boredom:

(a) people who are fresh and alert;

(b) people during their training period (e.g. a learner-driver has no time to be bored);

(c) people who are satisfied that their jobs are commensurate with their abilities.

An observation we made in Switzerland is of interest in this context. Jobs that were particularly dull and repetitive, without much demand for perception or precision, were most often given to female operatives coming from abroad. One is led to suppose that the educational level and potentialities of these foreign workers are relatively low, and that they are eager to earn as much money as possible in the shortest possible time, so as to be able to go back to their own country. Here may be an example of personal motives which help to resist boredom.

Extroversion

Several pieces of research concur in suggesting that extrovert persons[1] are very susceptible to boredom. On the other hand, the opinion often expressed that women are more resistant than men to boredom is scientifically questionable; equally, the supposed relationship between intelligence and susceptibility to boredom is still disputed.

Satiation

Many authors [17] [124] [129] draw a distinction between boredom itself and its emotional manifestations, which they call "satiation" (*Sättigung*). This means a state of irritation and aversion to activity which is provoking boredom. The person feels that he or she has "had enough". According to *Haider* [129] this is a state of actual conflict between a feeling of duty to work and the desire to have done with it, which puts the person involved under increasing internal tension.

Job satisfaction

A decline in work satisfaction can be looked upon as a precursor of psychical satiation. Several studies have shown [20] [128] [337] that in practice work satisfaction is lower where monotonous, repetitive work is concerned, than with jobs that allow a greater freedom of action. As an example of such a survey the results of *Wyatt* and *Marriott* [337] may be recalled. These authors questioned 340 workers in one car factory (Factory A) and 217 in another (Factory B) about their attitudes to their work. The authors compiled an "Index of satisfaction" from certain replies, and from self-assessments carried out by the workers themselves. *Table 24* summarises the most important of these:

The workers at the two factories agreed in rating working on the motorised production line as more rarely "interesting" and more often "boring" than work at either the non-motorised assembly line, or the free assembly. The indexes of job satisfaction worked out to the same effect. The same authors also compared this index with the length of shift of a selection of workers, and found that the shorter the shift, the greater the job satisfaction.

---

[1] Extroversion is a personal characteristic: the typical extrovert person is sociable, forthcoming, fond of taking risks and rather impulsive.

Table 24   Assessment of work, together with an index of satisfaction from 557 workers in two car factories which had the work organised differently. After *Wyatt* and *Marriott* [337].

Work	Factory	Production line (motorised)	Assembly line (non-motorised)	Free assembly
"Interesting"	A	35%	56%	67%
	B	34%	57%	94%
"Boring"	A	54%	42%	39%
	B	55%	37%	28%
Index of job satisfaction	A	0.53	0.92	0.96
	B	0.57	1.00	1.17

Contradictions in job satisfaction

The above results show, certainly, that about one third of the production line workers rate their job as "interesting", and only about half of them think it "boring". This confirms the experience of many factories that a proportion of the workers enjoy their monotonous, repetitive jobs and do not want one that is more variable or more demanding. A few older pieces of research, in which such conclusions are drawn, are cited by *Ulich* and others [312]. On the other hand works managers and personnel managers of other factories report that it is becoming more difficult to find workers to do the monotonous, repetitive jobs. We must therefore conclude that up to the present neither workshop practice nor scientific investigation has been able to demonstrate a clear relationship between monotonous, repetitive work and job satisfaction. Most likely neither the working conditions nor the personal factors have been sufficiently differentiated and evaluated in research work to date.

In this context we may quote again the comments made by *Ulich* and others [312] on this still unresolved problem:

"The contradictory results may perhaps be attributed to differences of attitude that really exist in practice. However, that would mean that for an indefinable proportion of the workers, male or female, working on a production line must actually be more relaxing than free assembly, since it allows them to express their personalities better by conversation, day-dreaming, and so on. For another, presumably larger, proportion of the workers, however, things are quite different. For them continuous work on a production line seems meaningless, and provides them with no opportunities to develop their personalities by exercising their brain power at their work."

The physiology of boredom

We have seen already that situations in which stimuli are few, or lacking in variety, induce a state of boredom, recognisable by weariness and somnolence, as well as by a decline in alertness.

Neurophysiological basis

This state of affairs is not difficult to explain in neurophysiological terms. It was shown in *Chapter 11*, under *"General fatigue"*, that a stream of impulses from the sensory organs, combined with feedback to the cerebral cortex, stimulate the reticular activating system, and that by this means the reactivity of the CNS (and hence of the whole body) is maintained in a high state of readiness. The level of activity of certain parts of the limbic system is maintained in the same way, with its consequent effects upon motivation and the emotions.

When stimuli are few the stream of sensory impulses dries up, bringing about a reduction in the level of activation of the cerebrum, and thereby of the functional state of the body as a whole (see the model of the neurophysiological mechanism of functional states set out in *Figure 124*).

Besides the reduced sensory inflow during quiet conditions there are two other physiological processes that should be noted, because they, too, are responsible for the decline in the level of stimulation, particularly in situations where the existing stimuli vary very little. These are *adaptation* and *habituation.*

Adaptation

Most sense organs have the peculiarity that *under a prolonged, steady stimulus the discharge from the receptor organ declines.* Obviously one function of this is to protect the CNS against prolonged overloading with impulses from the peripheral sense organs. Hence the term *adaptation*: the stream of sensory impulses is adapted to the needs of the organism.

In principle, all sense organs have this power of adaptation, even though they differ in the extent and speed of it.

Adaptation is particularly well-developed in the sensitivity of the skin to pressure (we soon get used to wearing a wrist watch), the stretch-receptors of the muscles, and the photoreceptors of the eyes.

Adaptation is not confined to the receptors of the peripheral sense organs, but also occurs in the synapses joining one nerve fibre to another.

What significance has adaptation in the problem of boredom? As we have said, the sense organs adapt themselves to external circumstances in such a way that they respond mainly to changes in stimuli and are relatively insensitive to a sustained level. Under uniform stimulation, therefore, the activating structures in the brain do not pass on any stimulus to those organs (reticular and limbic activating systems) responsible for the general level of activation of the body.

It is clear that *monotonous working conditions result in a reduced flow of afferent sensory impulses, not only because of the direct effect of lack of stimulation, but also through adaptation to what stimuli there are.*

Habituation

Habituation can be regarded as adaptation on a higher plane, which leads to a reduction of brain activation by repetitive stimuli, but which operates, not peripherally, but in the zone between cerebral cortex and limbic and reticular activating systems. The following may be quoted as an example of habituation. If a note of a regular pitch is sounded close to a sleeping cat, the cat will wake up the first time. If the same note is now sounded at regular intervals the effect on the cat will gradually diminish. If, however, the pitch of the note is changed, its original waking effect will be restored, whereas a note of the original pitch will still fail to wake the cat. This experiment shows that identical stimuli lose their effect with repetition, provided that the stimulus is meaningless, i.e. has no significance in the life of the animal. This is the essential nature of habituation, the elimination of reactions to meaningless stimuli.

Habituation is a filter

The mechanism of habituation may be compared to a filter which filters out stimuli that are meaningless in the life of the animal, allowing to pass only those that have a certain relevance.

Habituation and boredom among train drivers

A good example of human habituation is to be found in a report by *Endo* and *Kogi* [80]. Drivers of the famous Japanese high-speed train were studied over the section of the line from Tokyo to Osaka (515 km) and back. In three series of measurements (1965, 1966 and 1972) subjective flicker-fusion frequency was used as an index of the level of boredom[1] among six drivers measuring it every 30 minutes both on the outward and return runs.

The speed of the train had been substantially increased during the second and third years of the tests, so that each journey now occupied little over 3 hours against the 4 hours of the first series (1965). All the other conditions of the tests remained substantially the same. The average values for all three series are summarised in *Figure 132*.

In 1965 the flicker-fusion frequency rose slightly in the first hour, but then fell steadily; the rate of fall was most marked on the return journey. The Japanese scientists *Hashimoto, Kogi* and others [80] interpreted this fall in flicker-fusion frequency as a sign of either fatigue or boredom, induced by the monotony of the run. The situation was very different in 1966, since the much greater

[1] The Japanese authors *Hashimoto* [134] and *Endo* and *Kogi* [80] make no fundamental distinction between fatigue and boredom, both of these conditions indicating a lowered activation level of the brain.

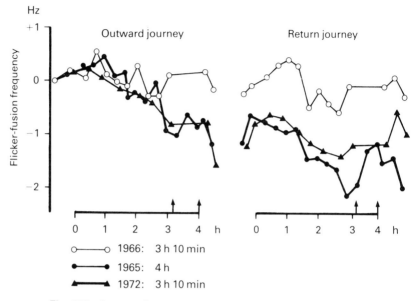

Hz

Outward journey   Return journey

○——○ 1966: 3 h 10 min
●——● 1965: 4 h
▲——▲ 1972: 3 h 10 min

Fig. 132 Curves of average subjective flicker-fusion frequency of the eyes of 6 engine-drivers for both outward and return journeys on the Shinkansen high speed train between Tokyo and Osaka. Arrows show the beginning and the end of the journey. Each series of measurements was taken on 22 or 24 days. The speeding up of the run in January 1966 caused the flicker-fusion frequency to revert almost to normal, but six years later (1972), "habituation" had set it back to the lower value.

accelerations during the run created new stimuli, and the symptoms of fatigue and boredom failed to appear. In subsequent years, however, the drivers became accustomed to the new conditions, and a process of habituation set in, until the original fall in flicker-fusion frequency was again observed.

The biological significance of habituation

Let us go back from this specific example to habituation in general. Its biological significance is the same as that of adaptation: the protection of the cerebral cortex (and thereby of the entire organism) against being inundated with irrelevant alerting or alarm stimuli. Without habituation, the organism would need to maintain itself constantly in a state of maximum alertness.

It is an obvious assumption that the process of habituation also has a part to play in monotonous situations, when it nullifies the effect of irrelevant and repetitive events.

It appears from these considerations that adaptation and habituation are neurophysiological mechanisms which are to be taken as indicating the existence of monotonous conditions. *Situations in industry and in transport which give rise to the phenomena of adaptation and habituation certainly involve an increased risk of monotony and boredom.*

The neuro-physiology of boredom (summary)	The physiological aspects of boredom may therefore be summarised as follows. Situations which are characterised by a low level of stimulation, or by a regular repetition of identical stimuli, or just by making few mental or physical demands upon the operator, lead to a diminution in the flow of afferent sensory impulses, as well as to a lower level of stimulation of the conscious spheres of the brain. The consequent fall in activation level of the reticular and limbic systems reveals itself in a reduced reactivity of the entire organism. It is hardly possible to distinguish between fatigue and boredom on physiological grounds, since both of these states are characterised by reduction in the level of cerebral activation. Nevertheless differences do exist, as will be shown later, on the evidence of results from certain experimental work on boredom.
Medico-biological aspects of boredom	Until a few decades ago, the science of work physiology was mainly interested in finding out how to relieve the worker of excessive physical load. *Increasing mechanisation and automation, as well as the tendency to divide up the work into as many simple operations as possible (Taylorism) has now led, in many occupations, to a new problem: insufficient demands on the physical and mental capacities. Unused physical and mental capacities characterise a state which we call "underload".*

Nearly all the organs of the human body have the important biological characteristic of being able to respond to stress by stepping up their performance. This is true, not only of the muscles, heart and lungs, but also of the brain. Human development from childhood onwards is heavily dependent upon this ability to adapt to the stresses of life.

Conversely, if an organ is not exercised, it atrophies. A good example is the wasting of muscle which becomes distinctly noticeable only a few weeks after a fracture of a limb. Cessation of development, followed by decline, takes place on a mental level as well as a physical one. Thus it is known from experiments on animals that the brain becomes better developed, functionally, morphologically and biochemically, when the animal is subjected to various mental demands and stresses than when it is allowed to grow up in a quiet situation with few external stimuli.

From these considerations it is therefore evident that underload, such as a person experiences from monotonous, repetitive work, is basically unhealthy from a medico-biological standpoint.

The relationship between stress and biological reactions can be broadly summarised as follows:

(a) *underload leads to atrophy;*
(b) *the right amount of load leads to healthy development;*
(c) *overload wears out the body.*

Medical research carried out by *Kornhauser* [182], by *Wisner* [332] and by *Laville* [190] may be mentioned in this context, showing that monotonous work is frequently accompanied by signs of damage to mental health, such as anxiety states, frustration, aggressiveness, irritability and a tendency to depression. The experimental results of *Caplan* and others [46] are also in general agreement with these findings. The following results of their surveys are of interest in relation to problems of boredom:

(a) compared with other operatives, those who worked on a production line were least often "content", and most often "bored";

(b) they also suffered most from anxiety states, depression, irritability and psychosomatic complaints.

These investigations made it possible to draw comparisons between two ways in which the assembly of heavy lorries was organised: assembly on a moving production line and assembly of each vehicle by a group of workers who carried out a variety of operations. The production line workers showed a higher level of boredom and greater tendencies to anxiety or depression. Nevertheless the interpretation of all these medical surveys is full of problems. It usually remains in doubt whether the working conditions are the cause of the effects that are noted, or whether those who choose to work on the production line may not be a socioeconomic group, who have an innately higher tendency to the complaints they report.

Field studies

Laboratory studies have the advantage that they are conducted under controlled conditions, and therefore it is usually possible to distinguish clearly between "cause and effect". On the other hand they have the drawback that the working conditions studied are usually simulated, and do not compare exactly with either industrial practice or everyday traffic. This reservation is particularly necessary when studying boredom. Even if it is possible to reproduce the physical working conditions very closely, the important psychological factors, e.g. the profit motive, or the role of social contacts, can be simulated only roughly in the laboratory. On the other hand it is usually possible to carry out field studies under everyday working conditions, though even so many factors are not under the investigator's control. Some results may be difficult to assess, or may be accepted only with reservations.

The following account concerns two field studies of the effects of monotonous work on the operators.

On the production line in factories

In 1961 *Haider* [128] [129] studied boredom among 337 female workers in various factories, 207 of whom worked on a moving production line, while the other 130 worked individually. He assessed the subjective feelings of these workers by using a self-assessment card, with 12 pairs of contrasting states. Comparison of the two groups gave, on average, the following differences: the production line workers were more "tense", "bored" and "dreamy" than those who worked at their own speed. *Haider* [128] writes of these: "Finally we may conclude that when the occupation is a long succession of simple, repetitive acts, we may expect to see "satiation phenomena", with increasing tension, restlessness, lack of incentive, and a declining performance of the boring work." Certainly the percentage of discontented and tense workers on the production line was as high as 20–25%.

Haider [130] also studied the psycho-physiological symptoms of boredom among a selection of these workers, who were engaged in machine-paced repetitive work. During their work, these women were subjected to light signals, which appeared at irregular intervals on the protective shield of their machine. Each time they noticed a flash of light they were required to press a pedal, and in this way it was possible to monitor their level of alertness throughout the day, as well as their speed of response. The author found that *the alertness of the 31 women workers declined during the working period, and their reaction time increased. Figure 133* shows part of these results.

Hours worked

Fig. 133 Vigilance and readiness for action during repetitive and paced work. The curves express average values for 29 female operatives. On the left, the percentage of signals omitted and on the right the simple motor reaction time in seconds. After *Haider* [130].

The results show that in the course of a working shift the women workers missed more and more signals, as well as taking longer to react to the ones they noticed. It seems fair to assume that this decline in performance is a result of the monotonous, repetitive nature of the job.

Checking bottles *Saito* and his colleagues [276] carried out a similar investigation in the food industry. The operatives studied were engaged in the visual control of bottles and their contents which moved so quickly that the job was rated as arduous, even though monotonous and repetitive. Even after a short time at work the number of rejected bottles became distinctly lower, a fact which the authors considered must indicate diminishing alertness. Concurrently a distinct fall in the flicker-fusion frequency could be recorded, accompanied by such subjective symptoms as increasing fatigue, sleepiness, headache, and a sense of "time dragging". The fall in flicker-fusion frequency was greatest at times when the operative was observed to talk less to the others and tended to doze, or even to fall asleep.

The progression of all effects was as follows:
(a) first hour: no change;
(b) 2nd–4th hours: marked impairments;
(c) final hour before lunch-break: operatives both felt and performed better.

The authors had the opportunity to make comparable studies of a group of operatives whose work was more varied and therefore less monotonous. This group showed less absenteeism and fewer subjective complaints.
Finally we may mention, a Swedish field study which investigated the problem of boredom from the standpoint of the action of hormones.

Adrenalin *Johansson* and his colleagues [165] were able to demonstrate that
and sawing work a group of sawmill workers, whose work was repetitive and at the same time responsible, secreted much more adrenalin than other groups of workers. They also had a higher incidence of psychosomatic ailments and more absenteeism. The authors interpreted these results as follows: the combination of monotonous, repetitive work with a higher level of mental stress called for a continuous mobilisation of biochemical reserves, which in the long term adversely affected the general health of the worker.

Critical final The above-mentioned field studies amply confirm the results of
conclusions the experiments under simulated conditions: *monotonous and repetitive work under industrial conditions gives rise to boredom, with the physiological and psychological symptoms associated with it.*
On the other hand field studies underline the fact that additional

mental and emotional factors often come into play at an actual working place. The causal connection between work and boredom seems to be much more complex than it appears to be in the theoretical picture. Industrial practice is much more complicated than the conditions of a laboratory experiment.

A further limitation of our theoretical model becomes apparent if the foregoing field studies are compared with those that were described under fatigue in *Chapter 11*. Both chapters described effects of working conditions which it was difficult to attribute with certainty to either fatigue or boredom. The fact is that both fatigue and boredom give rise to similar symptoms, and the causes are not easy to assess. Hence it is often impossible to say definitely whether either the symptoms or their underlying causes indicate a state of fatigue, or one of boredom.

Who can say, for example, whether the decline in performance, the rise in alpha waves in the brain, and the tired feelings of truck drivers are signs of a fatigue state, or the effects of boredom? Has the monotony of the job created boredom, or do the lowered alertness and impaired driving ability indicate that mental stress has resulted in an actual state of fatigue? Who can say whether the monotonous, repetitive job of the bottle-watchers is just boring, or whether excessive demands on their vigilance are actually fatiguing? There are many examples of this kind, which show that situations often arise, both in industrial practice and in traffic conditions, which can be simultaneously boring and fatiguing. In such cases the distinction between these two states is purely arbitrary, and the most credible research is that which makes no attempt to draw one.

Experiments concerning boredom

The following are a few selections from the very large number of important experiments that have been carried out to investigate boredom and its effects on people at work. Firstly we must mention the many vigilance studies, already discussed in *Chapter 10*. In practice nearly all these involved a monotonous task which nevertheless called for a constant state of alertness, and they have overwhelmingly shown that prolonged concentration on a monotonous task results in a steady decline in alertness. In the same chapter it was also pointed out that vigilance is dependent upon the functional state of the brain, the level of cerebral activation.

Bartenwerfer's simulated driving test

In 1957 *Bartenwerfer* [17] carried out an important experiment into boredom. His 39 test subjects were required to take part in a simulated night drive, during which they had to use a steering wheel to control a pointer on a projection screen and to keep it as closely as possible on a moving line which simulated a road along which they were driving. Their driving performance was assessed according to the number and duration of their departures from the line. The research room was sound insulated, and only dimly lighted. During the course of the experiment, which lasted 104 minutes, a buzzer sounded three times, and the drivers had to react by pressing foot pedals. Heart rate and cardiac rhythm were recorded throughout the experiment, and finally a note was taken of the drivers' fitness and of all their impressions during the test.

The most significant results of this experiment may be summarised as follows:

(a) a steady increase in feelings of weariness and sleepiness, and a growing indifference, accompanied by heaviness of the eyelids, double vision and difficulty in thinking;

(b) a marked decline in psychomotor performance, shown by more frequent and longer departures from the desired line, and an increase in reaction times;

(c) slower and more irregular heart beat;

(d) stimulating changes in the conditions of the experiment relieved the boredom, but only temporarily;

(e) the boredom took a sinusoidal course, with alertness alternately rising and falling.

Hashimoto's experiments [80]

The studies of engine-drivers [80] that were mentioned earlier had shown that during long trips they could not be relied upon to react to every signal without missing any. This observation was the starting point for a series of interesting experiments that were carried out at the Japanese Institute for Railway Science, under the leadership of *Hashimoto* [134] [135] [136]. To induce a state of boredom, these authors used simulated driving tests, in one series of which the task to be performed was easy, and in the other more difficult. The severity of the second test came from the fact that a prolongation of the feed-back control made it much more difficult to follow the "road". Subjective flicker-fusion frequency and reaction time were repeatedly measured during the 70-minute experiment. The results are shown in *Figure 134*.

The result was that the easier task produced a greater reduction in flicker-fusion frequency than the more arduous exercise. Similarly, reaction-times became longer during the easy test than during the harder one. *Hashimoto* gives the following interpretation for these results: the more difficult exercise demands a delicate sense of anticipation, associated with a complex processing of information in the brain. This process gives rise to stimulating impulses, which flow from the cerebral cortex to the reticular activating system. As a result, the level of cerebral activation is less damped down by boredom than it is during the easier exercises.

A similar observation about bus drivers was made by *Hashimoto*,

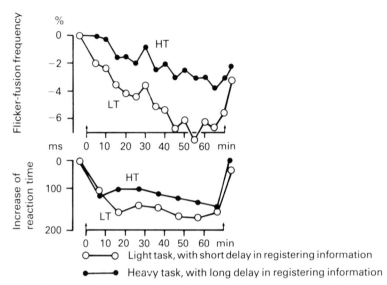

Fig. 134 Effects of a light (LT) and a heavy task (HT) during a simulated driving test on the subjective flicker-fusion frequency of the eyes and on the reaction time. Average values for four test subjects. After *Hashimoto* [135].

Kogi and others [136]. When the speed limit was raised from 80 km/h to 100 km/h, the flicker-fusion frequency remained at a high level throughout the 3½-hour journey. Travelling at higher speeds was more stimulating, and this reduced the incidence of boredom. In a further interesting experiment with the same simulated driving test, *Hashimoto* [134] not only measured the flicker-fusion frequency, but simultaneously recorded the electroencephalogram. One test took place under easy conditions, and this produced a distinct lowering of the flicker-fusion frequency and an increase in alpha rhythms as consequences of the boredom of the test. In a second test a set of chimes began to play every seven minutes and the "driver" had to wait one minute and then switch this off. Under these conditions of frequent stimulation, with a much increased inflow of sensory impulses the flicker-fusion frequency showed hardly any fall, and the alpha rhythms rose by only a small amount. In the first experiment, therefore, *boredom resulted in a fall in flicker-fusion frequency while a simultaneous increase in alpha rhythms was discernable on the electroencephalograph. The increased inflow of sensory impulses resulting from the periodic playing of the chimes was sufficient to ward off boredom for the duration of the test, 90 minutes. Figure 135* illustrates the relationship shown in this experiment between flicker-fusion frequency and alpha rhythms. The electroencephalograph uses comparison of the alpha and beta rhythms as an index of change.

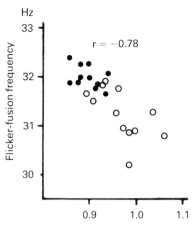

Relation between alpha and beta rhythms

● Values while chimes were being sounded

o Values during a state of boredom

Fig. 135 Flicker-fusion frequency of eyes and alpha rhythms in EEG during an experimental study of boredom. The reduction in flicker-fusion frequency is accompanied by a rise in alpha rhythms. After *Hashimoto* [134].

It appears from the figure that *the fall in flicker-fusion frequency is associated with a rise in alpha rhythms*, with a correlation coefficient of $r = -0.78$. A special physiological importance must be attached to this result, because it shows that during a condition of boredom both ways of measuring cerebral reactivity – the flicker-fusion frequency of the eyes, and the evidence of the electroencephalograph – show the same sort of effect.

Experiments with simulated repetitive tasks

Whereas the experiments described above were predominantly concerned with simulated traffic conditions, *Martin* and *Annette Weber* [215] [216] and *Baschera* and *Grandjean* [18] elected to produce their state of boredom by means of a very uniform and repetitive task. Over a period of several hours, their test persons were required to pick up nails singly, count them, and place a specified number into a series of envelopes. This task fulfils many of the conditions conducive to boredom: it is extremely repetitive, is undemanding, and does not require much alertness, yet it does not leave the mind entirely free for daydreaming.

In all these nail-counting experiments the work caused a lowering of the flicker-fusion frequency. Simultaneous checks, either with a questionnaire (*Nitsch* [248]) or a bipolar self-assessment chart [13], showed changes in the subjective well-being, either less efficient and less willing manipulation of the nails,[1] or a displace-

---

[1] These criteria come from *Nitsch* [248], and relate to corresponding changes in a variety of personal reactions: e.g. sleepiness, relaxation, self-confidence, and many others.

ment of the bipolar charts in the direction of increased sleepiness, fatigue, inattention and boredom.

Figure 136 reproduces some work of *Martin* and *Annette Weber* [216] in which the results of the nail-counting experiment were compared with a study of more stimulating conditions. This study required the research subjects to carry out a variety of psycho-motor tests, in the intervals of which they listened to music of their own choice.

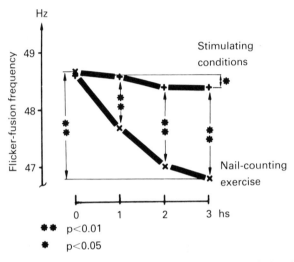

Fig. 136 **Average values of the subjective flicker-fusion frequency of the eyes of 25 test subjects in a nail-counting exercise, as compared with a stimulating situation.** After *Martin* and *Annette Weber* [216].

As the diagram shows, the monotonous counting of nails quickly provoked a distinct fall in the flicker-fusion frequency, which reached an average value of 1.7 Hz after 3 hours. In contrast, the more stimulating conditions of the second experiment produced only an insignificant fall in flicker-fusion frequency in the same length of time. Similar results were obtained from questioning the test subjects about how they felt. The nail-counting experiment produced an increased sense of strain (through loss of manual dexterity) and a loss of motivation (disinclination for the task): whereas almost all the effects of the more stimulating conditions were beneficial.

Research therefore shows that *a monotonous, repetitive job can quickly lead to boredom, as shown by a fall in flicker-fusion frequency and changes in subjective feelings.*

Both in driving tests [135] and in field studies among drivers of motor vehicles [133] [250] boredom could be avoided or at least postponed either by stimulating conditions, or by making the task more difficult. This state of affairs leads one to ask whether repetitive jobs could be made less boring simply by making them more difficult. For this reason we extended the nail-counting test by introducing a moderately difficult and a very difficult mental task [18]. The test subjects were required to perform the following three tasks, each lasting three hours:

Repetitive tasks with varying mental demands

(a) the nail-counting test as described above: *mental demand low*;
(b) select nails of different colours and with different ring-markings in such a way as to fill an envelope with the correct number, colour and ring-marking. We rated this exercise as making only a *moderate mental demand*;
(c) select nine nails, with a prescribed combination of colours and markings, and stick them into a board in an ascending order of rank according to their ring-markings. This was the exercise that made *heavy mental demands*.

Figure 137 shows the average fall in subjective flicker-fusion frequency of the 18 research subjects who were tested under the three sets of conditions.

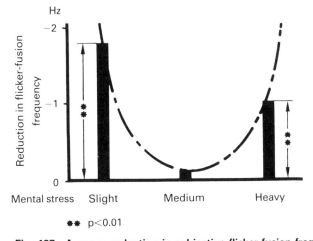

** p<0.01

Fig. 137 Average reduction in subjective flicker-fusion frequency of the eyes during repetitive tasks which involve various levels of mental loads. The vertical columns express the difference in the measurements before and after the 3½ hour experiments, involving 18 test subjects. The broken line is a hypothetical curve expressing the probable course of flicker-fusion frequency in relation to mental load. After *Baschera* and *Grandjean* [18].

The outstanding result of these three series of tests is the distinct relation between the flicker-fusion frequency and mental stress: the low stress level is accompanied by the greatest fall in flicker-fusion frequency; the high level of mental stress equally provokes a fall in this frequency, though it is somewhat smaller than before. On the other hand, the moderate mental stress leaves the flicker-fusion frequency almost unaltered. *We can draw a theoretical U-curve through the tops of the three columns in the block-*

diagram, which suggests that both understress and a high level of overstress cause a reduction in the level of cerebral activity. Between these two extremes lies a zone of moderate mental stress which has no effect worth mentioning on the functional state of the CNS.

This experiment shows a remarkable similarity to *Schmidtke's* results [270], which were discussed in *Chapter 10*, under "Vigilance": these, too, showed a U-shaped curve for the relationship between stress (= frequency of signals which required action) and performance (= frequency of signals acted upon) (see *Figure 119*). This U-curve leads one to suppose that few signals lead to understress, and, conversely, very many signals lead to overstress of the test subject. Here, too, we are forced to admit that there is an intermediate level of stress to which human beings react with a minimum of mistakes.

In the above-mentioned research of *Baschera* and *Grandjean* [18], analysis of the surveys relating to subjective feelings shows some interesting results. Several feelings, e.g. weariness and sleepiness show a parallel with flicker-fusion frequency. They were increased both for low and for high levels of mental stress. In contrast, "boredom" was predominantly associated with low level of stress. This shows that it is obviously possible to differentiate between the effects of under- and overstress by asking the appropriate questions. It is doubtful, however, whether it is yet possible to make this distinction by physiological methods.

Boredom and adrenalin

The significance of the performance hormone, adrenalin, produced by the adrenal cortex, was mentioned under "general fatigue" in *Chapter 11*. This hormone is responsible for the continuous maintenance of the reticular activity level, and all kinds of stress, physical, psychical and mental, which cause an increased flow of adrenalin. In recent years the excretion of adrenalin in the urine has been studied, particularly by Swedish workers, *Levi* [198] [199] and *Marianne Frankenhäuser* [85] [86] [87]. They found that the most diverse physical and emotional stresses led to measureable increases in the adrenalin excreted in the urine, which was interpreted as a mobilising of the performance reserves of the body. One study by *Marianne Frankenhäuser* and others [87] may be mentioned because it is relevant to our current problem of boredom. Their experiment with mental under- and overload produced the following results:

(a) *overload*, created by long lasting serial reaction time test, produced an increased flow of adrenalin (about 9.5 ng*/min);

(b) *moderate load*, in the form of reading a newspaper, gave only a small increase in adrenalin excretion (about 4 ng/min);

---

* 1 ng = 1 nanogramme = 10^{-9} g, (one millionth of a milligramme).

(c) *underload*, the effect of a uniform, repetitive operation also produced an increased flow of adrenalin amounting to about 5.7 ng/min, and so falling between the levels of "overload" and "moderate load".

Marianne Frankenhäuser [85] writes of this: "The results show that adrenalin production is increased not only when acting under pressure, against the clock, and with a high inflow of information, but also in conditions that are monotonous, and lacking in stimulation. This shows that the physiological reaction is produced by the mental and emotional stress, rather than by the physical effort as such."

Summarising it can be said that the functional state of boredom is accompanied by hormonal symptoms, which can be identified biochemically and which then give us an insight into the mental and emotional aspects of boredom.

Problems of monotonous, repetitive work

Criticism of Taylorism

It has been shown earlier in this chapter (boredom from a psychological point of view) that there is less job satisfaction in monotonous, repetitive work than in work of more flexibility [337]. For several years past there have been more and more critical objections on the part of industrial sociology and works psychology, against the Tayloristic principle of splitting the job into a large number of identical tasks which are repeated indefinitely. Working places organised on this principle are characterised by low cycles per piece, short training periods, and few demands on the operative. *The result of this fragmentation of the job is that individual freedom of action is severely curtailed, mental and physical systems lie fallow, and the human potentialities of the workers are not realised.*

Ulich [309] [312] sees operations of this repetitive kind as being a "dehumanisation of work by robbing it of any meaning".

Quality of life at work

Ulich [312] cites several surveys which favour the hypothesis that there is a link between the quality of one's working life and that of one's life in general. The restrictive effect of small, repetitive tasks creates an alienation not only from the job itself, but from society. Several comments on these social and ethical aspects of job satisfaction may be found in the work of *Friedmann* [88], *Ulich* [312] and in the *Report of the International Conference on Enhancing the Quality of Working Life* [256].

Adjusting to human capabilities

There are good reasons for believing that a job that takes account of a person's potentialities and inclinations will be carried out with interest, satisfaction and good motivation. Conversely, it is obvious that an undemanding job, which does not develop the potentialities of the worker will prove to be boring and lacking in motivation. At the other extreme, a job that requires more from the worker than he is capable of will be overtaxing. Hence the obvious requirement is that work should be so planned that it is matched to the capabilities of the operatives, without asking either too little or too much of them.

This basic recommendation embodies the idea of *Ulich* [309] [312] that the efficient performance of a complex or difficult task is at its optimum in the range between being under- and overdemanding. This concept appears in *Figure 138* as an inverted U-shape.

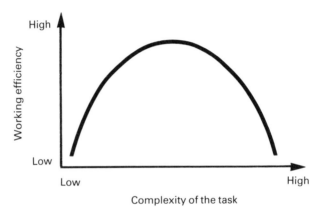

Fig. 138 Conjectured relationship between level of complexity of a job and working efficiency. The level of complexity is deduced from the number and variability of the operations involved. After *Ulich* [309] [312].

Similarly *Blumer* and *Naylor* [27] concluded that the level of frustration in relation to the level of complexity of the task was represented by a U-shaped curve, and that frustration was least when the demands of the job were most closely matched to the capabilities of the worker.

Summary of the consequences of extreme fragmentation of work

Table 25 summarises the most important critical objections, culled from various branches of science, against extreme Tayloristic fragmentation of human tasks in monotonous, repetitive work.

Table 25 Monotonous, repetitive jobs from the standpoint of various sciences.

As seen by:	Probable consequences:
Doctor	Atrophy of mental and physical powers
Work physiologist	Boredom; risk of errors and accidents
Work psychologist	Increasing discontent with the job
Student of ethics	Human potentialities not fully realised
Work scientist	Increased absenteeism; increasing difficulty in finding personnel to do the job

The various consequences that may follow from repetitive work, as set out in *Table 25*, have led in recent years to the development of different ways of organising assembly work and similar serial jobs.

The main objective of these efforts is to give the operator more freedom of action in the following two directions:

(a) *reduction of boredom*, with its concomitant feelings of fatigue and satiation;

(b) *making the work more worthwhile* by providing a meaningful job, and one which allows the operator to develop his full potentiality.

Basic to both these improvements is the assumption, already indicated above, that there will be a reduction in absenteeism, turnover of the work force, and social stress, and that the new conditions will attract more workers. Hence in the long run a higher productivity should result.

These desirable new forms of work organisation involve various improvements, ranging from simple variety of work, through various ways of broadening the scope of the job, to job-enrichment by giving the worker more information, and some degree of independence, with perhaps planning and control functions.

To facilitate a review of this field it is useful to divide it into two, as follows:

(a) *variety of work;*

(b) *broadening and enriching the job*.

Variety of work

This is a scheme by which each individual worker is entrusted with different jobs at different working places, which he carries out in rotation.

An example may be taken from the assembly of electronic calculators. The complete assembly is deployed round a work bench. There are eight places, but only six operatives, so that there are always two vacant seats. The resulting accumulation of components forces the female operatives to change their seats continually. It is an essential feature of this system that each operative must be trained to work at any of the eight places, and so the complete assembly is a function of the entire group. This is a particularly good example, even though it also shows traces of job enrichment in the free organisation of work within the group. The publication by *Ulich* [312] presents many examples from various countries, which show that variety can be introduced into the job in many different ways. The best organisations were judged to be those which at the same time introduced a certain group autonomy, allowing them to control the making of their own product.

If variety of work merely means moving to and fro between jobs that are equally monotonous or repetitive, the risk of boredom may be slightly reduced, but the desirable matching of the difficulty of the job to the capabilities of the worker is not being achieved. *Simply joining one monotonous, repetitive job to others is not going to lead to job enrichment.*

Broadening and enriching the job

For these reasons a special importance attaches to types of organisation which strive to enrich the work by broadening its scope, thus helping to develop the personalities and self-realisation of the employees. In organisations like these the tasks are so planned that a worker moves to a succession of different jobs, each of which makes different demands on his abilities. Additional responsibilities such as quality-control, or the installation and maintenance of machinery make an important contribution to the enrichment of the job. Many examples of such broadening and enrichment of employment can be found in the literature [256] [312]. An actual observation in a factory making electrical apparatus may serve as an example. A certain piece of equipment was originally assembled on an assembly line by six successive operations performed by six workers. On the new plan, one and the same worker carried out all six operations, being alone responsible for the quality of the entire assembly.

Autonomous working groups

The so-called autonomous working groups are certainly a further step towards job enrichment. The workers employed upon each production unit are organised into a group, and the organisation and subdivision of the work, as well as control of the end-product are delegated to them, the group therefore has planning and control functions.

Here is a Dutch example [138]. A firm making television sets introduced more attractive forms of work organisation, by stages, as it became a problem to find workers. At first the sets were assembled on a long assembly line, and the 120 workers pushed their set along once a minute, at a signal. The first stage of improvement was to have a new assembly line with 104 working places, divided into five groups. At the end of each group there was a barrier and a quality check. It was found that with this arrangement assembly times were shortened, and quality improved. Next the experiment was tried of allowing more time before the sets moved on, as well as a certain amount of job-switching. The final step was to introduce *autonomous working groups*. The groups were reduced in size to seven persons, and the range of each person's tasks broadened, so that intervals between moves rose from 4 to 20 minutes. The working groups undertook many responsibilities which previously had fallen upon the foreman or overseer. The groups performed their own quality control, and were self-administering. A certain amount of interchange within the group made them more flexible.

The authors declare that the first results include:
(a) more positive approach to the work;
(b) greater cooperation among members of a group;
(c) creation of a certain critical attitude to production levels;
(d) the groups were a nuisance to the management in one sense, since autonomous groups acted as a controlling force;
(e) there was less absenteeism;
(f) less waiting time;
(g) fewer "passengers";
(h) more working areas and new machines;
(i) more consultation;
(j) higher wages.

Taken as a whole, the economies predominated and the groups produced a television set more cheaply than did their predecessors.

All the efforts that have been enumerated above, to broaden and enrich people's work, must be regarded as experiments that are certainly not concluded, nor can their results yet be fully evaluated. *The search for new ways of organising monotonous, repetitive work is still going on.*

Interim balance-sheet

It is possible to draw up an interim balance-sheet from the publications and reports now in existence, and to show that positive advantages predominate. Generally, the most important results are:
(a) *reduction of absenteeism;*
(b) *more attractive working places;*
(c) *improvement of both quality and quantity of production;*
(d) *less boredom;*
(e) *more interest in the work;*
(f) *greater contentment with the job.*

These positive results must, in many cases, be set against higher costs, caused by longer periods of training, higher wages and sometimes new plant. According to *Ulich* [312] these higher costs should be fully compensated for by less time wasted in breakdowns and fluctuation of output.

Nevertheless a few experiments have been failures, due to insufficient planning and particularly to not giving the workers enough information and guidance.

Social contacts

This catalogue of organisational improvements may be concluded by emphasising the importance of social contacts at the working place. *The opportunity to talk to one's fellow-workers is an effective way of avoiding boredom.*

Conversely, social isolation brings monotony and increases the tendency to become bored with the work.

Sitting one after the other along a straight assembly line is bad: it is much better if the line follows a semi-circle, or is sinuous. Any arrangement is good so long as it brings several workers within conversational distance of each other.

Other ways of reducing the incidence of boredom include:

(a) more frequent short stops (see *Chapter 13*);

(b) opportunity to move about during these stops;

(c) a stimulating layout of the surroundings, making use of light, colour and music (see *Chapters 14, 15, 16*).

Supervisory work

The same principles apply to supervisory work as to driving a vehicle. An essentially dull situation, lacking in stimuli, must be enlivened just enough, so that it is neither soporific on the one hand, nor overstressed on the other.

In both types of job the problem is often to be able to recognise a critical situation at the right moment. The necessity to remain alert, which is often associated with a high level of responsibility, results in both fatigue and boredom, and increases the risk of mistakes and accidents. Hence preventive measures concentrate on making sure that relevant items of information are correctly assimilated.

Recommendations   To summarise, the following arrangements can be recommended for supervisory jobs at control panels, machines, projection screens and similar places of work:

(a) alarm signals and similar safety limits must be clear and decisive. A combination of a light signal and a noise (buzzer, gong or siren) is particularly effective;

(b) if a series of signals must be noticed without missing any of them, there should be between 100 and 300 of them per hour;

(c) the operator must be fresh, and must avoid getting tired beforehand. Night-shift workers are particularly liable to boredom until they have become night-adapted;

(d) the surroundings should be brightly lit. Music is helpful in the right circumstances. Room temperatures should vary only within comfortable limits;

(e) a change of work must be arranged as soon as boredom causes dangerous lapses of alertness. In extreme situations it may be necessary to consider hourly changes;

(f) short pauses — as far as these are possible — will help to avert boredom and improve alertness;

(g) in certain particularly critical situations it may be necessary to employ two people to keep a look-out together: e.g. in the driver's cab of a locomotive.

13
Working hours and eating habits

Daily working time

Many studies have shown that changes in the length of the working day often result in higher or lower output. Thus at one factory it was found that shortening the working day from 8¾ to 8 hours raised the output by between three and ten per cent, those whose work was predominantly manual showing a bigger increase than machine operators.

Figure 139 sets out the results of an old British survey [314]. These results confirm the general experience that shortening the working day results in a higher hourly output; the work is finished off more quickly, with fewer voluntary rest pauses. This change in working rhythm generally takes place within a few days, though occasionally it may be several months before the effect can be seen.

Conversely, making the working day longer causes the tempo of the work to slow down and the hourly output to fall. The relationship between a day's work and a day's total output has been worked out, to a first approximation, by *Lehmann* [194] (see *Figure 140*). From this it can be established that as a rule the total daily output does not increase in proportion to the daily working time (curve A, *Figure 140*); most often it behaves like curves B and C. In many cases it has even been found that increasing the daily hours beyond 10 results in a fall in total output, because the slowing down of the working tempo resulting from fatigue has more than offset the longer hours.

A longer working day or overtime working are common in wartime and in boom periods, but the results are often disappointing. Because of the relationship between hours and output just mentioned, productivity does not increase as much as is desired and expected, and it may even fall.

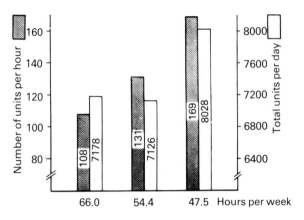

Turning a shell-case (hand operation)

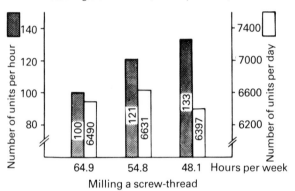

Milling a screw-thread

(mainly operating the machine)

Fig. 139 Working time and output. Above: for predominantly hand work, shorter hours (either weekly or daily) bring about an increase in both hourly and overall daily output. Below: for a job that is mostly operating a machine, shorter hours, while they increase hourly production, have little effect on the overall daily output.

All these observations point to the conclusion that the worker tends to maintain a particular daily output, and if the working day is varied, will to some extent adjust his working rhythm to compensate. This is only true, however, for work that is not linked to the speed of a machine. Workers on a conveyor belt, and others who must fit their work into the rhythm of a machine, cannot do much to compensate for changes in working times. This can be seen from *Figure 140*.

Obviously, of course, the extent to which working speed is varied to compensate for variations in working hours is also affected by rates of pay and other changes in motivation.

Effects on sickness rates

Many observations have provided evidence to show that excessive overtime not only reduces the output per hour, but is also accompanied by a characteristic increase in absences from sick-

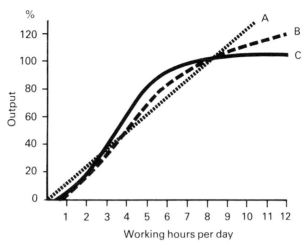

Fig. 140 Relation between working hours and output. 100% = output from an 8-hour day. A = curve of working hours and output. B = output in relation to working hours for moderately strenuous work. C = same for heavy manual work. After *Lehmann* [194].

ness and accident. A working time of eight hours per day, which makes the operator moderately, but not seriously fatigued, cannot be increased to nine hours or more without ill-effects. These include a perceptible reduction in the rate of working, and a significant increase in nervous symptoms of fatigue, often resulting in more illnesses and accidents.

The surveys made by *Behrens* [21] into the effects of overtime on absenteeism through sickness are shown graphically in *Figure 141*. These show distinctly how an increase in overtime in the summer months may be followed by an unseasonal increase in sickness and absenteeism.

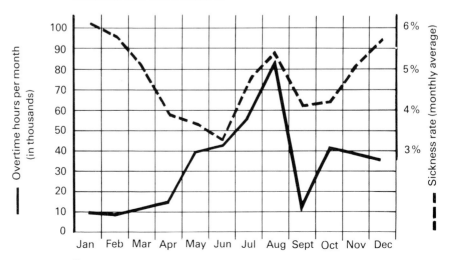

Fig. 141 Effects of overtime on sickness rate. After *Behrens* [21].

A similar observation was made by *Wyatt* [336] in seven British factories in the armaments industry. These results are summarised in *Table 26*.

Table 26 Working time and absenteeism (after *Wyatt* [336]. Statistical surveys in 7 British armament factories in 1944.

Men Working hours per week	% absenteeism	Women Working hours per week	% absenteeism
65.4	11.3	59.1	18.9
63.3	10.2	57.3	12.2
63.2	7.2	56.7	12.7
58.6	7.4	56.2	14.8
56.7	7.3	54.9	10.6
56.2	5.9		

Our physiological knowledge and present-day experience point to the conclusion that a working day of eight hours cannot be exceeded without detriment, particularly if the work is heavy. Modern firms, organised according to the principles of industrial science, usually arrange most of their work sensibly, whether the demands on their workers are heavy or only moderate. An extended working day is tolerable in lighter industrial work, or in jobs where the nature of the work provides plenty of rest pauses.

Arranging the working week

The duties of an employee are usually agreed as so many hours per week, which have become markedly shorter during the last hundred years.

Historical review

In Switzerland, the first federal Factory Act was accepted in a plebiscite in 1877, and stipulated 65 hours per week (11 each weekday and 10 on Saturday). An amendment in 1914 shortened this to a 48-hour week, and by 1960 the average working week had shrunk to about 44 hours.

Figure 142 shows the historical development of weekly working hours in the U.S.A.

It is evident from this diagram that the working week in the U.S.A. has tended to shorten almost continuously since 1850. Nowadays the 40-hour week is common, not only in the U.S.A., but in most industrial countries.

The five-day week

Nowadays the five-day week is common everywhere, and it presents little problem to combine this with a 40-hour week. Nevertheless the experience of introducing the five-day week is of

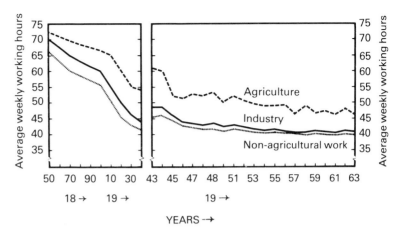

Fig. 142 The trend of weekly working hours in the U.S.A. from 1850 to 1963. After *Northrup* [249].

some general interest. So long as the total weekly hours were reduced at the same time, the introduction of the five-day week led to a relatively higher output per hour, and therefore to only a slight loss of overall weekly production. Occasionally factories found that the changeover from a six- to a five-day week led to less absenteeism. Experience shows that the work people in general, and especially the women, like the five-day week. It is preferred on social grounds, because the two-day weekend allows the women to get on with things at home. These social factors, combined with the increased opportunities for rest and relaxation are mostly responsible for the reduced absenteeism.

The introduction of the five-day week in Switzerland brought problems, because it was not accompanied by a corresponding reduction in total weekly working-hours. When the 44-hour week was divided among five days, it resulted in daily shifts of nearly nine hours (including rest pauses). Most certainly such a daily overload cannot be compensated for by the two-day weekend.

In our opinion, nine hours work per day is physiologically unhealthy, and the five-day week is an improvement only if the working day does not exceed eight hours.

The four-day week   A four-day week has come under discussion in the last few years. About 600 American firms and recently a few German and French ones too, are said to have had favourable experience of this [214] [273]. The total hours worked in four days are usually 36 in the U.S.A. and 40 in the European firms. *Rosenkranz* [273] thinks that the trend exists towards a four-day week of 36 hours, but that industrial conditions are not yet favourable to the idea. Advantages put forward are three days free at the weekend, and the possibility of increasing employment by taking on more workers.

From the point of view of industrial medicine, on the other hand, something must be said about "ruining the health" [224]. This criticism is certainly valid. It must be possible for a person to recuperate within each 24-hour period, not to go on exhausting himself (or herself) for four days and hoping to recover over the following three rest days. We now know that working days of nine or ten hours lead to excessive fatigue and increased absenteeism through sickness. *The four-day, 40-hour week must therefore be rejected on medical and physiological grounds.* Moreover there are organisational aspects to be considered. Compressing the working week into four days would throw up very severe problems, one of them that the recruiting of more workers would not suit every firm!

Maric [214] reported upon the American experience, which is said to have been very mixed. The results of a public opinion survey in Germany will be mentioned here.

In April 1971 the Allensbach Institute for Demoscopy (BRD) carried out a survey [4] of the introduction of the four-day week, among a representative selection of 900 employees in the Bundesrepublik. The relevant question, whether a four-day week of ten-hour days, or a five-day week of eight-hour days was preferred, received the following distribution of replies:

Four-day week   46 %
Five-day week   47 %
Uncertain   7 %

This distribution of preference shows no effect of education or sex, but there is a certain age effect. Of the overseers and clerks, 40 % were for the four-day week and 56 % for the five-day week.

Table 27 summarises these results:

Table 27   **Survey of the overall preference for a four-day week, among a representative selection of 900 employees in the Federal Republic of Germany, made by the Allensbach Institute for Demoscopy** [4].

Age group (years)	4 days of 10 hrs %	5 days of 8 hrs %	No opinion %
16–29	55	38	7
30–44	47	46	7
45–59	35	59	6
60 and over	21	69	10

It is very doubtful whether a similar result could be expected today, because in the meantime a trade recession has led to short-time working, and more unemployment, with consequent changes in working hours.

Working times *of the future*	Analysis of sociological trends has led many students of the future to believe that the working week will shorten still further. Thus, for example, the American forecaster *Kahn* [168] makes the prognosis that by the year 2000, weekly working hours, not only in America, but also in Europe, will have been reduced to 35 or even 30, and annual holidays extended to 2–3 months. The main arguments for this belief are as follows.

The employees want to profit from the expected boom not only in pay, but also by having more leisure. A higher income makes sense only if the money can be used to enhance the quality of life. The boom in tourism and the so-called leisure industries in the last 20 years reinforce this argument. Less certain is the view that as modern industry becomes more "rationalised" the work will be less satisfying and less stimulating, so that there will be more need to give a meaning to life in one's leisure hours.

Most of these prophecies were made before 1970, since when we have become more acutely aware that industry, the consumption of energy, and the destruction of natural resources cannot go on increasing at such a rate. Hence one of the most important arguments for shorter hours – that they would increase production and speed up industrial growth – is being called in question.

Flexible and continuous working schedules

Flexible working *hours*	Flexible working hours are a new form of work organisation which has suddenly found many adherents during the last decade. It is characterised by a fixed working period, called the block time or core time, and a flexible period at each end of this, the length of which is at the discretion of the individual worker. All the employees must be present at once during the core time, so obviously a certain number of hours must be worked per month, as a minimum. The weekly or monthly totals are recorded by various methods such as stamped cards, click-counters, or punched cards for data-processing machines.

A few examples of ways of regulating working hours are shown in *Figure 143*. In these examples the core time varies from 4½ to 7½ hours and the midday break may be part of this, or part of the flexi-time. The total flexible time may be reckoned weekly or monthly, and in most factories can be carried over into the next month, up to a limit usually between 10 and 20 hours.

A good review of experience up to date in various countries is to be found in the publication of the International Labour Office, by *Maric* [214]. On the whole it can be said that the new arrangement is popular with the employees.

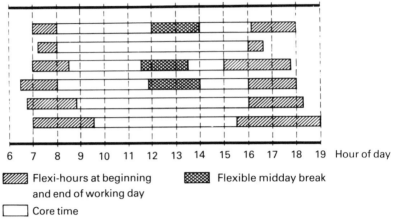

6 7 8 9 10 11 12 13 14 15 16 17 18 19 Hour of day

▨ Flexi-hours at beginning   ▥ Flexible midday break
   and end of working day
☐ Core time

Fig. 143 A few examples of flexible working hours.

Apparently not one firm has yet reverted from flexible hours to traditional working [214].

Advantages

The most important advantages of flexible working hours are the following:
(a) shorter commuting times by avoiding rush hours;
(b) working hours can be adjusted to suit personal convenience;
(c) eliminates punctuality problems;
(d) working day can be arranged to suit one's daily rhythm;
(e) a feeling of greater independence;
(f) tends to reduce absenteeism;
(g) tends to reduce overtime working;
(h) working hours can be adjusted to maintain production.

Drawbacks

The following objections are occasionally raised:
(a) higher administrative costs;
(b) direction of labour is more difficult because there is less time when everyone is available;
(c) people are tempted to work too long at a stretch in order to qualify for more free time later on;
(d) more difficult to maintain order and discipline;
(e) collections and deliveries are more difficult, and service to the customer suffers.

In reply to these objections, the last three are largely hypothetical, and not yet confirmed in practice. On the other hand it is a valid objection that not all kinds of work are suitable for flexible hours, and that in every factory there is one group of employees who must be excluded (for example workers at the information desk, telephonists, certain servicemen, switchboard operators, etc.). According to *Maric* [214] other groups for whom flexible hours are inappropriate are production line operatives, working in series, and other assembly workers organised in groups.

Looking over the very extensive literature on flexible working hours, one gets the distinct impression that the advantages greatly exceed the drawbacks, and that both on industrial and social grounds this new way of regulating working hours is a distinct step forward.

Continuous working hours

Before World War II a midday break of two hours was usual on the Continent, whereas in Britain, by long tradition, most factories had only a short break, not long enough for the employees to go home to lunch. Hence this type of organisation is called "continuous working", characterised by a midday break of 45–60 minutes, with a corresponding early shut-down in the evening.

After World War II many continental firms also adopted continuous working, which frequently raised problems, because some of the work force preferred the traditional long midday break. Social and physiological factors were involved. Continuous working interfered with family life, especially contact with one's children. On the other hand, the short midday break resulted in changes of eating habits: the midday meal was reduced to a small snack, and the evening meal became the principal meal of the day. This upheaval made trouble for not a few of the workers, because the heavy meal in the evening interfered with their sleep. Women with children also had the problem of finding someone to look after the children at midday, but this is now being tackled with nursery schools and school dinners.

These drawbacks must be set against the considerable advantages of less time spent in travelling (commuting), and having more free time in the evenings. By saving a 30-min journey each way the worker has one hour more in the evening.

To give as much consideration as possible to the great variation in travelling times, and to the wishes of individual workers, many Swiss firms have sought a compromise solution. Very often the midday break has been fixed at 1½ hours. *Flexible hours with a flexible break is often the best substitute for a previously unsatisfactory compromise working.*

The implications of flexible working hours for the daily eating habits of the workers concerned will be discussed later in the chapter under *"Snacks between meals"*.

Rest pauses

Biological importance

Every function of the human body can be seen as a rhythmical balance between energy consumption and energy replacement, or, more simply, between work and rest. This dual process is an integral part of the operation of muscles, of the heart, and, if we take all the biological functions into account, of the organism as a whole. Rest pauses are therefore indispensable as a physiological requirement if performance and efficiency are to be maintained. Military commanders have always known that a marching column should be halted once an hour, because the time lost will be more than compensated for by a better performance of his men at the end of the march. Rest pauses are essential, not only during manual work, but equally during work that taxes the nervous system, whether by requiring manual dexterity or by the need to monitor a great many incoming sensory signals.

Different kinds of rest pause

Work studies have shown that people at work take rest pauses of various kinds and under varying circumstances. Four types can be distinguished:

(a) spontaneous pauses;
(b) disguised pauses (i.e. switching to routine work for a time);
(c) pauses arising from the nature of the work;
(d) prescribed pauses, laid down by the management.

Spontaneous pauses are the obvious pauses for rest that the workers take on their own initiative. These are usually not very long, but may be frequent if the job is strenuous.

Disguised pauses are times when the worker occupies himself with some easier, routine task in order to relax from his concentration on the main job. Most jobs offer many opportunities for such disguised pauses, examples being the following: cleaning some part of the machine, tidying the work bench, sitting down more comfortably, blowing his nose, or even leaving his place on the pretext of consulting a workmate or the foreman. Such disguised pauses are justified from a physiological point of view, since nobody can do either manual or mental work continuously, without interruption.

Work-conditioned pauses are all those interruptions that arise either from the operation of the machine, or the organisation of the work, for example, waiting for the machine to complete a phase of its operation; for a tool to cool down; for a piece of equipment to warm up; for a component; for a machine or a tool to be repaired. Waiting periods are especially common in the service industries: waiting for customers, or orders. On a conveyor belt the length of work-conditioned pauses depends on the speed and dexterity of the operative. The faster he works, the longer he has to wait for the next piece to come along. Since speed and dexterity decline with age, the younger workers have long pauses, whereas the older operatives often have to work almost continuously to keep up. Hence on conveyor belts, the older operatives, as well as the less skilled often have to work hastily, and may be overstressed.

Prescribed pauses are breaks in the work that are laid down by the management: for example, the midday break, other pauses for snacks (e.g. coffee break).

Interrelationship between pauses

The four types of rest pause are to some extent interrelated, as was shown by *Graf* [94] with the help of a time study. In particular, this showed that introduction of a few prescribed pauses led to fewer spontaneous and disguised pauses being taken. *Figure 144* shows the results of a time study of a female operative in the electrical industry, who had to perform a highly skilled job on piece rates.

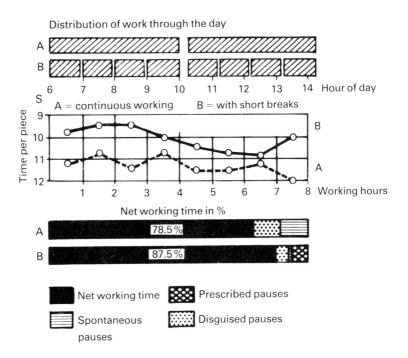

Distribution of work through the day

A = continuous working   B = with short breaks

Net working time in %

A — Net working time   Prescribed pauses
— Spontaneous pauses   Disguised pauses

Fig. 144 **Effects of short breaks (B) on the net working time (black block), on secondary tasks (disguised pauses) and on spontaneous pauses.** After *Graf* [94].

In a similar investigation *Graf* [94] showed that during tiring skilled work standing up (pressing fuse-holders out of china clay) both disguised and spontaneous pauses increased progressively during the course of an eight-hour study. There were about three times as many such pauses during the last three hours as during the first five, showing that the need for breaks in the work increased as the operator became more fatigued. To meet this need there was an increase of both disguised and spontaneous rest pauses.

In general, it can be said that all the different types of rest pauses (disguised, spontaneous, prescribed and work-conditioned) should amount to 15 % of the working time. Often a ratio of 20–30 % is allowed, and this is certainly necessary in some jobs.

Rest pauses and absenteeism

Shepherd and *Walker* [289] have investigated absenteeism among iron and steel workers, and found that if heavy work was combined with frequent work-conditioned pauses fewer shifts were lost than if the work was continuous. This did not apply if the work was light, thus suggesting that the increased shift-losses during uninterrupted heavy work were at least partly attributable to chronic fatigue, well-known as a factor in sickness rates. The results of this investigation are shown in *Table 28*.

Table 28   Yearly shift-losses in relation to severity of work to rest pauses, in two iron and steel works. After *Shepherd* and *Walker* [289].

Type of work	Shift losses in %		
	Heavy work	Moderately heavy work	Light work
Workers less than 45 years old			
Uninterrupted work	6.77	3.74	2.49
A few pauses	4.36	3.36	2.67
Frequent pauses	3.64	3.86	2.70
Workers more than 45 years old			
Uninterrupted work	7.55	5.99	4.60
A few pauses	5.19	4.22	3.94
Frequent pauses	3.51	4.74	3.39

Rest pauses and output

Many investigations into the effect of rest pauses on production have been recorded in the literature, and in general their results agree with those of research into working time and production. Introducing rest pauses actually speeds up the work, and this compensates for the time lost during prescribed pauses, as well as leading to fewer disguised and spontaneous pauses.

The hourly output of fatiguing work usually declines towards the end of the morning shift, and even more towards evening, as the rate of working slows down. Various studies have shown that if prescribed pauses are introduced the appearance of fatigue symptoms is postponed and the loss of production through fatigue is less. An example of the effect of various kinds of rest pauses on output is given in *Figure 145*.

In this example, as the results show, the onset of fatigue symptoms in piece work has been delayed by the introduction of short rest pauses. The collective effect of six short pauses has been an increase of production per shift (6.45 %) and this is much

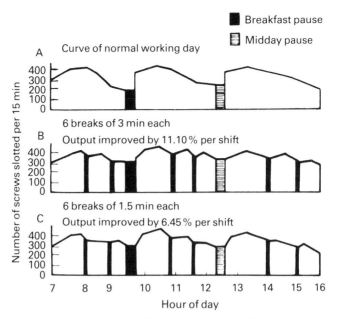

Fig. 145 Effects of prescribed pauses of two different lengths on output. The measure of output is the number of slots in screw heads cut in a period of 15 minutes. After *Hanhart* [132].

increased (11.10%) if the pauses are of three minutes each. The gain in output has more than compensated for the time lost. Such examples, with similar results, are frequent in all industrial countries. Even though not every investigation has been conducted on strictly scientific lines, they have indicated that *on the whole rest pauses tend to increase output rather than to decrease it. Ergonomics attributes these effects to the avoidance of excessive fatigue, or to the periodic relief of fatigue symptoms by an interval of relaxation.*

Number and length of pauses

For heavy work obligatory pauses should be laid down, evenly distributed throughout the eight working hours of the shift. If the pauses are only optional, workers tend to work continuously and save up all the permitted rest time until the end, so as to be able to leave work earlier. This leads to overstress, particularly among older workers. The arrangement of rest pauses for heavy work has already been discussed in *Chapter 6* under *"Limiting values and norms for energy consumption at work"*.

Moderately heavy work

For all other jobs in manufacturing industries, in offices and in administration the current recommendation is a rest pause of 10–15 minutes in the morning, and often a similar interval during the afternoon. These pauses serve the following purposes:
(a) *preventing fatigue;*

234

(b) *allowing opportunities for refreshment;*
(c) *time for social contacts.*
It is unthinkable not to have some rest pauses, which the workers value as much for social as for medical reasons.

Problem of time-linked jobs

Time-linked work on an assembly line poses a special problem. Many studies in the laboratory, as well as in factories [94], have shown that pauses of 3–5 minutes every hour reduce fatigue and improve concentration. They are specially necessary when the job is repetitive, has to be completed in a given time, and calls for constant alertness.

Under training

Rest pauses have a dramatic effect on the learning of a skilled operation, as we have already seen in *Chapter 8: "Skilled work"*. If training is interrupted frequently for short periods of relaxation, a new skill is acquired much more quickly than if training is continuous. *Rohmert, Rutenfranz* and *Ulich* [271] proved this in their monograph from much experience and many experiments. The results of one such experiment by *Iskander* [155] are set out in *Figure 146*.

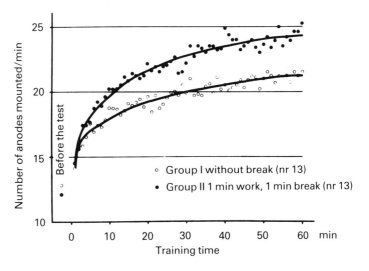

Fig. 146 Effects of short breaks on learning to mount anodes. After *Iskander* [155].

Two groups of 13 test subjects were learning to fit anodes into radio valves. They had to pick up two anodes at once and fit them into a holder, so that they could afterwards be grooved. The number of pairs completed per minute was taken to indicate how far the training had progressed. One group was trained to work 60 minutes without pause, whereas the other group alternated "one minute's work, one minute's rest".

The result confirmed many other similar results. The group with "one minute's work, one minute's rest" learned more quickly, and after 60 minutes achieved an end result distinctly better than that of the group who had worked continuously. *Iskander* obtained similar results during training for a punch operator's job, as well as for three different simulated driving tests.

Rest pauses during training do more than just prevent fatigue. *Ulich* [310] is very probably correct in his assumption that during rest pauses a trainee will look ahead and understand the process, so that it becomes easier to acquire the automatic skills required. So where a skilled operation is concerned, the rest pauses provide additional periods for mental training.

This aspect should be considered when dealing with the training of apprentices, who benefit from frequent rest pauses.

Recommendations To summarise, the following arrangement of rest periods is to be recommended:

(a) *where heavy work is concerned, or working in great heat, rest pauses should be such that the expected maximum hourly demands of the job should not be exceeded (see the relevant chapter on heavy work);*

(b) *for jobs demanding moderate physical or mental effort, a break of 10–15 minutes both morning and afternoon;*

(c) *a job making heavy mental demands, especially if it is timed work, or involves very little waiting time, should have, in addition to the morning and afternoon breaks, one or two shorter pauses of 3–5 minutes, both before and after midday;*

(d) *when learning a skill, or serving an apprenticeship, many pauses should be the rule, varying in frequency and duration to suit the difficulty of the job to be learned (see Rohmert, Rutenfranz and Ulich [271]).*

Nutrition and work

Comparison with a motor-car A motor-car requires three essentials for smooth running:

(a) *petrol as a source of energy;*

(b) *lubricating oils and greases to protect the moving parts;*

(c) *cooling water.*

Similarly the "human engine" requires:

(a) *foodstuffs (sugar, protein, fat) as source of energy;*

(b) *protective materials (vitamins, mineral salts, iron, iodine, unsaturated fatty acids, etc.) as "lubricants";*

(c) *liquids for cooling purposes.*

Figure 51 (in *Chapter 6*) demonstrated how chemical energy in the form of nutrients is taken in by the body and converted into heat and mechanical energy. This, again, is a comparable process to that which takes place in the engine of a car. A car can be driven only as long as the petrol lasts; a human being can go on working only as long as his food provides him with chemical energy. The more manual work done, the greater the demand for energy, which can be met by increasing the intake of food.

Food requirement and occupation

The energy content of foodstuffs can be measured, and is expressed in kilocalories (kcal). The same unit is used for the energy consumption of the human body, which is the higher, the more physically active is one's occupation. The average daily requirement of energy for men and for women in various occupations is summarised in *Table 10* (*Chapter 6*).

During recent years the overall pattern of the working population has changed in regard to physical labour. In every industrial country the proportion of workers with sedentary jobs has greatly increased, until we can expect these to account for about 70 % of employed persons. Conversely, the proportion of manual workers has fallen. A German example is given in *Table 29*, after *Wirths* [334].

Table 29 Distribution of employed persons, including housewives in the Federal Republic of Germany, in percentages.
*Figures from the old German Reich.

Severity of occupation	1882*	1925*	1950	1975
Light, sedentary work	21	24	58	70
Moderately heavy work	39	39	21	23
Heavy to severe work	40	37	21	7
Work force in millions	16.9	32.0	32.2	39.8

In general terms present-day adults can be divided into two categories according to their energy requirement at their occupation:

(a) sedentary workers and all female operatives: energy requirement 2000–3000 kcals per day;

(b) heavy workers whose average daily requirement is 3000–4000 kcals (ignoring the few whose exceptionally severe work calls for 4000–5000 kcals per day).

Table 30 summarises a proposal for distributing the calorie requirement among the five recommended mealtimes.

To illustrate the calorie content of some of the most important foodstuffs, *Table 31* gives the weight or quantity of each which is equivalent to 100 kcals.

Table 30 Distribution of daily food intake, in kilocalories.

	Sedentary workers and women	Manual workers
Breakfast	300–400	600–700
Morning break	25–50	150–250
Midday meal	800–900	900–1000
Afternoon break	25–50	150–250
Evening meal	1250–1400	1400–1600
Total	2400–2800	3200–3800

Table 31 Calorific equivalents of some important foodstuffs. The quantity of each that must be eaten to give 100 kcal of energy is:

Green vegetables	670 g
Turnips and swedes	400 g
Skimmed milk	3 dl (300 cm³)
Full-rich milk	2 dl (200 cm³)
Potatoes	150 g
Hens' eggs	60 g
Jam	50 g
Meat	50 g
Cheese	45 g
Bread	42.5 g
Legumes and pastas	30–40 g
Sugar	25 g
Butter or margarine	13.5 g

Needs of "sedentary workers"

For "sedentary workers" as well as for the great majority of female occupations, it is broadly true that *the quantity of food should be restrained in favour of high quality: in other words, fewer calories, more vitamins, minerals and trace elements.*

These categories of workers would also be well advised to cut down on energy-rich and highly refined foodstuffs, and give preference to natural foods containing protective elements: vegetables, salads, raw fruit, milk, brown bread and liver.

Under normal circumstances a person takes in just enough food to supply the energy he needs, regulating this according to his feelings of hunger, and so achieving an energy balance. Disturbances of this balance are fairly common among sedentary workers, who have a tendency to eat more than they need for their everyday life. Such people are often visibly overweight.

Needs of manual workers

Manual workers have quite different problems. They need a diet that is energy-rich, but not bulky, and so tend to prefer food that is protein-rich and fatty. Carbohydrate[1] foods are bulky and full of indigestible roughage.

---

[1] Carbohydrate foods include all kinds of sugar (sweet and non-sweet), which make up the greater part of flour, potatoes, pastas, and of course all sweet foods.

If a heavy worker wants to take in 3600 kcal in the form of potatoes, he will have to eat 5 kg of them. It would certainly be wrong for a manual worker to try to make up the balance of calories that he needs from carbohydrate foods alone. Such a diet would be too bulky and would overload the digestive organs. The recommended course is for manual workers to increase their intake of proteins and fats to approximately double the normal value. For a man weighing 70 kg this means about 100–110 g of protein per day, and about the same amount of fat.

Muscular work requires increase amounts of vitamin B_1 and phosphates. A heavy worker should have the benefits of brown bread (i.e. rye or wholemeal) and of milk products. On the whole the calorie-rich diet of manual workers will provide enough of the protective vitamins and minerals. To summarise, manual workers should have an energy-rich diet, giving preference to meat, eggs, milk, butter, cheese and brown bread.

The importance of proteins and fats

Since protein of animal origin is more valuable than plant protein for body building and muscular strength, half of the intake should come from meat, eggs and milk. A 70 kg man could find the necessary 50 g of animal protein in 1.5 l of milk, 300 g of meat, or seven eggs.

Fats are the foodstuffs that are richest in energy: 100 g of fat can replace about 300 g of bread or even 1 kg of potatoes. A further advantage of fat for heavy workers is that it remains longer in the digestive organs, and so postpones the onset of hunger.

Generally speaking, too much fat is eaten in civilised countries. Many studies have shown that ailments of the heart and circulation can be partly traced back to too much fat in the diet. The risk is less to manual workers than to "pen-pushers", who are strongly recommended to avoid fats and greasy foods as much as possible. The origin of the fats eaten is also of some significance to health, and one half of the fats should be plant oils, and the other half animal fats. Milk and milk products should be the preferred forms of animal fats, since they contain many vitamins and minerals.

Eating between meals (snacks)

Overall working time and nutrition

The traditional (continental) working hours, with a long two-hour break at midday, allowed the worker to go home to take his midday dinner with his family, and still have time for a rest, if he wished. There was time for a good deal of the meal to be digested during the lunch hour, so that readiness for work in the early afternoon was only moderately affected. Such an arrangement is good from the point of view of health as well as of social life.

The undoubted advantages of a long midday break become nullified and illusory, as soon as the distance from work to home becomes so great that half or more of the time is taken up with travelling to and fro. Then there is no time for relaxation, and there is the additional fatigue of the two journeys. This happens more frequently as towns and cities grow bigger, and causes more and more factories to change over to continuous working, with a shorter midday break and an earlier release in the evening.

A shorter midday break usually means eating in a works canteen or a nearby restaurant. A meal away from home is more expensive and there is less time for eating, and a worker instinctively feels that it would be unwise to go back to work immediately after a large meal; hence a change in eating habits, transferring the main meal to the evening, and making the midday meal little more than a large snack. This has the advantage that the afternoon's work is less affected by digestive problems, as is partly illustrated in *Figure 149*.

From a medical point of view it should also be noted that *a midday break of 45–60 minutes is usually enough for relaxation, provided that there is also another rest pause of 10–15 minutes, both morning and afternoon, for relaxation and eating a snack.*

Food intake and biological rhythm

In a Swedish gas works the reading errors of several inspectors were tabulated daily over a period of 19 years. *Figure 147* shows these reading errors, according to the time of day at which they occurred. It is clear, among other things, that they reached a

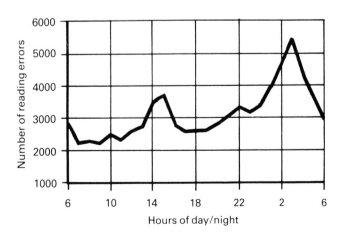

Fig. 147 Reading errors of gas meter inspectors, according to the time of day. Evaluation of 175,000 records, with a total of 75,000 errors, and extending from 1912 to 1931. Besides the distinct peak after the midday meal, mentioned in the text, there is a much higher peak in the early hours of the morning, when readings are taken at night. After *Bjerner, Holm* and *Svensson* [25].

maximum immediately after the midday break, that is while digestion was still going on.

This observation confirms a biological law, by which loading the digestive system damps down the state of readiness of the entire organism. This effect of large meals has been known for a long time, and finds expression in the familiar proverb: "A full stomach makes an unwilling student."
It is known, however, that the converse, i.e. an empty stomach, has an equally bad effect on efficiency. This is particularly well known to sportsmen, who call it "hunger anxiety".

Distribution
of meal times

The American physiologists *Haggard* and *Greenberg* [127] measured the blood sugar level and the respiratory quotient of volunteers throughout a whole day. They found that the blood sugar and the respiratory quotient, as well as the muscular efficiency, increased after each meal, falling away again steadily as time passed. The blood sugar level reached its lowest level about 3–4 hours after breakfast, when symptoms of fatigue and loss of efficiency often made their appearance. If, however, the test subjects were given small snacks or a meal every two hours, the anticipated low point was not reached: the blood sugar and efficiency remained more or less at their high level throughout the working day. Some of these results are set out in *Figure 148*.

The same investigators carried out a later experiment under practical, everyday conditions in a factory making tennis shoes. Their results are reproduced in *Table 32*.

Table 32 Effect of food on output in a tennis-shoe factory. After *Haggard* and *Greenberg* [127].

| Feeding | Number of items produced | | |
	During 1st hour	During 4th hour	Daily total
Without breakfast	144	156	1379
3 main meals	192	168	1455
3 main meals, plus 2 snacks	193	186	1521

Without breakfast, the output during the first hour was considerably less than with three main meals; furthermore, the addition of two snacks raised the output during the fourth hour, at the end of the morning shift. These results have been confirmed in recent years by several other experiments, and lead to the conclusion that *the intake of food five times daily (three mealtimes and two snacks) is good for both health and efficiency, and is to be recommended.*
This recommendation applies particularly when continuous work-

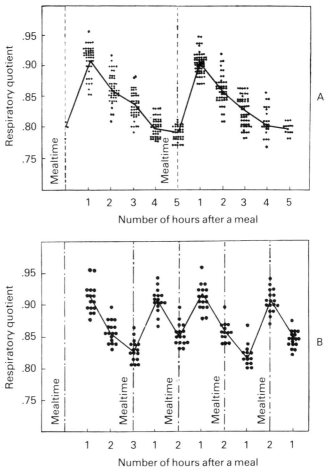

Fig. 148 Effects of two meals (above) and of five smaller meals (below) on the respiratory quotient, as an index of bodily efficiency. The RQ (ordinate) follows the fluctuations of blood sugar level and can be considered as an indicator of efficiency. The RQ does not fall as steeply after five meals as after two, so that efficiency remains at a higher level throughout the working day. After *Haggard* and *Greenberg* [127].

ing, with a short midday break is in operation, since then the need for extra periods to relax and eat is even greater. *Figure 149*, based on current scientific knowledge, and general experience, shows a "theoretical" curve of eagerness for work, both during traditional working hours and during a working day that is planned according to physiologically sound principles.

Intake of fluids

A person needs not only food, to provide energy, but also water, to maintain a correct water balance. The average requirement is 35 g water for every kilogram of body weight per 24 hours (= 2–2½ l*

---

* 1 litre = 1 dm³.

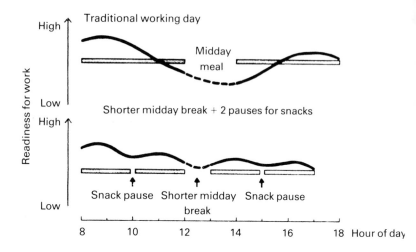

Fig. 149 Working time, food and readiness for work. The dotted blocks cover the working periods, and the curves indicate the conjectural rise and fall of readiness for work. During the working day there should be two breaks for snacks, in addition to a short midday mealtime (below) when the curve does not fall as low as it does with a larger midday break (above).

per day). The water that is drunk and that which is obtained from the food eaten is being continuously excreted through the kidneys and the sweat glands. The liquid excreted is not pure water, but a fluid rich in waste materials (urea, sodium chloride and many other metabolic end-products). Although foodstuffs have a high fluid content (e.g. meat 70–80%; bread 43%; fruit 85%; potatoes 78%; pastas 19%), we still need more liquid. The amount varies between individuals, but is generally about 0.5–1 litre, rising to 1.5–2 litres and more in summer.

The amount of water we drink is governed by feelings of thirst, which in turn depend mainly on the concentration of salts in the blood. An increase in salt concentration makes one more thirsty.
In summer much of the water is lost by sweating. Since sweating is essential to maintaining body temperature, like the cooling of a motor, the consequent water loss must be replaced in summer, in tropical countries, and in hot jobs in industry. This is best done by drinking easily assimilated liquids such as tea, coffee, or soft drinks. The loss of minerals during heavy sweating may be offset by taking salt tablets.

Summary of snacks Snacks between meals are important both for sedentary and manual workers, to meet a significant part of their daily fluid requirements, as well as of additional energy-producing food, according to their bodily needs. *Table 30* shows the necessary calorific values of such between-meal snacks, but people vary greatly in their eating habits, and hence in their choice of snacks.

The range of choice lies between a refreshing drink that is low in calories and, at the opposite extreme, a heavy snack equivalent to some 400 calories, with many intermediates. A few suggested snacks between meals are listed in *Table 33*, the choice depending primarily on the fluid and energy requirements of the person concerned.

Table 33   Suggestions for between-meal snacks, and their calorie content.

Type of snack	Number of kcal
1 cup of mineral water	–
1 cup of soup	10–15
1 cup of tea with two lumps of sugar	35
1 cup of coffee with milk and two lumps of sugar	37
1 cup of apple juice	65
1 cup of milk or yoghourt	66
1 cup of ovaltine in milk	130
Bread (50 g)	120
Bread with fruit	about 240
Bread with cheese	about 300
Bread and sausage	about 300

A drink with few calories (soup, tea, or coffee) and no solid food is recommended for desk-workers, whereas manual workers need a calorie-rich drink such as ovaltine, apple juice, milk or yoghourt, supplemented by bread and cheese, sausage or fruit.

Coffee and tea are specially popular drinks for snack-time because they have an immediate stimulating effect, though this is slight and does not last very long. The worker feels a need for frequent stimulants of this kind, and there is no objection on medical grounds, provided that they are not drunk to excess. A certain amount of stimulation is a good thing where the work is monotonous, yet responsible, as often happens nowadays at such jobs as switchboard operator, or controller of a machine that is mainly automatic.

Effects of snacks on teeth

An important aspect of between-meal snacks is the modern concern with dental health, because the link between daily sugar intake and the incidence of dental caries is undeniable. The stickiness and physical consistency of sugar foods is bad for the teeth, especially pastries, nutty crunch, and many kinds of sweet biscuit, chocolate, bananas and crystallised fruit. On the other hand bread and fruit are rarely harmful, in spite of their starchiness. From a dental point of view the following are recommended items for between-meal snacks: *apples, nuts, fresh fruit, mineral water, or plain milk, bread and butter, cheese, yoghourt, sausage, meat.*

Planning of between-meal snacks

The recommended between-meal snacks can be a problem for the management. Introducing rest periods when snacks can be eaten easily leads to overstaying the permitted time and to a slackening of discipline. The service must be quick, so as to waste as little as possible of the rest period on obtaining food. A works canteen of average size can serve about 100—150 persons in 10—15 minutes. If we allow a few minutes walking to and from the canteen, and a few minutes for the last served to eat their snack, it will be seen that only 60–80 persons can be served and back at work within the time limit (15 minutes). If the breaks of different sections are staggered over a period of one hour, then the canteen can serve 250–300 persons.

In larger factories it is more convenient to have decentralised snack-bars for different departments. A serving counter operated by a single person can meet the needs of 40–60 persons during a break period, and with two persons serving this number is doubled, and doubled again if four break periods are staggered.

There have been many recent attempts to solve the problem of between-meal snacks by providing vending machines for both food and drinks. Service from a vending machine is slower than by hand, each machine serving not more than 20–30 people during a break period. The machines have the advantage of taking up little space, and not needing an attendant, but the disadvantage of offering a limited range of foods and drinks, and little opportunity for personal preferences. Servicing the machines (filling up, cleaning, etc.) may be carried out by employees of the firm, or by special canteen staff, who may belong to an outside catering firm which specialises in such machines. A well-organised repair service is necessary so that breakdowns can be quickly remedied. Sometimes snacks are distributed through the factory by serving trolleys. The worker loses no time going to and from a canteen, and the passage of the trolley automatically dictates the rest pause. On the other hand the interruption may not always be convenient for a break in the work. The amount of food and drink that can be carried is limited, and time is wasted when it goes back to the canteen to refill. It may be difficult and tiring to manoeuvre the trolley through some departments, so the system is most suitable for servicing small and isolated sections.

Cold or warm snacks?

Cold drinks and food cause the blood vessels of the stomach to contract and reduce the flow of digestive juices, impairing digestion, and causing much discomfort to susceptible persons. Hence neither meals nor between-meal snacks should consist exclusively of cold food.

Even in summer, or during hot work, lukewarm to warm drinks are better than cold, since they place less strain on the stomach, and pass more quickly into the blood, where they have their recuperative effect.

14
Night work and shift work[1]

The problem

In recent times all industrial countries have turned over more and more to continuous production, an example being the following from France. The proportion of shift workers to total work force rose from 12% in 1957 to 21% in 1974 [47]. Nowadays the proportion is more than 20% in most industrial countries. This explains why shift working is no longer a fringe problem, but of ever-increasing importance.

The main reasons for going over to continuous production are economic ones. Many manufacturing processes are said to be economic only with this form of working. In other cases expensive machinery needs to be used 24 hours a day to be profitable.

We have already said repeatedly that the human organism is in its ergotropic phase (geared to performance) in the daytime, and in its trophotropic phase (occupied with recuperation and replacement of energy) during the night. Hence the night worker must approach his work, not in a mood for performance, but in the relaxed phase of his cycle. Herein lies the essential physiological and medical problem of night work. Another aspect is the burden it lays on family life, and the social isolation. Ergonomics is therefore faced with the problem of planning work schedules in such a way that shift work does as little harm as possible to health and to social life.

Comprehensive surveys of the problem of night work and shift work may be found in the publications of *Ulich* [311], *Mott* [230], *Swensson* [299], *Rutenfranz* and *Knauth* [274], *Andlauer, Carpentier* and *Cazamion* [8], *Colquhoun* [56], *Conroy* and *Mills* [57].

---

[1] "Shift work" in this context includes some night work.

Circadian rhythm and night work

The various bodily functions of both man and animals fluctuate in a 24-hour cycle, called the circadian rhythm (*circa dies* = approximately a day).

Even if the normal influences of day and night are excluded, e.g. in the Arctic, or in a closed room with unchanging artificial lighting, a kind of internal clock comes into play, the so-called endogenous rhythm. This varies in different individuals, but usually operates a cycle of between 22 and 25 hours.

Under normal conditions endogenous circadian rhythms are synchronised into a 24-hour cycle by various "*time-keepers*":

(a) changes from light to dark and vice versa;
(b) social contacts;
(c) work;
(d) knowledge of clock time.

The bodily functions that are most markedly circadian are sleep, readiness for work, and many of the autonomic, vegetative processes such as metabolism, bodily temperature, heart rate and blood pressure. It may be said that practically all the human functions that have yet been studied show a regular daily cycle. *Table 34* summarises a few characteristic day/night changes.

Table 34   Circadian rhythm.

The following bodily functions increase by day and decrease by night:

Body temperature
Heart rate
Blood pressure
Respiratory volume
Adrenalin production
Excretion of 17-keto-steroids
Mental abilities
Flicker-fusion frequency of eyes
Physical capacity

The bodily functions listed above certainly show these trends throughout the 24 hours, but they do not all reach their maxima and minima at the same time. There is a distinct phase-difference between one and another. Taken as a whole they confirm the rule that was mentioned above:

(a) during daytime all organs and functions are ready for action (ergotropic phase);
(b) at night most of these are damped down, and the organism is occupied with recuperation and renewal of its energy reserves (trophotropic phase).

Normal sleep

The most important function that is geared to circadian rhythm is sleep. While it is still not possible to say just what is the scientific function of sleep, it can certainly be said that sleep that is undisturbed either in quantity or in quality is a prerequisite for health, well-being and efficiency.

An adult human being requires about eight hours' sleep per night, though there are considerable individual variations. While there are some people who need ten hours' sleep, if they are to be fresh and alert, others need only six hours, or even less. It was said of *Thomas Alva Edison* that he needed only about three hours, and that he himself dismissed even these three hours as merely a bad habit.

Length of sleep is mainly a matter of age. A new-born child needs 15–17 hours daily during its first six months, whereas old people sleep less and less, and often this in broken periods.

The quality of sleep is not uniform, but cyclic in form, and has various stages of different depth (see *Figure 150*). The following five stages may be distinguished.

Stage 1 The electroencephalogram (EEG) shows low amplitudes, with many theta waves. This is the stage of going to sleep, sleeping lightly. Duration 1–7 minutes.

Stage 2 EEG shows low amplitudes. Besides the theta waves there are also the so-called "sleep spindles", strong peaks between 12 and 14 Hz, following in quick succession. Stage 2 is a condition of light sleep and its duration is about 50% of total sleeping time.

Stage 3 EEG shows increase of amplitudes and decrease in frequencies, up to 50% of waves being below 2 Hz. Many delta rhythms, interspersed with sleep spindles. Deeper sleep.

Stage 4 In EEG more than 50% of the waves are below 2 Hz. Maximum synchronisation and deepest phase of sleep.

Stage 5 Rapid eye movements (REM). In EEG similar to Stage 1, with a mixture of frequencies. REM, frequent salvoes of quick movements of the eyes are characteristic, this is the stage during which dreams are especially common. Despite the picture on the EEG, the REM stage is characterised by maximum relaxation of the muscles and resistance to being awakened; hence Stage 5 is known as the "paradoxical stage".

Quality of sleep

Although little is yet known about the significance of the five stages of sleep, it can be said in general terms that *Stages 3 and 4, and the REM stage 5 are the ones that have particular recuperative properties*. These stages determine the quality of one's sleep.

As mentioned above, cyclical changes take place during sleep, with about four descents into deep sleep, linked by intervening shallow periods. *Figure 150* shows the cyclical course of an ordinary night's sleep.

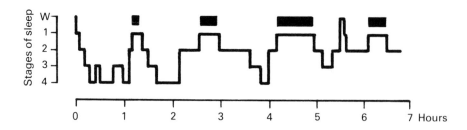

Fig. 150 The cyclic course of a night's sleep. W = being awake. Black bars = periods of rapid eye-movements (REM).

A single sleep cycle lasts between 90 and 100 minutes. The deepest level (Stage 4) is usually reached during the first cycle, as all experience confirms, with the result that most of one's deep sleep comes during the first half of the night.

Daytime sleep of night workers

For a long time past, works doctors have recorded frequent cases of disturbed daytime sleep among night workers. The results of three such surveys are shown in *Table 35*.

Table 35 Frequency of disturbed daytime sleep among night workers.

Authors and circumstances	% disturbed Night shift workers	Day shift workers
Andersen [6] 600 3-shift workers and 300 day shift workers	66	11
Ulich [311] 115 3-shift workers and 152 workers without a night shift	63	5
Thiis-Evensen [304]; *Aanonsen* [1] Among workers who have voluntarily undertaken night work	84–97	–

Part of this disturbed sleep must be attributed to noise, which is usually greater round a residential area during the day than at night, but many night workers say in addition that they feel a certain restlessness during the day, and their daytime sleep is not refreshing enough [274] [311].

Length and quality of daytime sleep

Some recent EEG studies of length and quality of sleep among night shift workers are of interest. *Figure 151* shows the results of a study by *Françoise Lille* [200], who analysed in detail the daytime sleep of 15 regular night shift workers.

It appeared that daytime sleep was distinctly shorter than night sleep the workers took on their rest day. The average length of

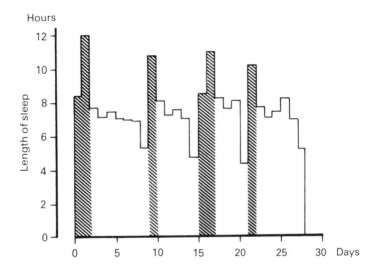

Hours

Length of sleep

Fig. 151 Length of daytime sleep of night-shift workers. White columns = total of daytime sleep. Shaded columns = night sleep on rest days. Average of 15 workers. After *Françoise Lille* [200].

sleep in the daytime was six hours, whereas on the rest day the average varied between eight and twelve hours, with longer sleep on the second of the two rest days than on the first. *Françoise Lille* concluded that *the night worker accumulated a "sleep debt" which he "paid back" on his two rest days.* Evidently a singly day's rest was not enough for this purpose.

Daytime sleep on the EEG

Detailed analysis of the EEG showed that the quality, as well as the duration, of daytime sleep was impaired, as evidenced by a greater number of periods of light sleep, and more body movements.

Later on these findings were confirmed by other authors. For example, *Knauth* and *Rutenfranz* [177] studied five persons who worked at night from time to time during a 40-day period, and observed a shortening of daytime sleep, particularly of Stage 5 of REM. Comparison between sleepers in noisy surroundings and in a soundproof room showed that the disturbance was not caused by noise, but was an integral feature of daytime sleep.

Finally we may mention the work of *Pternitis* [258] on 3-shift workers in a thermal power station. The results showed the following average sleeping times:

(a) after the night shift: 6 h 03 min
(b) after the late day shift: 7 h 35 min
(c) after the early shift: 6 h 47 min

The daytime sleep after the night shift, in comparison with the other two shifts, showed a shortening of the deep sleep stages 2 and 3, as well as of the REM stage. On the other hand Stage 1, the initial sleep phase, almost doubled. Comparison of all three shifts showed that sleep after the late day shift was very refreshing, with least disturbance, while sleep after the early shift was intermediate in quality between the other two.

To summarise, all these studies show that sleep following a night shift is curtailed and of little restorative value.

Capacity for work at night

Both mental and physical working capacity show a characteristic circadian rhythm. As an ergonomic example the reading errors of the Swedish gas inspectors may be mentioned once again [25]. As is evident from *Figure 149*, psychophysiological readiness for work is at a maximum in the forenoon, and in the second half of the afternoon, whereas it is poor immediately after the midday break and declines even more at night.

Two more examples may be given. The study of air traffic controllers, mentioned earlier in *Chapter 11* [100] [115] [116] allowed the level of fatigue to be monitored all round the clock, for a period of three weeks. The results are shown in *Figure 152*.

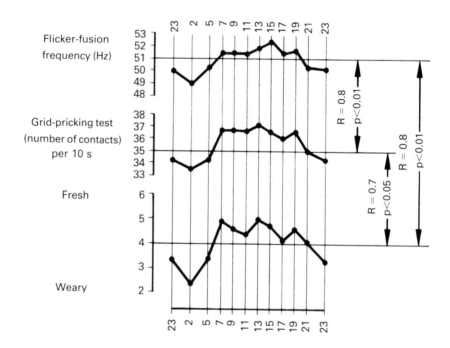

Fig. 152 **Circadian rhythm of the indicators of fatigue among air-traffic controllers over a period of three weeks.** Spearman's R factor between the three curves was 0.8 and 0.9 with p<0.01. Each point on the graph is a mean value of the individual averages for that day. Usually there was one night shift each week. After *Grandjean* [100].

From this diagram it will be seen that the subjective flicker-fusion frequency, the psychomotor performance in a grid-pricking test, as well as the self-assessment of subjective feelings of fatigue, all showed a markedly parallel circadian rhythm. The indicator showed lower readings during the night (pointing to a higher level of fatigue), and higher readings in the daytime (which could be interpreted as showing a state of increased readiness for action).

The second example comes from some work by *Prokop* and *Prokop* [257], during which 500 truck drivers were asked at what times of day they had fallen asleep at the wheel at least once. The numbers of replies for each of the hours is shown in *Figure 153*.

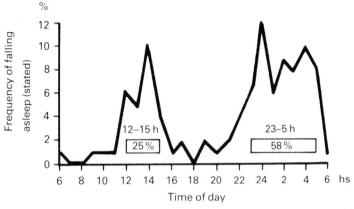

Fig. 153 **Frequency with which 500 truck drivers fell asleep at the wheel, in relation to the time of day.** After *Prokop* and *Prokop* [257].

The statements about falling asleep show a clear daily cycle, with one peak in early afternoon and an even more pronounced peak during the night. These three examples, to which others could be added, show that readiness for action is high during the daytime, and declines at night. These results reflect the rule that was formulated in the previous chapter, according to which the human organism is performance-orientated during the daytime, and damped down at night.

Productivity and frequency of accidents

These facts lead us to suppose that night work would be conducive, not only to lower output, but also to more frequent accidents. Several authors have recorded these [8] [338], yet the hypothesis is still not confirmed. Often the accident rate at night seems scarcely altered, or even reduced [9]. This contradiction between theory and practice is hardly to be wondered at, if we think the conditions surrounding the night worker (fewer disturbances from other people, higher wages, etc.) can be compared with those of the day worker only with certain reservations. Moreover there is much evidence that the night worker has made a "positive choice" of particularly rewarding work.

252

Reversal of
circadian rhythm

It was pointed out earlier that circadian rhythm is affected by a variety of other time-keeping factors. It is reasonable to assume that the circadian rhythm of the night worker may be reversed by the factor "work". So far this assumption has only been partially confirmed. Thus, for example, it seems likely that after several successive night shifts, body temperature, heart rate and adrenalin secretion will experience a reversal of their previous circadian rhythm.

Nevertheless most research workers stress that this reversal is not complete even after several weeks, that the curves become flattened, yet the maxima can hardly be said to be interchanged in position.

Bonjer [28], for example, observed that shift workers on a weekly rotation showed a decline of circadian rhythm of body temperature and heart rate during their night shifts, but that the maxima always appeared during the daytime. Normal circadian rhythms were soon restored on rest days. A characteristic curve of body temperature during normal day working, and during a night shift, is shown in *Figure 154*.

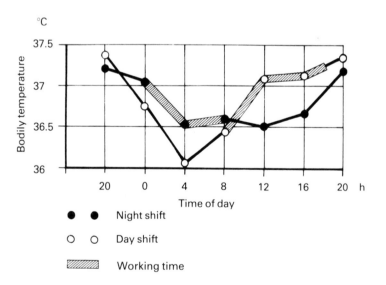

Fig. 154 **Body temperature during normal daytime work, and during the night shift.** Average values for 3 workes. After *Bonjer* [28].

Most research workers have obtained comparable results in similar studies [56] [57] [89] [178] [202]. So we can conclude that *biological circadian rhythms show first signs of reversal after several night shifts, but the reversal is not usually total, even after several weeks.*

Night work and health

It has been known for a long time that night workers commonly have bad health, and that more and more workers are forced to give up shift work for this reason. When shift work increased after World War II these matters of adjustment became a problem in industrial medicine, which was analysed in detail in several large field studies.

Sickness rate

The first big surveys came from Scandinavia. In Norway *Thiis-Evensen* [304] and *Aanonsen* [1] studied overall sickness rates among 6000 and 1100 workers, respectively. See *Table 36*.

Table 36 **Sickness rates among shift workers in Norwegian factories.** Studies made between 1948 and 1959 by *Thiis-Evensen* [304] and *Aanonsen* [1]. The percentages quoted relate to the total workers in each group studied.

| Ailments | Thiis-Evensen | | Aanonsen | | |
	Day work	Night work	Day work	Night work	Former night workers
Stomach troubles	10.8	35	7.5	6	19
Ulcers	7.7	13.4	6.6	10	32.5
Intestinal disorders	9	30	11.6	10.2	10.6
Nervous disorders	25	64	13	10	32.5
Heart troubles	–	–	2.6	1.1	0.8

"Positive choice" of night workers

Thiis-Evensen's survey showed that shift workers had significantly more digestive ailments and nervous disorders, and Aanonsen's work revealed an interesting corollary. Among the day workers investigated there were many who had abandoned shift work, either on grounds of health, or because they did not like it. This group of former night workers, who had certainly exercised a "negative choice", showed a distinct increase of digestive and nervous troubles. This discovery proved that comparisons with so-called normal groups must be carried out with caution, and that up to a point, night shift workers must be regarded as being a "positive selection" of particularly tough workers. This might also account for the contradictory results of surveys of sickness rates [304] [338].

In spite of these statistical difficulties, an ever increasing sickness rate has been observed during the past 20 years among "active" as well as former night shift workers [6] [8] [96] [299] [311]. Several reports should be mentioned here [8], according to which night shift workers often misuse drugs, taking stimulants during the night and sleeping tablets during the day. The reasons for the

increased liability to nervous disorders and ailments of the stomach and intestines are primarily:
(a) *chronic fatigue;*
(b) *unhealthy eating habits.*

Occupational sickness among night workers

Nowadays it is justifiable to talk about *occupational sickness* among night workers, the dominant symptoms being those of *chronic fatigue* (see also *Chapter 11: "Fatigue in industrial practice"*), thus:
(a) *weariness, even after a period of sleep;*
(b) *psychic irritability;*
(c) *moods of depression;*
(d) *general loss of vitality, and disinclination for work.*

The state of chronic fatigue is accompanied by an increased liability to psychosomatic disorders, which in night workers commonly takes the form of:
(a) *loss of appetite;*
(b) *disturbance of sleep;*
(c) *digestive troubles;*
(d) *stomach and duodenal ulcers.*
Thus the nervous disorders observed among night workers are no more than the symptoms of chronic fatigue, which, combined with unhealthy eating habits, is the cause of the increased liability to digestive troubles.

The causes

What, then, are the actual causes of occupational sickness among night workers? The answer to this question lies in what we have already said about circadian rhythm and the disturbances arising from the change from day work to the night shift. A conflict is generated in the body of a night worker by "desynchronisation" of his time-keeping mechanism; the working cycle is opposed to the "light–dark" and "social contact" cycles. No one of these mechanisms seems to be fully dominant, so that the functional unity of the body is lost, and the harmonic correlation between the separate biorhythms is impaired.

Symptoms

Expressed in other words: *since a complete adjustment to night work does not take place quickly enough, the night worker's bodily system is only partly switched over to "working" at night, and "sleeping and resting" by day.*
The result is insufficient sleep, both in quantity and quality, with less adequate recuperation, resulting in chronic fatigue with its associated symptoms.
Since at the same time eating habits are unhealthy (meals at unfamiliar times and so on), the psychosomatic symptoms tend to show themselves mainly in digestive disorders.

The nature of occupation sickness of night workers, with its causes and symptoms, is set out diagrammatically in *Figure 155*.

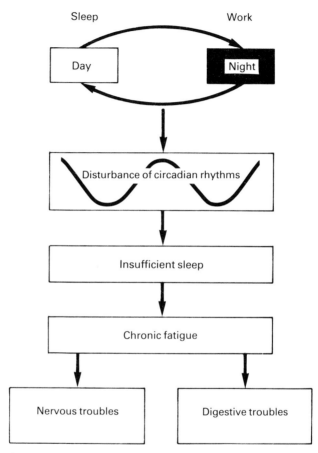

Fig. 155 Diagram illustrating causes and symptoms of occupational ailments among shift workers who periodically work at night.

Individual susceptibility

These ailments do not afflict every worker in the same way, and even if they show the same symptoms, the extent of the disorders varies very much from one person to another. *It is broadly true that about two thirds of shift workers suffer some degree of ill-health, and about one quarter sooner or later abandon shift work because of major health problems.*

Effect of age

Resistance to the special stresses of night work declines with age. Most burdensome is the need to change over to being active during the sluggish period of the night. The older worker is less adaptable, and tires more easily. On the other hand the older worker cannot enjoy the benefit of longer sleep, since the sleep of older people is very easily disturbed. *Hence the older worker suffers both from greater stress and fewer opportunities to recuperate from them.* In fact many surveys have shown that shift workers in the age groups over 40 are distinctly more prone to disturbed sleep and complain of their ill-health [9] [52]. It may be concluded that *during the course of the year people do not usually become accustomed to night work, but on the contrary they become increasingly liable to occupational ailments.*

Social aspects of shift work

Since "social well-being" is closely related to physical health, we must now briefly consider the social effects of shift work.

In the forefront of these is the dislocation of family life, interference with wider social contacts among friends, and fewer opportunities of participating in group activities. Particular investigations into these social effects were made in France by *Maurice* and *Monteil* [217], and by *Chazalette* [52]; in Germany by *Ulich* [311]; in the U.S.A. by *Mott* [230], and in England by *Brown* [40].

Family life

All these investigations show that most of the people questioned give priority to complaints about fewer meal times among their families. As a result, the father of the family is forced to leave the upbringing of the children to his wife.

Free time

Another complaint that is frequently expressed concerns dislocation of social life outside the family circle. Active participation in group activities, whether in sport or in politics is so limited that the night worker often feels that he is excluded from society altogether. There are similar impediments to the cultivation of friendships, especially if there are not many other shift workers living nearby. This state of affairs also dictates what the shift worker can do in his free time, so that he often pursues solitary hobbies. The French authors [52] [217] talk in this context of a tendency for shift workers to "feel themselves to be on the fringe of society", or of "social isolation".

Some of *Ulich*'s results [311] are given in *Table 37* as evidence of the impact of social factors.

Table 37 Social disruptions among 438 industrial workers (= 100%).
After *Ulich* [311].

Disruption of family life	59%
Disruption of private life outside	31%

These effects are particularly irksome if the rest days do not come at weekends.

| *Opinion polls* | Opinion polls have shown that many people are of two minds |
| *of shift workers* | about shift work. On the one hand they are opposed to it on |

Opinion polls of shift workers

Opinion polls have shown that many people are of two minds about shift work. On the one hand they are opposed to it on grounds of health and social life, but on the other they see certain advantages in it, such as more pay, or more freedom to plan their leisure. On the whole, however, the drawbacks predominate. Thus *Ulich* [311] found that of 495 workers:
(a) 53 % were "against shift work";
(b) 23 % were "completely decided against it".

The three shifts

Each of the three shifts has its advantages and drawbacks.
The early shift makes a communal evening meal possible. Leisure activities are possible, either in the afternoon or evening. If the shift begins very early (e.g. 4 a.m.), it is tiring and night sleep is cut short.
The late day shift is particularly bad for social life, since there are no opportunities for family life, either round the table, or in leisure activities. On the other hand sleep is good after this shift.
The night shift is bad from all angles. Family life is often limited to taking the evening meal together. Leisure activities are usually possible only in the second half of the afternoon. Sleeping habits vary: one group of night workers interrupt their daytime sleep for a midday meal, and then lie down again afterwards; the others sleep through until early afternoon. The night shift is bad for social life, and very tiring because all sleeping is carried on in the daytime.

Organisation of shift work

Distribution of shifts during the day

On the 3-shift system the day is divided into three equal periods of eight hours each. A common system is this:
(a) early shift: 6–14 hours
(b) late day shift: 14–22 hours
(c) night shift: 22– 6 hours
but there are many variants. In the U.S.A, for example, 8–16–24 hours is commonly worked, and this arrangement seems to have advantages, both physiological and social. Each shift allows at least one mealtime in the family circle, and at the same time provides good opportunities for sleeping, especially on the early and late day shifts.
A few firms work a system of two shifts of 12 hours each, but a 12-hour working day cannot be recommended from the standpoint of either industrial medicine or ergonomics (see *Chapter 13: "Arrangement of weekly work"*). At most, exceptions might be made for undemanding jobs, with long built-in pauses. When shifts are as long as this, each shift, whether day or night, is followed by two rest-days, and many workers like this.

Rotation of shifts　　In Europe periodic rotation of shifts is the rule, but in the U.S.A it is not uncommon to work the same shift all the year round. *Mott* [230] sees certain social advantages in this arrangement, but in the long run continuous night work is not acceptable, either on social or medical grounds, at least in big gangs.

Shift-rotation cycle　　Until 1960, many experts had been of the opinion that the intervals between shift rotation should be as long as possible. Recommendations for rotation every 3–4 weeks were based on the idea that

Mon	Tue	Wed	Thur	Fri	Sat	Sun
N	—	E	L	N	—	—
—	E	L	N	—	E	E
E	L	N	—	E	L	L
L	N	—	E	L	N	N

Weekend patterns and frequency per year			
Saturday	Sunday	Monday	Frequency per year
—	—	—	13
E	E	E	13
L	L	L	13
N	N	N	13

Fig. 156 Above: an example of a shift rota, in which the night shifts are widely scattered. Below: summary of free shifts (rest periods) over the year. E = early shift; L = late day shift; N = night shift. After *Graf* [95].

people need several days to change their biological rhythm, and adaptation to the new shift can take place only if several weeks are allowed. Nowadays we know that this interpretation is misleading. Even after several weeks adaptation is not complete, especially in regard to sleep, one of the most important bodily functions. The daytime sleep of workers on the night shift remains inadequate, both quantitatively and qualitatively for a long time.

The latest information leads to the recommendation that rotation of shifts should be short-term.

Criteria for shift rotation

As a start, it may be helpful to consider what criteria apply to shift systems.

The following may be considered the most important requirements:

(a) loss of sleep should be as little as possible, so as to minimise fatigue;

1. Week	M T W Th F S Su	E E L L N N —	5. Week	M T W Th F S Su	N N — — E E L
2. Week	M T W Th F S Su	— E E L L N N	6. Week	M T W Th F S Su	L N N — — E E
3. Week	M T W Th F S Su	— — E E L L N	7. Week	M T W Th F S Su	L L N N — — E
4. Week	M T W Th F S Su	N — — E E L L	8. Week	M T W Th F S Su	E L L N N — —

Fig. 157 The 2-2-2 shift system ("metropolitan rota"). E = early shift; L = late day shift; N = night shift.

(b) there should be as much time as possible for family life and other social contacts.

The best shift plans to meet these requirements are those with single, isolated night shifts, each followed immediately by a full 24-hours rest.

Figure 156 shows a shift plan devised by *Graf* [95] which meets most of the requirements.

From this it can be seen that over a period of four weeks there is only one set of three consecutive night shifts. All the other night shifts are scattered singly, and each is followed immediately by a rest day. A very good feature of this plan is the distribution of free shifts, which, throughout the year, include 13 complete weekends, Saturday to Monday inclusive.

Two plans widely used in England are the 2–2–2 system (the so-called "metropolitan rota") and a 2–2–3 system that is called the "continental rota". Both of them are short rotations, which comply with the current ergonomic recommendations. The two are shown in *Figures 157* and *158*.

It will be seen that on one system the free days follow two nights' work, and on the other three nights. The 2–2–2 system is slightly the less favourable, because a free weekend (Saturday/Sunday) comes only once in eight weeks. The 2–2–3 system is more advantageous in this respect because a free weekend occurs every four weeks.

1. Week	M T W Th F S Su	E E L L N N N	3. Week	M T W Th F S Su	N N – – E E E
2. Week	M T W Th F S Su	– – E E L L L	4. Week	M T W Th F S Su	L L N N – – –

Fig. 158 The 2-2-3 shift system ("continental rota"). E = early shift; L = late day shift; N = night shift.

Short-term rotations are made more difficult because they sometimes bring production to a halt at weekends, but it should be possible to reach a compromise over this, as suggested by *Rutenfranz* and *Knauth* [274].

Recommendations Shift work that includes night shifts is burdensome, and often leads to ill-health which can rightly be classified as occupational. *Night work is therefore a danger to health.*

Since there is no way of planning shift work that significantly reduces this occupational risk, it should be introduced only with the greatest hesitation.

Rutenfranz [274] has written in this context: "In my opinion, and from the standpoint of medical safeguards in industry, continuous production is permissible only where it is unquestionably essential to the manufacturing process. Its introduction simply to increase profits is to be deplored." *If a night shift is unavoidable, then the following nine recommendations should be considered:*

(a) *night shift workers should not be engaged when they are below 25 years old, or over 50;*

(b) *workers with a tendency to ailments of the stomach and intestine, who are emotionally unstable, prone to psychosomatic symptoms and to sleeplessness, should not be employed on night work;*

(c) *workers who look after themselves, who live far away from their work or in noisy neighbourhoods, are unsuitable for night work;*

(d) *the usual 3-shift system, changing over at 6–14–22 hours would be better altered to 7–15–23 or 8–16–24 hours;*

(e) *short-term rotations are better than long-term ones, and continuous night work without change should be avoided;*

(f) *a good shift rotation either calls for scattered single nights at work, or else the 2–2–2 or 2–2–3 rotation (Figures 157, 158);*

(g) *whether 1, 2, or 3 nights are worked in a row, they should be followed immediately by at least 24 hours' rest;*

(h) *any shift plan should include some weekends with at least two consecutive rest days;*

(i) *every shift should include one break for a hot meal, to ensure adequate nourishment.*

15
Light and colour in surroundings

Human visual apparatus, accommodation, adaptation, glare and the faculty of sight have all been dealt with under *"Visual perception"* in *Chapter Nine*. The present chapter will be concerned primarily with how to make the light and colour of one's surroundings meet the physiological and psychological requirements of visual perception. First, however, a few facts about optical technology.

Light measurement and light sources

It is necessary to define two of the many units employed in the study of illumination, in order to understand what follows after.

Intensity of
illumination

The *intensity of illumination* is a measure of the stream of light falling on a surface. The unit of measurement is the lux, defined as follows:
1 lux (lx) = 1 lumen (lm) per square metre.
(An obsolete unit formerly used throughout the English-speaking world was the "footcandle", which is self-explanatory).
The human eye is sensitive to a very wide range of light intensities, from a few lx in a darkened room to 100 000 lx in the open air under the midday sun. Light intensities in the open vary throughout the day, between 2000 and 100 000 lx, whereas at night artificial light of 50–500 lx is customary.

Brightness
(luminance)

Luminance is a measure of the brightness of a surface, and the impression of brightness of a surface is proportional to its luminance. Since luminance is a function of the light that is emitted or reflected from a surface, that of walls, furniture and other objects is greatly affected by the reflective power of their surface. The luminance of lamps is a direct measure of the light they emit. The units of luminance are the *apostilb* (asb) and the *stilb* (sb).

1 asb = 0.32 candela (cd) per square metre;

1 sb[1] = 10 000 cd per square metre = 31 416 asb.

The apostilb is usually used as a unit of measurement for the light reflected from walls, .furniture and other objects that do not themselves emit light, while the stilb is the unit for light sources. Some examples of the approximate luminance of some common source of light, in stilbs:

Moon:	0.25
Clear sky:	0.40
Lighted candle:	0.7–0.8
Oil lamp:	0.6—1.5
Electric filament lamp:	70–1000
Fluorescent tubes:	0.45–0.65.

The luminance in asb and the unit of light intensity (lx) are related as follows:

Luminance (asb) = reflectivity (%) × light intensity (lx). A simple example is this: if a white wall has a reflectivity of 80 %, and the light intensity is 100 lx, the luminance of the wall will be 80 asb. If, however, the reflectivity was 100 %, then the luminance would become 100 asb. Since the visual impression of brightness depends mainly on the luminance within the visual field, the reflectivity of surfaces within a room is very important; indeed physiologically it is as important as the lights themselves.

Light sources

There are four kinds of light source that are specially distinctive.

1. *Direct radiants*. These direct 90 % or more of their light towards the object in the form of a cone of light. Light sources of this kind throw hard shadows, with a contrast between light and shadow of 1:10 or much more. They are useful in offices, shop windows and showcases, and switchrooms, but the excessive contrast tends to produce relative glare. They can be recommended as working lights only where the general illumination is high enough to modify this contrast.

2. *Mixed direct and indirect lighting*. A translucent shade allows a substantial part (about 40 %) of the light to radiate in all directions, while the rest is thrown directly or indirectly on to the ceiling and walls. This type of lighting throws only moderate shadows, with soft edges, and is useful as general

---

[1] Today another unit is often used: The *nit* (1 nit = 1 candela/m²). In the U.S.A. luminance is expressed in Foot Lamberts (fL); the relation to the nit is: 1 nit = 0.292 fL or 1 fL = 3.43 nits.

illumination in the home, in shops, in offices, etc. It is suitable for moderately close work, but not for work requiring great precision. Its best application is for even lighting over the whole room, including objects on the walls.

3. *Opalescent globes*. These, and similar free radiants, give out light equally in all directions, and throw slight to moderate shadows. Because they are bright, they often cause glare, and so they should not be used in living rooms or work rooms. They are suitable for store-rooms, corridors, entrance halls, vestibules, lavatories, etc.

4. *Indirect lighting*. These throw 90 % or more of their light on to the ceiling and walls, which reflect it back into the room. This system requires the ceiling and walls to be light-coloured. The light is diffused and practically without shadows. Architects sometimes prefer this type of lighting because it can be used to emphasise architectural features, as well as produce a pleasant aesthetic effect. It is not suitable for workrooms unless it is supplemented by extra lamps over the work places; in that case the indirect lighting has the advantage that it does not cause glare. Indirect lighting is particularly suitable for displays, salerooms, anywhere where the eye is drawn to the walls.

Light sources

Modern light sources are broadly of two kinds: electric filament lamps and fluorescent tubes. The following are important physiological points.

Filament lamps

The light from filament lamps is relatively rich in red and yellow rays. It changes the apparent colours of things, and so is unsuitable when correct assessment of colour is important. When used over a work place, they have the further drawback that they give off heat. Lampshades can reach temperatures of 60°C and more, and if they are close to the head, can cause discomfort and headaches. On the other hand their warm glow is welcome because of its association with evening light and a cosy atmosphere.

Fluorescent tubes

Fluorescent lighting is produced by passing electricity through a gas (usually argon), or through mercury vapour. This method converts electricity into light much more efficiently than a heated filament, so that fluorescent tubes give off three or four times as much light as filament lamps. The inside of the tube is covered with a fluorescent substance, which converts the ultraviolet rays of the discharge into visible light, the colour of which can be controlled by modifying the chemical composition of the lining. Hence fluorescent tubes can be manufactured to give the warm light of a filament lamp, the white light of a cloudy sky, or the bluish light of bright daylight. Since the use of fluorescent tubes is so widespread, their advantages and drawbacks should be clearly understood.

Advantages

(a) *High output of light, and long life* – but not if they are frequently switched on and off, when their life is reduced to little more than that of a filament lamp.

(b) *Low luminance, so that there is little glare.* The luminance of fluorescent tubes is 0.45–0.65 sb, compared with the 70–1000 sb of a filament lamp.

(c) *Ability to match the light to daylight*, thus avoiding the tiresome colour difference between the light from the lamp and daylight in the room, as well as making colour-matching at work more accurate.

Drawbacks

(a) *Visible and invisible flicker.* Because they operate from alternating current, fluorescent tubes produce a flicker at the main frequency: 50 Hz in England, 100 Hz in some other countries. This is above the normal flicker-fusion frequency of the human eye, and so is not usually visible, but it can become noticeable as a stroboscopic effect on moving, reflective parts of a machine. It is greater with daylight tubes than with white or warm tones. When tubes become old or defective they develop a slow perceptible flicker, especially at the ends.

Both visible and invisible flicker is bad for the eyes. After long exposure, people often complain of headache and painful irritation of the eyes, accompanied by reddening and a flow of tears. Our own studies [109] have shown that invisible flicker from a fluorescent tube can increase the physiological signs of fatigue, and measurably reduce performance.

(b) Invisible flicker, and the stroboscopic effect can be largely avoided by using two or more fluorescent tubes out of phase with each other, using suitable equipment (duo apparatus, 120° switching, or three-phase switching). In this way the fluctuations in light intensity can be substantially reduced; with three-phase switching the flicker becomes little more than that of a filament lamp. *Rooms in which people normally work for more than a few minutes at a time should never be lighted by a single fluorescent tube, but always with two or more tubes out of phase.*

The invisible flicker of light sources is designated by the so-called uniformity figure, which expresses the range between the brightest and dimmest parts of the cycle.

For fluorescent tubes this number ranges from 0.2 to 0.6 (according to type), whereas filament lamps often have values in excess of 0.9. A flickerless light has a uniformity figure of 1. *Figure 159* shows uniformity figures for various groups of fluorescent tubes.

3 fluorescent tubes:

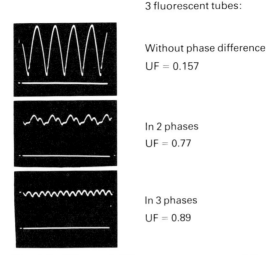

Without phase difference
UF = 0.157

In 2 phases
UF = 0.77

In 3 phases
UF = 0.89

Fig. 159 Effects of different arrangements of fluorescent tubes on the uniformity of the light. The curves express the invisible flicker, as registered by a photoelectric cell, with the horizontal lines marking the zero level of light intensity. UF (Uniformity Figure) = min/max light intensity.

Fluorescent tubes that flicker visibly should be replaced promptly. New tubes should be carefully checked, and rejected if necessary. It is desirable to cover the two ends of the tube, so as to mask the end-flicker, and the tube should be rejected as soon as flicker becomes visible.

(c) *"Cold, pale light".* Fluorescent tubes are often said to create a cold, unfriendly atmosphere. This is true of white, or daylight tubes, and the lower the general level of illumination, the paler and colder the effect: if they are brighter than 1000 lx the "atmosphere" becomes more like true daylight, and the chilly effect disappears. So daylight tubes should be more powerful (over 800 lx for shops and salerooms, 500 lx or more for workrooms). When there is no need to match the illumination to daylight (e.g. in livingrooms, restaurants, etc.), warm tone fluorescent tubes may be used, to avoid creating this "chilly atmosphere".

It is clear that appropriate installation will avoid most of the drawbacks of fluorescent tubes, so that their undoubted advantages can be utilised.

Physiological requirements of artificial lighting

For visual comfort and good optical performance, the following conditions should be met:
(a) suitable level of illumination;

(b) balanced arrangement of the lights;
(c) matched phasing of lights;
(d) avoidance of glare.

The physiological requirements under these four headings are just as valid for artificial light as for natural daylight, but since the practical problems are somewhat different, the requirements for artificial lights will be considered first.

The following publications may be recommended for several aspects of the subject:

Jacob and *Scholz* [158], *Duboic-Poulsen* [71], *IES Lighting Handbook* [154], *Hopkinson* and *Collins* [147], *Handbuch für Beleuchtung* [131] and the *DIN Norm 5035 of 1972* [69].

Intensity of lighting

Until 40 years ago recommended light intensities were between 10 and 50 lx, but since then the figure has been steadily increased, as a result of the greater efficiency of fluorescent tubes, and better performance of work under brighter lighting. Since 1960 *Blackwell* [26] [154] has used an optical test which takes into account the ambient lighting, the size of the object, the contrast, and the speed of perception. The results of these studies are embodied in the U.S. Norms of the *IES* [154].

Table 38, after *Weston* [328], gives guidelines based on contrast between the object of the visual task and the ambient lighting.

Table 38 Guidelines for intensity of lighting.

Visual tasks	Light intensities in Lux		
Contrast between the object of the visual task and its immediate surroundings	*high*	*medium*	*low*
Extremely precise	1000	3000	10 000
	700	2000	7000
	500	1500	5000
		1000	
Very precise	300		3000
		700	
	200		2000
		500	
Precise	150		1500
Moderate precision	100	300	1000
	70	200	700
	50	150	500
		100	
Less precise	30		300
		70	
	20		200
		50	
Not precise	15		150
	10	30	100

These recommendations should always be followed, on ergonomic grounds. If the current recommendations are compared, it will be seen that the norms of the American "Illuminating Engineering Society" (IES) prescribe significantly higher levels of lighting than are embodied in the European guidelines.

A few of these proposals are quoted in *Table 39*.

Table 39 Comparison between German (DIN) and U.S. (IES) Norms for intensity of lighting, each in lx.

	DIN [69]	IES [154]
Rough assembly work	250	320
Precise assembly work	1000	5400
Most delicate assembly work	1500	10800
Rough work on toolmaking machine	250	540
Fine work on toolmaking machine	500	5400
Most precise work on toolmaking machine	1000	10800
Technical drawing	1000	2200
Book-keeping; office work	500	1600

Excessive brightness is bad

A very high level of illumination is often undesirable in practice. Levels above 1000 lx increase the risk of troublesome reflections, of too heavy shadows, and excessive contrast. In one of our own studies in 15 open-plan offices [243] we measured the intensity of illumination, among other things as well as using a questionnaire to record the impressions and general complaints of the office workers. 23% of the 519 employees questioned stated that they were disturbed by either reflections or glare. The sources of glare were distributed as follows, expressed as percentages of the number of people affected.

(a) window: 36%
(b) lamps: 25%
(c) polished table tops: 18%
(d) at dusk: 17%
(e) other causes: 4%

The question relating to eye troubles is comparatively often answered in the affirmative. *Figure 160* sets off the frequency of complaints about eye troubles against the average lighting level in the office.

It is clear from this figure that complaints about eye troubles were more frequent in open-plan offices with lighting levels of 1000 lx or over than in other offices with 200–800 lx, and the difference was statistically highly significant ($p < 0.001$). The direct question of what was the best lighting for these offices gave the answer that on the whole it lay between 400 and 850 lx. Obviously we cannot interpret these results as meaning that there is a direct causal relationship between level of illumination and eye troubles, but we

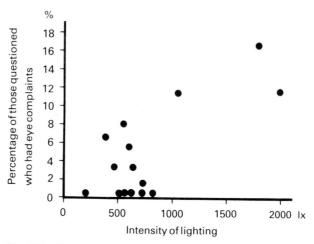

Fig. 160 Frequency of eye complaints (open question) in relation to the intensity of lighting in each of 15 open-plan offices. The number of people questioned in each office = 100%. After *Nemecek* and *Grandjean* [243].

have grounds for thinking that brightly lighted open plan offices more often create reflections, deep shadows and relative glare, and that these possibly contribute to the eye troubles recorded.

Our observations conflict with several studies carried out in test rooms, where a level of 1000–4000 lx was often rated as best by the test subjects. Presumably the test rooms had been carefully designed from an optical point of view, so as to avoid contrast and glare. This may account for the contradictory results.

As far as current results go, the values given in *Table 40* may be recommended as a basis for comparison for work rooms for different purposes.

Table 40 Examples of suitable lighting levels in work rooms.

Kind of work	Examples	Recommended lighting (lx)
Rough	Storeroom	80–170
Moderately precise	Packing; despatch	200–250
	Works laboratory; simple assembly; winding thick wire on to spools; work on carpenter's bench; turning; boring; milling; locksmith's work	250–300
Fine work	Reading; writing; book-keeping; laboratory technician; assembly of fine equipment; winding fine wire; woodworking by machine; fine work on toolmaking gig	500–700

Very fine to extreme work	Technical drawing; colour proofing; adjusting and testing electrical equipment; assembling delicate electronics; watchmaking; invisible mending	1000–2000

If a strong light is necessary this is best achieved by the use of working lights, but these should always be used in conjunction with a good general illumination, to avoid creating too much contrast. Guidelines might be as follows:

Working light(s)	General illumination
500 lx	150 lx
1000 lx	300 lx

Specifications for lighting levels can be no more than general guidelines, and other circumstances must be taken into account in any particular situation. For example:
(a) the reflectivity (colour and material) of the working materials and of the surroundings;
(b) the extent of difference from natural lighting;
(c) whether it is necessary to use artificial lighting during the daytime;
(d) the age of the people concerned.

Age

Blackwell [26] studied the effect of age with his test procedure. If the degree of contrast necessary to satisfy people in the age-group 20–25 was taken as unity (1), then for older people this must be multiplied by the following factors:

40 year olds:	1.17
50 year olds:	1.58
65 year olds:	2.66

According to *Fortuin* [83], the lighting level required for reading a well-printed book at various ages followed the same factors as those recorded by *Blackwell.*

Spatial uniformity of illumination

Physiological studies have shown that the optimum conditions for seeing things comfortably are decisively dependent on the distribution and relative contrast of the biggest surfaces in the visual field. *Figure 161* shows the results of an investigation by *Guth* [126].

According to *Guth*, relative contrast levels of 1:5 in the middle of the visual field significantly impair the efficiency of the eye (assessed by measuring perception of the smallest bright spots), as well as visual comfort (assessed by measuring the frequency of blinking the eyelids). Surfaces that are brighter than the object being looked at are more disturbing than darker surfaces. The

272

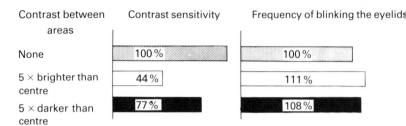

Contrast between areas	Contrast sensitivity	Frequency of blinking the eyelids
None	100%	100%
5 × brighter than centre	44%	111%
5 × darker than centre	77%	108%

Fig. 161 Physiological effects of contrasting areas in the middle of the visual field. The contrast is measured between the 15% of the visual field in the centre and the area immediately adjacent to this. After *Guth* [126].

explanation lies in the uneven illumination of the retina, and its effect on adaptation; the strong contrasts produce relative glare.

Modern knowledge and experience suggest the following seven rules.

1. All the objects and major surfaces in the visual field should as far as possible be equally bright.
2. Surfaces in the middle of the visual field should not have a brightness-contrast of more than 3:1 (see *Figure 162*).
3. Contrast between the middle field and the edge of the visual field, or its vicinity, should not exceed 10:1 (see *Figure 162*).
4. The working field should be brightest in the middle, and darker towards the edges.
5. Excessive contrast is more troublesome if it occurs at the sides of the visual field, and below, than at the top of the field.
6. Light sources should not contrast with their background by more than 20:1.
7. Maximum permissible range of brightness within the entire room is 40:1.

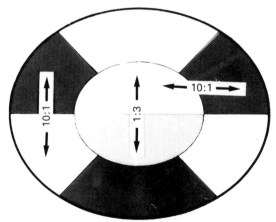

Fig. 162 Acceptable contrasts between the brightness of various parts of the visual field. Within the middle field, 3:1; within the outer field 10:1; between middle and outer field 10:1.

These rules are often not carried out. In everyday practice the following should be avoided:

(a) bright windows;
(b) dazzling white walls and dark floorings;
(c) blackboards on a white wall;
(d) reflective tabletops;
(e) black typewriters on bright underlays;
(f) polished machine parts.

The choice of colour and material is of great importance in the design of walls, furniture and larger objects in a room, because of their varying reflectivity.

The following reflectivities are recommended:

Ceiling:	80–90 %
Walls:	40–60 %
Furniture:	25–45 %
Machines and equipment:	30–50 %
Flooring:	20–40 %.

Windows must always be equipped either with adjustable venetian blinds or with translucent curtains, so that excessive contrast can be avoided on sunny days. Work places should be at right-angles to the window, as shown in *Figure 163*. This applies equally to schoolrooms, meeting rooms, conference halls, libraries, etc.

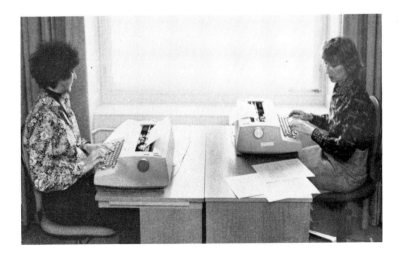

Fig. 163 Correct and incorrect arrangement of the work place. Left: the secretary, when reading, has a bright window surface in her visual field; the contrast between the window and other surfaces is more than the ratio 10:1. Right: the bright window is not in the visual field of the secretary; the luminance contrasts have been decreased and the ratio corresponds to the recommendations of *Fig. 162*.

Work places where most delicate visual work is being done are exceptions to this rule. Here the light must come from in front, so the bench is often placed across the window, and then the operator, to avoid glare, must bend his head so far forward that he is looking almost vertically down on his work. Hence this frontal lighting is often a cause of bad posture of the neck and body.

Steady lighting

Even more disturbing than static contrast are rhythmically fluctuating bright areas in the visual field. These occur if the work requires the operator to glance alternately at a bright and a darker object; if bright and dark objects pass by on a conveyor belt; if moving parts of a machine are bright and reflective; or if a lamp itself flickers.

As we have already seen, the pupil and the retina of the eye can cope with changes in brightness only after a certain delay, so that fluctuating brightness leaves the eyes either under- or over-exposed for much of the time. Hence such lighting conditions are particularly disturbing. Physiological research has shown that if two brightness levels in the ratio 1:5 fluctuate rhythmically, visual performance is reduced as much as if the level of illumination had been lowered from 1000 lx to 30 lx.

To avoid fluctuating levels of brightness as far as possible:

(a) *cover moving machinery with an appropriate housing;*
(b) *equalise brightness and colour along the main axes of sight;*
(c) *take the precautions mentioned earlier to avoid flickering light sources.*

Light as a source of glare

The special importance of glare has already been discussed in *Chapter 9* "Visual perception". Glare, or "dazzle", is a gross disturbance of the state of adaptation of the retina, comparable to the over-exposure of a photographic film. All forms of glare – even the comparatively weak relative glare – make seeing difficult and uncomfortable. *Avoiding glare inside a room is one of the most important ergonomic considerations when designing work places. Figure 164* sets out the results of a classic piece of research by *Luckiesh* and *Moss* [203]. Their test subjects carried out a visual test, in which a light source of 100 watts was moved by steps closer and closer to the optical axis. Visual performance was gradually impaired.

As the experiment shows, the lighting in a workroom must be arranged carefully. The artificial lights in workroom and offices often create both absolute and relative glare, while reflections from polished surfaces may also dazzle.

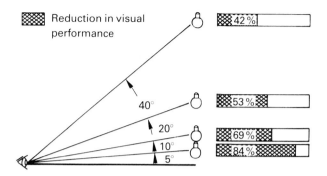

Reduction in visual performance

40°

20°

10°

5°

42%

53%

69%

84%

Fig. 164 Effect of source of glare (dazzle) on visual performance. The black blocks indicate the reduction in visual performance as a percentage of the normal performance when there is no glare present. Visual performance becomes worse, the closer the light source is to the optical axis. After *Luckiesh* and *Moss* [203].

A bad example

Figure 165 shows a very unsatisfactory arrangement of lights. Opalescent globes are being used in a drawing office, where they will often come into the visual fields of the draughtsmen, as well as being reflected back from the polished floor covering. The result is very strong contrasts, far exceeding the recommended maximum of 10:1.

The following recommendations should be considered, in order to arrive at a good arrangement of lights, and appropriate overall distribution of light:

(a) *no source of light should appear in the visual field of any worker* during working operations;

Fig. 165 Unsuitable lighting in a drawing office. The free radiants are sources of much glare. The dark floor contrasts too strongly with the white working surfaces (relative glare), as well as throwing back strong reflections of the lamps.

(b) *all lights should be provided with shades*, of such a kind that the average luminosity does not exceed 0.3 sb overall, and 0.2 sb at the work place;

(c) *the line from eye to light source must make an angle of more than 30° with the horizontal (see Figure 166)*. If a smaller angle cannot be avoided, e.g. in large workrooms, then the lamp must be effectively shaded;

(d) *fluorescent tubes should be aligned at right angles to the line of sight;*

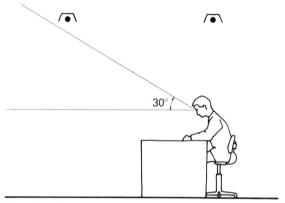

30°

Fig. 166   The angle between the horizontal and the direction from eye to overhead lamp should be more than 30°.

(e) *it is better to use more lamps, each of lower power, than a few high-powered lamps;*

(f) *to avoid glare from reflection the lines from work place to lamps should be such that they do not coincide with any of the directions in which the operator normally needs to look.* No reflection giving a contrast greater than 10:1 should come within the visual field;

(g) *the use of reflective colours and materials on machines, apparatus, table tops, switch panels, etc. should be avoided.*

A good and a bad arrangement are shown in *Figure 167.*

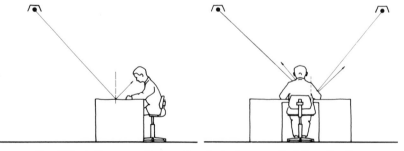

Fig. 167   Left: poor placing of a single lamp so that its reflection comes into the line of sight of the operator, with the risk of direct glare.
Right: the reflection from two lamps placed to the sides are not in the line of sight, so reflected glare is avoided.

Lighting for the work place

Fine and delicate work	Very precise work, say from the size of normal print down to fractions of a millimetre, needs special lighting to supplement the general illumination. Examples of such work include:

(a) colour testing in chemical works, paper factories and the textile industry;
(b) delicate assembly work; adjusting and testing electronic equipment; making watches and clocks; precision mechanics;
(c) grinding, etching, polishing and engraving glass;
(d) weaving, sewing, knitting, colour printing, invisible mending and quality testing in the textile industry.

Very small objects may need to be magnified, and lenses, magnifying glasses and other optical aids should be provided.

Requirements for good vision	The following considerations are important in this context:

(a) level of illumination at the work place;
(b) distribution of bright surfaces with the visual field;
(c) size of the objects to be handled;
(d) how much light is reflected from these objects;
(e) contrast between objects and surroundings; cast shadows;
(f) how much time is available for seeing whatever is necessary;
(g) the age of the persons concerned.

As already mentioned the best level of illumination for visual activity is 5000 asb, and that for contrast sensitivity should lie between 200 and 10 000 asb. The maximum contrast sensitivity is obtained with 1200–1500 in the middle of the test field, and 200–300 asb peripherally. High levels of illumination are usually demanded by operators who need to concentrate them on very tiny objects, or wish to throw these objects into relief by creating strong contrast.

So we can deduce the general principle that work demanding high visual acuity (recognition of the tiniest shapes or objects) and contrast sensitivity (control of colour or pattern in textile, checking X-ray pictures, etc.) calls for a high level of illumination. The values given in *Tables 38 and 40* (pp. 268, 270) can be recommended as guidelines.

Lights that are too bright can be damaging	Occasionally, however, the lighting can be too bright and harmful. Reflections from metallic surfaces may impair vision. Secondly, fine structures in materials, or surface irregularities in metal, may actually be seen more easily with moderate lighting than if over illuminated.

Lighting systems	In one of our own investigations [109], we studied the effects of different lighting systems on the performances and subjective impressions of four test subjects engaged in a precision task,

characteristic of the watch-making industry: they were placing tiny cog-wheels into holes in shiny metal casings. The most important results are shown graphically in *Figure 168*.

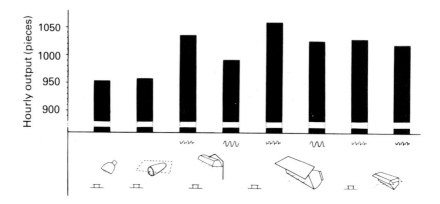

Fig. 168 Effects of different lamps and shades on precision work. The columns indicate the average hourly output in each of 15 experiments. The light sources tested, reading from left to right, were: filament lamp without reflector; filament lamp with reflector; fluorescent tubes with and without phase shift; broad and deep lighting over the work place from fluorescent tubes; fluorescent tubes with diffusing screens in front of them.

The fluorescent tubes were rated better than the filament lamps (left of picture) for the same output of light, and these gave better performances with a phase shift than without. *The best results, both for performance and for subjective impression, were obtained from fluorescent tubes with deep diffusing screens, big radiating surfaces, and reflectors.*
Measurement of light intensities at the working place showed that filament lamps caused significantly more surface contrast than the other lamps.
For example, *Table 41* shows the light intensities we measured at one watchmaking factory.

Table 41 Light intensities (in sb) reflected from various parts of a watch, when the general lighting was 1000 lx.

Part of watch	Filament lamps sb	Fluorescent tubes with diffusers sb
Bridge of gear-train	0.12	0.05
Shining ratchet wheel	0.27	0.04
Ruby bearing	0.02	0.006
In shadow	0.015	0.005
Maximum value	0.27	0.07
Minimum value	0.014	0.005

Hence the ratio of darkest/lightest areas was 1:20 with the filament lamps and 1:14 with the diffused fluorescent lighting.

Figure 169 shows the two forms of illumination, and their different effects on the visibility of the individual parts of the watch can be clearly seen.

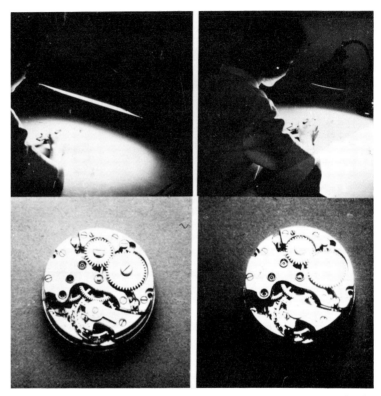

Fig. 169 The effects of two different types of lighting on illumination in the working field (watch repairing).
Left: two fluorescent tubes, with 2-phase shift give a large evenly illuminated area, in which details of the watch show clearly.
Right: a simple filament lamp gives strong contrasts, and dazzling reflections from the watch casing.

Contrasts in light intensity

In contrast to large areas, very small objects are better for being lighted with strong contrasts; thus dark markings or objects on a light background are easier to see than bright objects on a dark background. For this reason, when working with very small objects, it is better to have the work lighted from in front rather than from the side, since then the back of the object is in deep shadow, and the object stands out against the bright, reflective working surface. Our own research, mentioned above, confirmed the correctness of this reasoning, performance as well as subjective assessment showed that frontal lighting was better than any other arrangement.

Thus the incidence of the light and the casting of shadows can make a considerable difference to the recognition of objects and to the interpretation of their surface structure.

Very diffused light, without shadows, makes everything look flat and featureless, whereas lighting that casts shadows makes things more solid.

Current knowledge leads to the conclusion that for very fine work in industry, neither a completely diffused light, nor one full of deep shadows, is entirely suitable. As our example shows, a strong beam of light from a lamp with reflector can create deep contrasts which make visual work difficult. We had the same experience in relation to the checking of metal parts for uneven areas and spots of rust. This visual task was easier with a lamp that was half diffused than with one that gave full, direct light. *Figure 170* shows this work place diagrammatically.

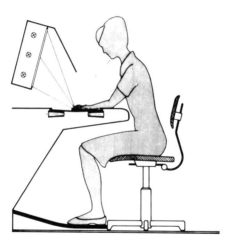

Fig. 170   Diagram of lighting at a work place where metal parts passing on a conveyor belt are inspected, and defective parts discarded. The light comes from three fluorescent tubes, out of phase, giving a large light source, with a diffuser in front of it, and a shade to protect the eyes from direct light. The frontal lighting enables defective parts to be quickly recognised.

Different kinds and arrangements of light for very fine work should always be tested with experienced workers. Many kinds of precise work pose special visibility problems, which cannot be solved by stereotyped methods.

General recommendations

In spite of this reservation we can formulate a general rule that is specially valid for precise assembly work, or delicate mechanical tasks:

(a)   frontal lighting;

(b)   screening of the lamps from being directly visible;

(c) lamps should have diffusive screens of ribbed or frosted glass. This half diffused light is reflected less and creates less contrast;

(d) the light source should emit from a large area;

(e) the diffusing screens should be broad and deep so as to make the illumination of the work bench as uniform as possible;

(f) fluorescent tubes with phase shift are preferable to filament lamps, since the latter give off more heat.

Daylight

The physiological demands of daylight are essentially similar to those of artificial light, but there are some special problems, which will now be briefly discussed.

Besides providing illumination, natural daylight penetrating into a room establishes contact with the world outside, giving a view of the surroundings and indicating the time of day and the state of the weather.

Daylight quotient The starting point for measuring daylight is light from an overcast sky, or in a shaded area (diffused daylight).

An important unit of measurement for daylight inside a room is *the daylight quotient, the ratio between light intensity inside and that outside;*

expressed as a percentage: $DQ = \dfrac{Ep}{Ea} \times 100$

where

DQ = daylight quotient;

Ep = light intensity at a given measuring point inside the room;

Ea = light intensity falling on a horizontal surface when the sky is uniformly overcast.

The desirable level of daylighting within a room is often expressed as the required daylight quotient. Assuming a minimal external light intensity (Ea) of 5000 lx, various daylight quotients at the work place would provide the following illumination in lux.

Dq at work place (%)	lx at work place	Level of illumination
3	150	Low
6	300	Moderate
10	500	Average
20	1000	High

It therefore seems appropriate to assess the light level at a work place, not by the daylight quotient, but by the values given in *Tables 38 and 40*.

On the other hand, daylight quotients can be used to determine the distribution of light within a room, as well as the dimensions of windows, sunblinds, and other features which affect daylight indoors.

It is physiologically desirable to have as much light, as evenly distributed, as possible. The higher the daylight quotient, the less need there is for artificial lighting, especially in winter. On a cloudy day in December, the daylight intensity indoors, with a DQ of 10%, will reach 500 lx only for the four hours from 10.00–14.00 hours.

Distribution of light indoors

The distribution of DQ values within a room depends mainly on the position and type of the windows, as shown in Table 42.

Table 42   Comparison of the effects of windows of two different heights on the distribution of light inside a model room.

Distance from window to measuring point in metres	DQ values in % Window height 1 m	Window height 2.7 m
0	25	25
1	12	20
2	4.5	16
3	2	11
5	1	5.5
8	0.5	2.5

One reservation must be made. Big, high windows certainly help to distribute daylight within a room, but they have the drawback that they admit a great deal of summer heat, especially if they face south or south-west. Furthermore, in winter they act as cold surfaces, with an adverse effect on the temperature of the room. The sizes of windows must not be decided solely in relation to daylight; there must be a "balance sheet" of all the pros and cons.

Recommendations

The following are nine rules of thumb relating to daylight indoors.
1. *High windows are more effective than broad ones,* since the light penetrates further into the room. The lintel should not be deeper than 30 cm.
2. *Window sills should be at table-height.* If the window extends below the tabletop it is cold in winter and may cause glare.
3. *The distance from window to work place should not be more than twice the height of the window.*
4. *For workrooms the window area should be about one fifth of the floor area.* This is only a general rule which is very flexible according to circumstances.
5. *It is important that the glass should transmit plenty of light.* Clear glass has a transparency of more than 90%, whereas

frosted glass, glass bricks, or special heat-insulating glass may have transparencies from 70 % down to only 30 %.

6. *Effective protections against the glare of direct sunlight, and against radiant heat*, are important in securing good visibility and comfort indoors. *The most efficient method is an adjustable external sunshade*, either venetian blinds or curtains. Venetian blinds inside the window, or between the panes of double-glazing, are a mistake, because they afford no protection against radiant heat.

Insulating window panes are not adjustable; they reduce the amount of light transmitted, which is undesirable in winter, while in summer they trap heat inside the room (greenhouse effect). Balconies, overhanging eaves and other projections also create problems, by cutting off some of the light in dull weather. In our latitudes they should be restricted to south-facing walls.

7. *Each window should receive direct light from the sky, and it is desirable that a portion of sky should be visible from every work place.*

8. The nearest building should be at least twice as far away as its own height.

9. Pale colours should be used, both in the room itself, and in any courtyard outside, so as to reflect as much of the incident daylight as possible.

Glass houses and windowless factories

The trend in modern architecture has been to increase the window area of new buildings, sometimes ending up with a glass house. As we have already said, glass walls, or very large window areas, *radiate heat away in winter and admit excessive heat in summer*. On the other hand they admit more daylight and allow the occupants to see more of the outside world. The problems of regulating the room climate are both difficult and expensive to solve.

Equally problematical are factory buildings that are not provided with windows at all, on grounds either of economy or increased production. The arguments in favour of such buildings are that they can be given a uniform internal climate (through air-conditioning), and that the internal lighting, being entirely artificial, can be controlled by modern techniques. The more complete insulation of the walls reduces the cost of air-conditioning. Against these methods it is argued that the absence of windows makes the inmates feel "imprisoned" and "cut off", and that people need some contact with the outside world when they are at work. Since the validity of these various arguments is not yet proved it would seem sensible to proceed slowly with building practices that are so unnatural, and so psychologically questionable. The requirement of some authorities that at least some windows should be provided seems a sensible interim provision.

Skylights in the roof and fanlights in the walls or door are often useful accessories in single storey buildings, in lofts and attics, and in any other rooms where insufficient daylight penetrates.

The following are the most important types, from the standpoint of industrial physiology:

(a) *pitched roofs*. If the axis of the skylight is no further away from the work place than the height up to the ridge, the lighting is bright and even. It is particularly good if the work places are so arranged that each lies between two skylights at equal angles, and so is lighted from both sides;

(b) *skylights in ridge of roof*. This gives glass above the longitudinal axis of the building. It is particularly suitable for high work shops, where the operatives in the middle need the best light;

(c) *mansard roofs*. Here the skylights are placed in the sloping part of the mansard roof, above the windows, giving a relatively high level of daylight all along the outer walls. However, they increase the risk of glare and do not give much extra light to the people in the middle of the room;

(d) *zig-zag shed roofs*. The whole working area has clear glass above it, in panels which usually face north and slope upwards at an angle of 60°. The reverse face of each zig-zag is opaque, faces south, and slopes at an angle of 30°. This arrangement gives the highest and most uniform values of daylight quotient, and is specially recommended for large workshops.

Colour in the workroom

The accepted colours of the spectrum cover the following bands of wavelength, in nanometres (nm = 10^{-9} m = one millionth of a mm).

Violet:	380–436
Blue:	436–495
Green:	495–566
Yellow:	566–589
Orange:	589–627
Red:	627–780

Electromagnetic waves longer than 780 nm belong to the infra-red (radiant heat); shorter than 380 nm, they are ultraviolet rays which are of critical importance for the synthesis of vitamin D from its precursor ergosterol and for normal organic growth.

The colours that we see arise because the molecular structure of the surface of objects reflects only part of the light that falls on it. This is what we perceive. For example a green-painted machine absorbs all the incident light except the green (495–566 nm). The colour receptors of the retina are the cones, which are capable of distinguishing more than 100 000 shades of colour.

"Colour mixtures"

The many different shades or "tones" of colour arise from the mixing of the primary colours, i.e. the combination of different wavelengths.
(a) red and green in different proportions give all the intermediate colours such as orange, yellow and yellow-green;
(b) red and violet give different shades of purple;
(c) green and violet give all the intermediate colours, mainly various shades of blue.

Any desired colour can be obtained by mixing two of the three primary colours, red, green and violet.
A mixture of all the wavelengths gives the impression of white light, as in sunlight, but "white" can also be obtained by mixing separate pigments. It is then called a complementary colour.

Reflected colour

When deciding upon the colours at and around a work place, it is necessary to consider reflectivity. *Table 43* gives a few examples.

Table 43 Reflectivity in percentage of the incident light.

Colour and materials	Reflectivity (%)
White	100
Aluminium; white paper	80–85
Ivory; deep lemon yellow	70–75
Deep yellow; light ochre; light green; pastel blue; pale pink; cream	60–65
Lime green; pale grey; pink; deep orange; bluegrey	50–55
Powdered chalk; pale wood; sky blue	40–45
Pale oakwood; dry cement	30–35
Deep red; grass green; wood; pale leaf green; olive green; brown	20–25
Dark blue; purple red; reddish brown; slate grey; dark brown	10–15
Black	0

Colours in and around work places have the following functions:
(a) orderliness, and as an aid to identification;
(b) to indicate safety devices;
(c) colour contrasts to make work easier;
(d) their psychological effect on the operator.

Orderliness

Certain rooms, or floors, or sections of the factory can be given a colour code, which helps to keep the whole works on an orderly plan. In very big, scattered industrial plants, or those which are under multiple direction, different colours can be used to introduce a certain degree of orderliness and to facilitate supply and servicing.

Safety colours

If the same colour is always used to indicate a particular danger, the correct reaction to it becomes automatic, so this practice is now followed in most countries. More details will be found in *DIN Norms 4844* and *5381* [68], or in *Hettinger, Kaminsky* and *Schmale* [143].

These are some of the common colour codes:
(a) *Red* is the *"danger colour"*: Halt, Stop, Prohibited. Red is also the warning colour for FIRE; on extinguishers and other equipment;
(b) *Yellow*, usually in contrast with Black, means *"danger of collision"*, *"look out"*, *"risk of tripping"*. Yellow and Black are much used as warning colours in transport;
(c) *Green* means *rescue services, safety exit*, etc. It is used to indicate all forms of rescue equipment and first aid;
(d) *Blue* is not actually a safety colour, but is used for *giving directions*, advice, signs, etc.

Colour contrasts of large areas

When deciding upon colour contrasts large areas such as walls and furniture must be considered separately from small areas such as splashes of colour intended to attract attention on knobs, handles, levers, etc.
The colours of large areas should be chosen so that they have similar reflectivities (see *Table 43*), so as to have colour contrast without differences in brightness. The avoidance of large areas of contrasting brightness, close together, is an important factor in ensuring good visual acuity. Large areas and big objects should never be painted in pure colours, nor with fluorescent paint, since these cause local overloading of the retina, and lead to the production of after-images. So walls, partitions, table tops, etc. should be painted with unsaturated colours, in a matt finish.
It is easier to lay out working materials and to select the one required, if they are coloured differently from their immediate surroundings, and this matter should be borne in mind when the work place is being designed. At the same time contrasts in brightness and intensity of colour should be avoided. For example, if the working materials are made of leather, wood, or similar materials of an ochre yellow or brown colour, a suitable colour for the background would be matt green, nile green, or matt bluish. Greyish-blue materials such as steel and other metals show up well against a background of dark ivory or light beige. The area surrounding the machine, or the work table, might be painted in cool, neutral colours, from yellow green to pastel blue.

Eye-catching
colours

Eye-catching colours are the little spots of strongly contrasting colours that are used to attract attention, to "catch the eye". Colour is used in this way in nature: a red strawberry among green foliage; brilliant flowers which attract insects and other creatures by their colour contrast. This effect is as important to the animals as it is to the plants. On the other hand nature also uses colours for concealment. Defenceless wild animals are often neutral in colour, and merge into their background, so that they are almost invisible. Similarly, it is a good idea to provide a few eye-catchers at a work place, marking such things as the more important handles, levers, control-wheels, knobs, and so on. If the eye-catchers are small, not more than a few square centimetres in area, they should contrast strongly, not only in colour, but also in brightness. Eye-catchers make the controls easier to find, reducing the time taken up in searching for them, and hence the diversion of attention from the work itself.

The human eye sees the greatest contrast between yellow and black, because the brain adds together the effect of colour and intensity. So yellow–black contrast is recommended for indicator instruments as well as for switch panels.

The greatest danger in colour planning, and especially in the planning of eye-catchers, is excess. If there are too many eye-catchers, in too many different colours, then the whole work place becomes restless and distracting. Colour does not mean bunting! *The most important physiological requirement in the use of colour is restraint, with three, or at most five eye-catchers to each work place.* This applies, too, to colour in schoolrooms, restaurants, homes – everywhere, in fact, where people either work or relax. Less restraint is appropriate in shop-windows, display cases and exhibitions, where the customer is meant to be stimulated by eye-catchers.

Psychological
effects

By the "psychological effects" of colour we mean optical illusions and other psychical phenomena that are triggered off by colour. In part these are caused by subconscious associations with previous sights or experiences, and partly by hereditary factors. They influence the psychical affectivity, and thereby the whole of a person's behaviour. ("Events" in art are phenomena of this kind. Modern abstract painting strives to produce such effects by colour and form alone, which, to the "initiated", are at least as stimulating emotionally as representational pictures.)

Psychical effects can be induced, too, by colours in a room, arousing strong feelings of like or dislike. Since, however, rooms have to serve particular functions, their colours do not only have aesthetic consequences: their physiological and psychological effects must also be taken into account. Yet there is always a good deal of latitude for aesthetic considerations.

Illusory effects   Particular colours have their special psychological effects, which are all more or less similar in character, though with great individual variation. The most important chromatic illusions concern distance, temperatures and the effects of the general psychical affectivity. *Table 44* summarises these illusory effects of individual colours.

Table 44   **Psychological effects of colour.**

Colour	Distance effect	Temperature effect	Psychical effect
Blue	Further away	Cold	Restful
Green	Further away	Cold to neutral	Very restful
Red	Close	Warm	Very stimulating, not restful
Orange	Very close	Very warm	Exciting
Yellow	Close	Very warm	Exciting
Brown	Very close, claustrophobic	Neutral	Exciting
Violet	Very close	Cold	Aggressive; unrestful, tiring

Broadly speaking, all dark colours are oppressive and tiring; they absorb the light and are difficult to keep clean. All light colours are bright, friendly and cheerful; they scatter more light, brighten up the room, and encourage greater cleanliness.

Colours in   Before starting to plan the colour for a room, there must be a
a room   careful consideration of its functions, and who is going to use it. After that it will be possible to plan its colours in relation to psychological and physiological factors.

The principles mentioned above must be adhered to. Consideration must be given to the work to be carried on in it, whether this is likely to be monotonous, or whether it will make heavy demands on the concentration. If the work is monotonous, it is advisable to include a few areas of exciting colour, but not big areas such as the main walls or ceilings; merely a few items such as a pillar or column, a door, or a partition wall.
If the workroom is very large, it can be divided up by the use of different colours, thereby making it less anonymous.
If the work going on in the room demands close concentration, the colours should be chosen carefully so as to avoid unnecessary distractions and unrestful items. In this case walls, ceilings and other structural elements should as far as possible be painted in light colours that do not attract attention.
Walls and ceilings painted yellow, red or blue may be very

attractive at first glance, but as time goes on they become a strain on the eyes. Hence such rooms often become unpleasant after a while.

More intense colours can safely be used in rooms that are mainly used briefly, e.g. entrance halls, corridors, lavatories, storerooms, etc. Here strong colours may help to brighten up the room, and make it structurally more pleasing. Colour plans for schools, hospitals, and administrative blocks should be based on similar principles.

Field studies

Excessive eyestrain can have two main effects: tiring the eyes, and adding to general fatigue.

Visual fatigue

Visual fatigue comprises all those symptoms that arise after excessive stress on any of the functions of the eye. Among the most important of these are straining the ciliary muscle of accommodation by looking too closely at very small objects, and the effects of strong local contrasts on the retina. Visual fatigue manifests itself as:

(a) painful irritation (burning), accompanied by lachrymation, reddening of the eyelids, and conjunctivitis;
(b) double vision;
(c) headaches;
(d) reduced powers of accommodation and convergence;
(e) reduced visual acuity, sensitivity to contrast, and speed of perception.

These comparatively severe defects are brought about in particular by inadequate lighting, by optical aberrations such as hypermetropia (long sight) or the effects of age, in so far as these cannot be corrected by suitable spectacles.

Obviously all types of visual work contribute to the general fatigue discussed earlier, since every job that calls for more rapid and precise eye movements will make heavier demands on perception, concentration, and motor control of the hands. So whenever the eyes are overstressed for long periods the symptoms of eyestrain (sore eyes and headaches) will be added to those of general fatigue.

The effects of visual fatigue on a person's occupation may include:

(a) *loss of production;*
(b) *lowering of quality;*
(c) *more mistakes;*
(d) *increased accident rate.*

Accident rate

In a report of the American "National Safety Council", the experts reckoned that bad lighting was the cause of 5 % of all industrial accidents, and, together with the optical fatigue it engendered, contributed to as many as 20 % of them.

The experience of an American heavy industry (Allis Chalmers) may be mentioned as an example [92]. After the level of illumination over an assembly line had been increased to 200 lx, there was a fall of 32 % in the accident rate. As a further step the walls and ceilings were painted in light colours, to reduce contrast.and provide a more uniform illumination, and the accident rate fell by another 16.5 %. Similar surveys in England and France showed drastic cuts in accident rates, especially in shipyards, foundries, big assembly lines and mechanics' shops.

Lighting and production

There are also many reports of increased production after the lighting was improved. These increases are partly a direct effect (through more rapid visual assessment of the work), and partly indirect, through the reduction in fatigue. *McCormick* [220] has given a table summarising the results of 15 industrial studies, all of which showed increases in output, ranging from 4–35 %, after increasing the level of illumination. The original level had been very low, however, less than 100 lx. *McCormick* himself [220] had reservations, because of the existence of other, uncontrollable factors that were always present in such situations. Yet in spite of this valid criticism, there is no doubt that the increase was partly due to the previous inadequate lighting.

Less waste

A typical example, from an American cotton spinners, is given in *Figure 171*. When the level of illumination was raised from 170–340 lx, production rose about 4.6 %, while simultaneously the

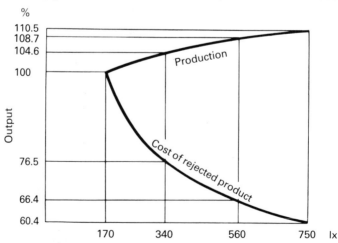

Fig. 171 **Effect of level of illumination on output and on cost of rejected product in an American cotton-spinning factory.**

amount of rejected product was sharply reduced. As a result, the total costs fell by 24.5%. This result encouraged the management to increase the illumination still further up to 750 lx, whereupon production rose to 10.5% above the original level, and the reduction in wastage brought costs down by almost 40%.

Similar results were obtained in England, France, Germany and other countries, often showing increases in production, reduction in rejected products, and fewer accidents, as the level of illumination was increased.

Uniform lighting   Brighter lighting is not the only improvement recommended by modern industrial physiologists; the correct distribution of light over the working field, and the avoidance of glare, are equally important. The effects of reducing contrast between large areas are illustrated in the following example.

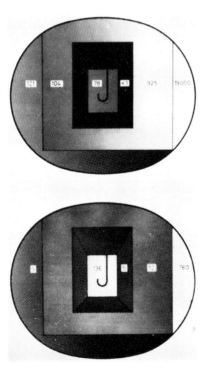

Fig. 172 Surface brightnesses in the visual field of female needle adjusters before and after the lighting has been improved. The numbers indicate the measured light intensities in asb in the following areas: 19 000 and 760 = side windows; 925, 104 and 12 = bench surfaces; 121 and 9 = floor; <1 and 5 = interior of box on bench; 78 and 136 = projection screen on which shadow of needle appears. For further explanation, see text. After *Grandjean, Bloch, Egli* and *Gfeller* [102].

Adjusting knitting needles

A group of women working in a Swiss factory were employed in adjusting needles on a machine for knitting stockings. The needle was fixed in position, and bent into shape with pliers, the enlarged shadow of the needle being projected on to a screen, on which a dotted line indicated the correct shape to be achieved. This projection screen was deeply recessed into a black box built into the work bench. The black box, the bright screen, the floor, the surface of the bend, and the window all contributed to excessive contrasts in the visual field, far beyond the recommended values of between 1:3 and 1:10. *Figure 172* gives these contrasts in asb.

Fitting a curtain over the window, and brightening up the dark box with a coat of green paint reduced the contrast considerably, though still not as much as could be wished. The brighter box also made it possible to increase the contrast between the shadow of the needle and the screen, thus improving visual acuity, yet without creating relative glare. *Figure 173* shows that these improvements in lighting produced a significant increase in hourly output, and a corresponding reduction in fatigue at the end of the day, as measured by the flicker-fusion frequency. This example shows that attention to lighting can both improve performance and reduce stress.

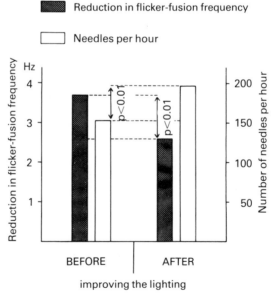

Fig. 173 Effects of reducing surface contrasts upon fatigue levels and output among female needle adjusters. The level of fatigue was assessed from the reduction in flicker-fusion frequency each day in Hz (ordinate). The shaded columns indicate the average amount of daily reduction of flicker-fusion frequency during 11 weeks before the lighting was improved, and 4 weeks afterwards. The blank columns give the output in number of needles adjusted per hour for the same periods of time.

16
Noise and vibration

The nature of sound, the units of measurement of sound pressure, and the physiological phenomena of acoustical perception were all explained earlier in *Chapter 9*. We saw that human hearing has two basic functions:

(a) the transmission of acoustic information;

(b) alarm signals, ranging from waking up to the highest level of alertness.

The present chapter will be concerned with the effects of noise on work and output, and with the principles of protection against noise.

Definition

The simplest definition is that *noise is any disturbing sound*. In practice we call it simply "sound" when we find it not unpleasant, and "noise" when it annoys us. *Hawel* [137] defined it more precisely as follows: "Sound only becomes burdensome if the person affected feels that it is discordant, i.e. if it does not harmonise with his intentions at that particular moment." This definition is particularly apt when applied to noise at work.

The various aspects of noise have been described in detail by *Broadbent* [37], *Jansen* [159], *Furrer* and *Lauber* [90], *Kurtze* [189], *Chavasse, Pilon* and *Wisner* [51], *Kryter* [187], *Klosterkötter* [175], in the *VDI Guidelines 2058* [316], in the *ISO Publication* [156], in a *Report of the Swiss Institute for Accident Prevention* [285], and by *Ward* [319].

Measurements and sources of noise

Measurements of noise load

The *noise load* is the extent of the noise, as measured in physical units, taking account of all the acoustical factors over a given time. Various pieces of research work have shown that the *noise level* is not the only factor operating, but that the frequency with which the noise occurs and other quantities contribute to the total noise load.

These studies have led to the development of units of measurement which combine various components of the noise load into one quantity, so that it is then possible to characterise the noise load at a particular point by a single figure.

Two such units are important in assessing noise problems at the work place:

(a) *the equivalent level of sustained noise (continuous sound level) (L_{eq});*

(b) *the summated frequency level.*

The equivalent level of sustained noise L_{eq}

The equivalent level of sustained noise (L_{eq}) expresses the average level of sound energy during a given period of time (i.e. the energy level). This quantity is an integration of all the sound levels which vary during this time, and so compares the disturbing effect of the fluctuating noises with a continuous noise of steady intensity.

The summated frequency level

The summated frequency level is measured with a sound level indicator and a frequency counter, operating over a given time. Commonly used units of sound measurement include:

(a) L_{50} *(average noise level);*

(b) L_1 *(peak noise level).*

"L_{50} = 60 dB" means that the level of 60 dB was reached or exceeded during 50 % of the relevant time.

"L_1 = 70 dB" means that the level of 70 dB was reached or exceeded for 1 % of the time.

These two levels, L_{50} and L_1, are related to the equivalent level of sustained noise by the following approximation:

$$L_{eq} = L_{50} + 0.43 (L_1 - L_{50}) \approx \frac{L_{50} + L_1}{2}$$

Sources of noise

Disturbing noise may be either *external*, coming from outside the building, or *internal*, generated within the building itself. The most important sources of external noise are traffic, industry, building and neighbours.

The most important source of internal noise in factories are machines, motors, compresed air, milling machines, stamping machines, looms, sawmills and many other pieces of noisy machinery. In addition to this, the office has its own internal noise which comes from telephones, typewriters, calculating machines (mechanical), and people walking about and talking.

Street noise

External noise is a disturbing factor in offices, drawing offices, conference rooms and schools. *Table 45* lists some of the noise levels to be expected in premises alongside a road:

Table 45 Noise levels from street traffic, given as L_{eq} in dB(A). Measuring point in front of the window: if this is opened a decrease of 5–10 dB(A) can be expected indoors.

Traffic density	L_{eq} in dB(A) By day	By night
Heavy (main road with through-traffic)	65–75	55–65
Moderate	60–65	50–55
Light (local street)	50–55	40–45

Industrial noise

Internal noise in factories is always a very variable factor. It may be continuous or intermittent, and may take the form of banging, clattering, rattling or whistling. *Table 46* shows some of the peak levels that may be attained.

Table 46 Peak noise levels in dB(A).

Source	Noise level in dB(A)
Rifle-shot; motor test bench	130
Pneumatic bore-hammer	120
Pneumatic chisel	115–120
Rocking sieve; chain saw; compressed air rivetter; electric cutter; compressed air hammer	105–115
Milling or weaving machine; crosscut saw; stamping machine; boiler-house; weaving shed	100–105
Electric motor; rotary press; wire-drawer; sawmill; composing room; bottle-filling machine	90–95
Toolmaking machine (running light)	80
Typewriter	65–75

To assess the extent of the danger to hearing from such noise we need to consider the average level of L_{eq} during the 8-hour shift. According to *ISO/R 1969–1971* [156] the measurement should be in dB(A), and equivalent level of sustained noise (L_{eq}) calculated for intermittent noise.

Table 47 gives a few examples of the L_{eq} for an 8-hour day.

Table 47 Equivalent level of sustained noise L$_{eq}$ in dB(A) at various work places, calculated over an 8-hour shift [285].

Department or machine	L$_{eq}$ dB(A)
In a flexible-tube factory:	
At yarn-spinning machine	95
In the spinning room	90
At the weaving machine	95
In the weaving shed	95
In the soft-drinks industry:	
At the mixing plant	95
At the washing check point	100
At the automatic sealing machine	100
At a can-filling machine	90

Noise in offices

Concentrated mental work, or jobs at which understanding of speech is important are "noise-sensitive" occupations, and even if the noise level is comparatively low it can be disturbing. Noise levels to be expected in offices are listed in *Table 48.*

Table 48 Noise levels usual in offices.

Type of office	Average L$_{eq}$ in dB(A)
Very quiet small office and drawing offices	40–45
Large, quiet offices	46–52
Large, noisy offices	53–60

The peak noise level L$_1$ lies in the zone between 56 and 65 dB(A), which is about 5–10 dB(A) above the average noise level L$_{50}$. Individual peak noise levels are often produced by the following:

	dB(A)
(a) telephone bell, within 2 m:	75
(b) normal typewriter, within 2 m:	70
(c) "silent" typewriter, within 2 m:	60
(d) conversation, within 1 m:	60–65

Damage to hearing through noise

Damage to hearing

Strong and repeated stimulation from noise can lead to loss of hearing, which is only temporary at first, but after being "deafened" repeatedly some permanent damage may occur. This is called *noise deafness*, and is brought about by slow but progressive degeneration of the sound sensitive cells of the inner ear. The louder the noise, and the more often it is repeated, the greater the damage to hearing. Moreover it is well known that noise consisting of predominantly high frequencies is more harmful than low-frequency noise. Intermittent noise, such as hammering, is more harmful than continuous noise, and a single very loud noise – a detonation, or an explosion – can damage the ears immediately. Individual sensitivity to noise varies greatly from one person to another. Some, who are particularly sensitive, may suffer permanent deafness after only a few months, whereas less sensitive persons may not show the first symptoms until after many years' exposure. Noise deafness starts with the higher frequencies, of about 4000 Hz, and extends only gradually to the lower frequencies. At first the worker is unaware of it, and only gradually notices his loss of hearing when it begins to involve the lower frequencies. Noise deafness is progressive, and often combines with the deafness of increasing age, or is mistaken for the early onset of the latter. In most industrial countries, noise deafness ranks among the occupational hazards of working life.

Audiometry

Nowadays the extent of loss of hearing is usually measured by means of so-called pure tone audiometry, which determines the threshold of hearing of pure tones of various frequencies. The result is shown on an audiogram, which shows in dB how much the threshold of hearing has been raised for each frequency. Such an audiogram of impairment of hearing by noise is shown in *Figure 174*.

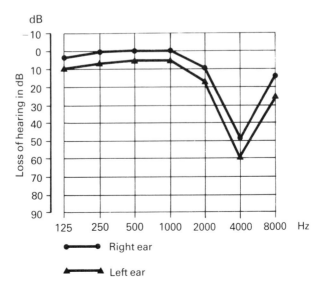

Fig. 174 Pure-tone audiogram showing impairment of hearing by noise. The zero line is the normal threshold of hearing. The loss of 50–60 dB at 4000 Hz is characteristic of hearing damage. After [285].

Temporary loss of hearing

In this context we must return to temporary noise deafness, already mentioned above. This phenomenon is characterised by the fact that *hearing returns to normal:* what is known as the *"temporary threshold shift"*. Reversible deafness has been the subject of intensive study in recent years, by *Kryter* [188], *Ward* [319] and others. These authors have found that there is a very close relationship between temporary and permanent deafness, and that the results for temporary loss of hearing allow many conclusions to be drawn that are of more general application. The most important of these may be listed as follows:

(a) noise up to 80 or 90 dB causes only slight shifts in the threshold of hearing, eight or ten decibels only, but if the noise is increased to 100 dB, the threshold goes up by 50 to 60 dB;

(b) the temporary shift in the threshold of audibility is proportional to the duration of the noise. For example, a 100 dB noise for 10 minutes produces a shift of 16 dB, and after 100 minutes, one of 32 dB;

(c) the time taken for hearing to return to normal is also proportional to the intensity and duration of the preceding noise. The time of restitution is about 10 % longer than the duration of the noise;

(d) alternate noisy and quiet periods produce less temporary deafness.

Age deafness

The threshold of hearing rises progressively with age, and loss of hearing is greatest in the higher ranges of frequency and more pronounced in men than in women. Taking a frequency of 3000 Hz as standard, the loss of hearing to be expected at various ages is as follows:

50 years:	10 dB
60 years:	25 dB
70 years:	35 dB.

The audiogram of age deafness differs from that of noise deafness in that the loss of hearing increases progressively as the frequency is raised, so that the highest frequency still audible shows the greatest shift of its threshold. The characteristic dip in the curve at 4000 Hz is not evident in cases of age deafness.

Older workers often show the combined effects of age deafness and noise deafness, and it is very difficult to distinguish the two.

The risk of loss of hearing

From the evidence of many comparisons between exposure to noise and the frequency of impaired hearing, it is now possible to estimate the risk of hearing damage in noisy factories. The publication of the *ISO* [156] sets out in a comprehensive table this risk in relation to age, duration of exposure, and the intensity of the noise (expressed as L_{eq} for a 40-hour week). A simplified extract from this report is given in *Table 49*.

Table 49   The risk of damage to hearing, as a presumptive percentage of the work force. These percentages will be increased by several units with age.

L_{eq} dB(A)	Length of exposure in years		
	5	10	20
	%	%	%
80	0	0	0
90	4	10	16
100	12	29	42
110	26	55	78

These figures show that the risk of damage increases both with sound intensity and duration of exposure, *the damaging intensities being those above approximately 90 dB(A)*.

Factory workers are often exposed to noise that varies widely, and it has been shown that interruptions, or periods of relative quiet, reduce the risk of damage to hearing.

To assess the extent of such risk, the equivalent level of sustained noise over the 8-hour working day must be calculated. The relation between length of exposure and intensity of sound to create the same degree of risk is as follows:

hours	dB(A)
8	90
6	92
3	97
1½	102
½	110

As far as current knowledge of noise damage goes, *the following may be proposed as a limiting value: L_{eq} for an 8-hour day not to exceed 85 dB(A)*. Most of the limits in use today approximate to this, though they are often more precisely defined and formulated [156] [285] [316].

Physiological and psychological effects of noise

Figure 93 (Chapter 9) showed the auditory tracts and how they are linked with the activating and alarm-sensitive structures of the brain. Here lies the explanation of many of the effects of noise and the spread of a state of activation or alarm throughout the consciousness.

This may result in:

(a) *impaired alertness;*
(b) *disturbance of sleep;*
(c) *a feeling of stress.*

At the same time this activation affects the autonomic centres and produces the so-called *vegetative effects* in the internal organs. Finally, a special problem of noise is that it makes it *difficult to understand what people say.* The ergonomic aspect of these various problems will now be discussed.

Understanding of speech

We all know from our own experience that the sensitivity of the ear to one particular sound – say the voice of a colleague – becomes less and less as the ambient noise increases. The ability to pick out one particular sound from the rest depends on its auditory threshold, which rises linearly with the sound intensity up to 80 dB. Where human speech is concerned, however, it is not enough to hear the pure tones. Their message must also be understood, and to do this requires a very special discriminating ability in the ear. A critical factor is correct hearing of the consonants, which are much softer sounds than the vowels.

Since the understanding of words and sentences depends greatly on individual intelligence, as well as on familiarity with the test language, research workers into the damaging effects of noise have used *syllables* as the criteria. A speaker utters a succession of meaningless syllables and records what proportion of the total number are correctly understood. This has become the standard method for research in this field.

It has been shown that a considerable understanding of sentences and their meaning is possible without understanding all the separate syllables. *Figure 175* shows the ratio between syllable comprehension (S) and sentence comprehension (W).

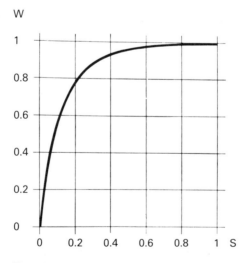

W

Fig. 175 Comprehension of syllables (S) and complete sentences, or sense (W). Each is plotted as a decimal fraction of the total number of syllables or sentences offered for comprehension.

It will be gathered from this graph that with a syllable comprehension of only 20 % it is still possible to understand nearly 80 % of the sentences; if half of the syllables are understood (S = 0.5), then about 95 % of the sentences are intelligible.

Speech comprehension in a workroom depends very largely on the loudness of the voice concerned and the level of background noise. *Figure 176* shows the relationship between syllable comprehension (S), noise level (N) and voice level (P) (solid lines).

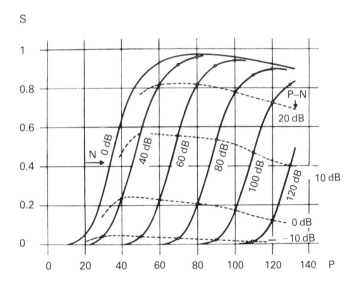

Fig. 176 Syllable-comprehension (S) in relation to the sound-pressure (P) of speech and the noise level (N) in the room. The dotted lines connect all those points at which P-N is 20, 10, 0 or –10 dB.

In addition, the dotted lines join together points at which there are equal differences between sound pressure of noise and sound pressure of speech: P-N = 20, 10, 0 and – 10 dB respectively. It is apparent from this graph that a syllable comprehension of 40–56% is possible, as long as P-N = 10 dB. According to *Figure 175*, this implies 93–97 % comprehension of sentences, or sense (W). *Experience shows that this level of comprehension is good enough for most factories and offices, and therefore speech comprehension is considered to be unimpaired, as long as the background noise level is at least 10 dB below the level of the speaking voice.*

If, however, we are concerned with an exchange of verbal information, on a subject that is unfamiliar, with difficult new words, then a higher level of syllable comprehension is necessary. It has been demonstrated that in these circumstances syllable comprehension must be as high as 80%, and this requires a difference of 20 dB between the pressure levels of voice and background noise (*Figure 176*).

The normal speaking voice, indoors, at a distance of 1 metre, operates at the following pressure levels in dB:

Quiet conversation:	60–65
Dictation:	65–70
Speaker at a conference:	65–75
Delivery of a lecture:	70–80
Loud shouting:	80–85

If the voice has to be used frequently to dictate, or to convey information in the course of a person's employment, it should not exceed 65–70 dB at a distance of 1 m. If this is to be understood clearly and without strain, the background noise level must not exceed 55–60 dB, and if the verbal communication is more difficult to understand, e.g. contains many strange words, or unfamiliar names, then the background noise must not exceed 45–50 dB.

If offices or workrooms with these requirements are situated immediately on roads with moderate or heavy traffic densities, then a maximum permitted background noise of 55–60 dB cannot usually be adhered to, especially in summer when windows are open. In these circumstances noise levels of 70–75 dB are often reached in offices. Offices in summer need more effective protection against traffic noise, in the form of air-conditioning, which allows the windows to be kept permanently closed. In towns, traffic noise is the most important reason for installing air-conditioning.

Effects on performance

Exposure to noise has little effect on manual work, whereas we all know from experience that thought and reflection are more fatiguing in a noisy environment than in a quiet one. Sports managers, in particular, know that discipline and all movements that call for intense nervous concentration are adversely affected by noise. Many everyday examples show that noise impairs concentration, and one is led to assume that under such conditions performance and output will also suffer. It is interesting, though, that what seems to be a truism in everyday life is only partly confirmed by either experiments or field studies. Research into the effects of noise on either mental or psychomotor performance has given very contradictory results: noise is just as likely to improve performance as to make it worse.

Contradictions

We must conclude that *a decline in performance can be attributed to the effects of noise, only with important reservations.*

In practice, analysis of results up to the present show that the following considerations are important in assessing the effects of noise:

(a) the intensity of the noise;
(b) the nature of the background noise: is it steady or intermittent, predictable or unexpected?
(c) the information content of the noise: is it merely mechanical, or does it include speech that must be understood?
(d) the nature of the task being performed: is it boring or stimulating?

Improving performance

Noise can even be positively stimulating in the right circumstances. Performance may be *improved* if the work is boring, and perhaps also in situations where there are many other distractions. Sometimes, if these distractions can be concentrated upon one dominant noise, even mental performance may be improved.

Loss of performance

However, these examples of improved performance must be set against many more pieces of research that have shown performance to be adversely affected by noise. For example, *Broadbent* [37] [38] and *Jerison* [163] have observed a decline in performance during exacting tests of sustained vigilance.

To summarise, we can say of the deleterious effects of noise in performance that:

(a) *background noise often interferes with complex mental activities, as well as certain kinds of performance that make heavy demands on skill and on the interpretation of information;*
(b) *noise can make it more difficult to learn certain kinds of dexterity;*
(c) *many studies have shown that high levels of noise (over 90 dB as a rule), either discontinuous or unexpected, can impair mental performance.*

Field studies

The effects of noise have occasionally been studied under industrial conditions. Thus *Wisner* [330] reports the following experiences:

(a) in one machine shop a reduction of about 25 dB in the noise level led to 50% fewer rejected pieces;
(b) in an assembly shop, a reduction of noise level by 20 dB raised production by about 30%;
(c) in a typing pool, a reduction of noise level by about 25 dB was accompanied by a 30% reduction in typing errors.

Such results must always be interpreted with caution, however. *Wisner* himself hints that the new situation in the factory may have a psychological effect on the workers.

Effects of noise
in offices

An interesting example of this was revealed during a study in 15 open-plan offices, already mentioned in *Chapter 15* [243]. On the one hand the summated frequency level was measured in each of the offices, while on the other hand 519 office workers were questioned about their experiences and opinions. About 8000 measurements of noise levels gave average values (L_{50}) between 38 and 57 dB(A), and peak noise levels (L_1) between 49 and 65 dB(A). The distribution of replies to the question "were you disturbed by noise?" is shown in *Figure 177*.

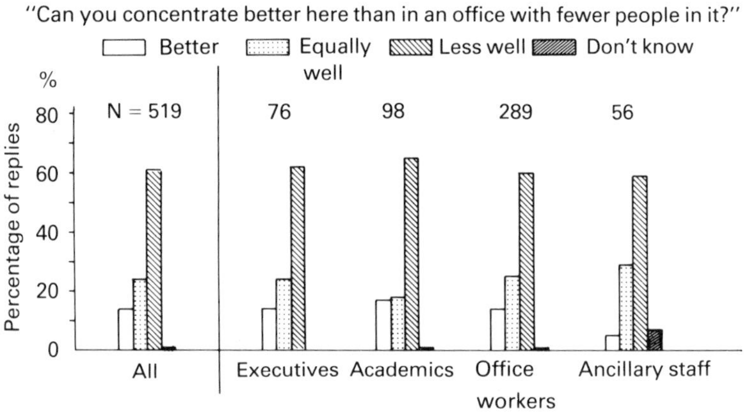

Fig. 177 **Frequency distribution of the answers to questions about disturbance from noise in 15 open-plan offices.** Total 519 questioned, 79% men and 21% women. The differences between the social groups are statistically significant. ($p < 0.01$, chi^2 test). After *Nemecek* and *Grandjean* [243].

From this the following conclusions may be drawn:
(a) 35% of those questioned were "greatly disturbed" by noise;
(b) those classified as "office workers" and "assistants" were less disturbed by noise than "overseers" and "graduate staff".
Among other results may be mentioned the following:
(a) 69% of those questioned mentioned "disturbance to concentration" as one of the drawbacks of noisy offices;
(b) about 2/3 of those questioned listed talking, the telephone and business machines as the commonest causes of distraction;
(c) 37% of those questioned would have preferred a smaller, more conventional office. A statistical analysis (tetrachoral correlation) shows that the following reasons for rejecting open-plan offices are significant:

(i) "annoyance from noise"
(ii) "disturbance of concentration"
(iii) "occupational ailments"
(iv) "lack of privacy".

Disturbance from conversation

The distribution of replies to the open question "What kinds of noise disturb you?" is shown graphically in *Figure 178*.

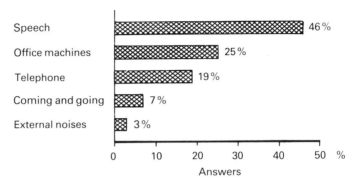

Fig. 178 **Frequency distribution of replies to a question about the sources of noise.** 411 people questioned, some mentioning more than one source, so a total of 762 replies = 100%. After *Nemecek* and *Grandjean* [243].

It is clear that *"talking" or "conversation" is the disturbance most often complained of.* Many of those questioned added that it was not the loudness of the conversation that disturbed them, but its content! These indications were confirmed by the results of correlations calculated between summated noise levels and the frequency with which the noise was rated as "very disturbing". In practice, no correlation could be found between the degree of disturbance and the summated noise levels, L_{50}, L_{10} or L_1. The graph for L_{50} and degree of disturbance is shown in Figure 179.

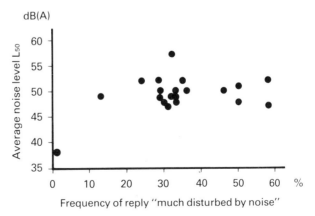

Fig. 179 **Frequency of complaints of disturbance for various average noise levels (L_{50}), in each of 15 open-plan offices.** After *Nemecek* and *Grandjean* [243].

It is clear from the figure that the extent of disturbance, measured as a percentage of "severe disturbance", is almost independent of the sound level in decibels. We must conclude from the results as a whole that *conversation, which is rated as the most disturbing source of noise, distracts from concentration, not so much by its mere loudness as by its "information content"*. These results show how complex are the effects of noise upon mental activities. Later on [244] it was possible to confirm these results for offices of a more traditional kind, which were no exceptions to the rule.

Studies in factories and in the laboratory have agreed in showing that people feel noise to be a burden and a discomfort, and that to carry out difficult tasks in noisy surroundings always brings about feelings of strain and requires an effort of will. This has been found even by authors who were unable to record any measurable objective interference with the subject's work. It follows, therefore, that if we must carry out work which makes heavy demands on thought, concentration and skill, in a noisy atmosphere where it will necessarily require a certain nervous expenditure and a certain mental stress, we must be prepared to isolate the noise, so to speak, so that we are no longer conscious of it.

The effects of noise on mental work may be summarised as follows:

(a) an intermittent noise, especially if unexpected, is more disturbing than a continuous one;
(b) high frequencies are more disturbing than low frequencies;
(c) work which calls for sustained alertness over a long period, and without pause, is particularly sensitive to noise;
(d) disturbance is greater during a learning stage, than when the work has become more automatic;
(e) sounds which have a certain information content are more disturbing than meaningless noise.

Vegetative irritation from noise

Many physiological studies have shown that exposure to noise produces:

(a) raising of the blood pressure;
(b) acceleration of heart rate;
(c) contraction of the blood vessels of the skin;
(d) increase in metabolism;
(e) slowing down of the digestive organs;
(f) increased muscular tension.

All these reactions are symptomatic of a spreading state of alarm, which is generated and controlled by a state of increased stimulation of the autonomic nervous system. This is actually a defensive mechanism which prepares the whole body for facing possible danger, by being ready for fight, flight, or defence. It should not be forgotten that throughout the animal kingdom the sense of hearing is primarily an alarm system, and this basic function still remains even in the human organism.

Waking effects

It is essential for the maintenance of good health that the stresses of the day should alternate with the restorative powers of sleep. During sleep the activities of the muscles, brain and many other organs go on at a reduced rate, while only those organs which contribute to restoration of physical powers and the assimilation of food (digestive and metabolic organs) continue to function without restraint. If sleep is curtailed, or frequently disturbed, its restorative functions are impaired, and if this continues it leads to a loss of efficiency and of well-being, which become evident during the daytime.

The acoustic sense is the most effective awakener from sleep. When the eyelids are closed, optical stimuli are largely excluded, whereas the sense of hearing is only slightly muffled during sleep. It still retains its primary function as an alarm system.

Experience shows, in fact, that familiar noises are less likely to awaken a sleeper than unfamiliar ones. People who live close to a railway are not awakened by passing trains, whereas an anxious mother awakes if her child coughs or breathes strangely. Obviously the human brain can "tune in" to certain sounds and react to those by awakening, while ignoring others, but has little defence against totally unpredictable sounds, when neither their nature nor their timing can be foreseen. Such noises have a powerful waking effect.

Noise may either awaken people completely, or rouse them into a twilight sleep. This is particularly likely if the noise is repeated. Since twilight sleep is not as restful as deep sleep, it is undesirable, and so intermittent noises may be said to impair the quality of sleep even if the subject does not wake up completely.

Noise and sleep

The effects of noise on the duration and quality of sleep under experimental conditions has been studied in detail with the aid of the electroencephalograph. An evaluation of this "literature of sleep" has been published by *Barbara Griefahn, Jansen* and *Klosterkötter* [120], as well as by *Lukas* [207]. Studies up to now have shown that a noisy environment:
(a) seriously curtails the total time asleep;
(b) cuts down the amount of deep sleep;
(c) increases the time spent awake, or in light sleep;
(d) increases the number of waking reactions;
(e) prolongs the time of falling asleep.

Waking experiments

The sources of noise that produce awakening vary enormously, not only from one person to another, but even for the same individual under different circumstances. The background noise at the moment of taking the measurement is important, as well as the various physiological factors that have been mentioned.

A classic experiment involving the waking effect of noise was that of *Steinicke* [296]. He carried out waking experiments on 343 persons in their bedrooms, using an automatic apparatus. All the experiments took place between 2 am and 7 am. The waking noise lasted three minutes on each occasion, and started at 30 phon, increasing by steps of 5 phon until the subject was awakened. The sound was a mixture of frequencies between 60 and 5000 Hz.

The results of the 343 experiments are shown in *Figure 180*, in the form of a summation curve.

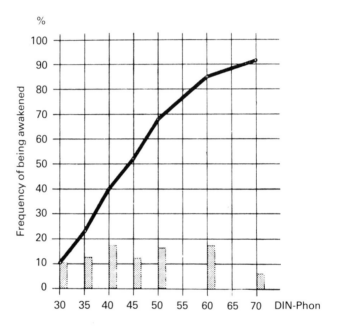

Fig. 180 Waking threshold of noise. The values are quoted in DIN-phons, but these are approximately the same as the dB(A) values currently in use. Total: 343 test subjects. The curve is a summation curve, and the vertical columns indicate the fraction of the total subjects who are awakened by that particular level of noise. After *Steinicke* [296].

In *Steinicke*'s experiments the test subjects were given long periods of noise (3 minutes) at a time of day when they were most likely to wake up anyway, so the threshold values obtained for them cannot be applied to realistic circumstances without further consideration. In fact later research [120] [207] has given thresholds that were about 20 dB higher than those in *Figure 180*. On the other hand, these experiments confirm *Steinicke*'s findings about the kind of relationship between awakening and the intensity of the sound, and about the wide variation between individuals.

Field studies

Questionnaires are more reliable than experiments when investigating the effects of noise during the night, especially if large populations are investigated. *Figure 181* shows the results of such a field study carried out in the Basel area by *Graf, Meier* and *Müller* [97].

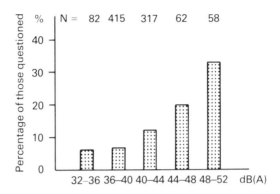

Fig. 181 **Disturbance of sleep by traffic noise**. Ordinate: percentage of those questioned who were "often prevented from going to sleep, or awakened from sleep", by traffic noise.
Abscissa: summated frequency level L_{50} in dB(A) measured in front of the window. Total of 934 persons questioned. After *Graf, Meier* and *Müller* [97].

As the figure shows, increasing noise significantly raised the percentage of persons who reported that they were "frequently disturbed, either when falling asleep, or when sleeping". The increase began when L_{50} exceeded 40 dB(A), and when $L_{50} = 50$ dB(A) about one third of the persons questioned were affected. As a final corollary to this section, let us recall the sleeping problems of night workers. The disturbance of their biological rhythm was added to the noisier environment when they were sleeping in the daytime, and the result was a serious interference with their quota of sleep.

The burden of noise

Everyday experience teaches us that many noises have effects of an emotive nature on people. These arouse feelings and sensations, and so are strongly subjective. They are reckoned among the psychological effects of noise. Not all noises or sounds are burdensome. Natural sounds, such as the rustling of leaves or the murmur of a stream, are pleasant. Noise may also be agreeable in itself, if it drowns other sounds that are not pleasant. On the other hand there are many kinds of noise, or noisy situations, which people regard as being subjectively unpleasant and burdensome. The nature and extent of the burden depends on a number of subjective and objective factors, the most important of which are as follows.

1. The louder the noise, and the more high frequencies it contains, the more people are affected by it.

2. Unfamiliar and intermittent noises are more troublesome than familiar or continuous sounds.

3. A decisive factor is a person's previous experience of the particular noise involved. A noise which often disturbs one's sleep, which excites anxiety, or which interferes with what one is doing, is particularly burdensome.

4. A person's attitude to the source of the noise is often specially important. A motor-cyclist, a workman, a child or a musician is not disturbed by the noise generated by his own activities, whereas a bystander or someone not participating is disturbed to an extent which depends on how much he dislikes either the sounds being produced, or the person producing them.

5. The extent of the disturbance by noise often depends upon what the person affected is doing, and what time of day it is. A working housewife is less disturbed by traffic noise and noise from the neighbours, all day long, than is her husband during his short midday break, when he wants to rest and relax. The rustling of papers is disturbing during a lecture, whereas out in the street it would pass unnoticed.

The burdensome feelings generated by noise are among its most important effects, because they are so widespread, and they must be regarded as the decisive factor in developed techniques for combating noise, and formulating regulations against it.

Becoming accustomed to noise

It is still not clear how far people can become accustomed to noise. Experience shows that there may be some degree of adaptation in certain circumstances, yet in others there is either no adaptation at all, or even the converse, an increasing sensitivity to noise. These phenomena depend on so many external circumstances, and so many psychological factors, that it is still not possible to generalise. We can only infer from the continued increase in noise and its burdens that no adaptation will be possible. On the contrary, the limits of adaptability will be overreached more and more often.

Noise and health

The recuperative processes that are essential to health take place, on the one hand, during night sleep, and on the other, during pauses of all kinds, interruptions of work, and during a person's leisure time.

If the irritating effects of noise on the vegetative nervous system are not confined to working hours, but extend into the hours of sleep, then this will upset the balance between stress and recuperation. Noise will then become a causative factor in states of chronic fatigue, with all its ill-effects on well-being, efficiency and the incidence of ailments.

According to the definition of the World Health Organisation (WHO), health is a state of physical and mental well-being. If we take this definition as a basis for discussion, then not only must deafness, but also the frequent disturbances to sleep, delayed recuperation, and the daily repetition of the burdens of noise be reckoned among the hazards to health.

Protection against noise

Noise in factories can be countered in the following ways:
(a) *protective planning;*
(b) *reduction of noise at source;*
(c) *insulation against reflection and scattering;*
(d) *personal sound protection.*

Planning

The most important technological step in the battle against disturbing noise lies in the choice of building materials and in planning of the subdivisions of the building. *Hence noise protection begins on the architect's drawing board.*

It is fair to assume that the noise level will decrease with increasing distance from its source, so it is advantageous that offices, drawing offices, and any other places where mental work is carried on should be sited as far away as possible from the noise of traffic. If the factory itself is noisy, the noisy sections, too, should be as far away as possible from places where work is carried on that calls for concentration and skill; intervening rooms, used for packing and storing, will act as "buffers".

When considering the division between two rooms, account must be taken of the damping effect of walls, doors, windows and hatches; and *Table 50* gives some examples of these.

Table 50 Sound-deadening effect of various building items.

Item	Damping effect dB	Remarks
Normal single doors	21–29	Speech clearly understandable
Normal double doors	30–39	Loud speech still understandable
Heavy special doors	40–46	Loud speech still audible
Window, single glazing	20–24	
Window, double glazing	24–28	
Double glazing with felt packing	30–34	
Dividing wall, 6–12 cm brick	37–42	
Dividing wall, 25–38 cm brick	50–55	
Double wall, 2 × 12 cm brick	60–65	

Tackling noise at source

The most effective and rational way to deal with noise is usually to tackle it at source. Certain jobs, and certain pieces of machinery, are noisy because heavy, hard surfaces clash together. Sometimes this noise can be reduced by replacing the hard material with something softer, e.g. vulcaniser rubber, ordinary rubber or felt. For the same reason transport vehicles are quieter with rubber tyres than with metal wheels. All forms of transmission should be examined to see if they are still in good condition, since old and worn components create unnecessary noise. Belt drives, using rubber, leather or fabric belts, are quieter than a series of cog-wheels. Toothed belts, engaging with toothed wheels are the quietest of all, especially if the toothed wheels are made of some synthetic material.

Noise radiated from vibrating plates can be reduced by stiffening them, loading them with weights, making them curved, or using non-resonant materials. Moving machinery, and motors in operation not only send out sound waves, but also set the structure of the building in vibration. Such vibrations and resonances, and the secondary noises that they set up, can be disturbing throughout the whole building. For this reason very heavy machines should be rigidly set in concrete or iron mountings. If necessary they can be mounted in special concrete troughs, with intervening layers of sound-insulating material. Such sound-insulating layers can be of spring steel, rubber, felt or cork, according to the weight of the offending machine.

Enclosing the source of noise

An especially effective way of reducing noise is to enclose the source. A housing of suitable material may reduce the radiated noise by 20–30 dB. The inside wall of such a housing should be lined with sound-absorbing material, while the wall itself should be as heavy and as airtight as possible. The housing should enclose the source of noise, with as few gaps as possible, to avoid the transmission of structural vibrations. Of course some apertures are usually necessary, for the passage of leads, or for operating the machine; such apertures impair the soundproofing of the machine, and it can be stated as a general guideline that their total area should not exceed 10 % of the area of the housing. Doors or sliding panels may be provided to allow the machine to be controlled and serviced.

Sound insulation in a room

When all the possibilities of sound-damping at source and by enclosure have been exhausted, it is sometimes possible, in the right circumstances, to give further protection by covering the walls and ceilings with sound-absorbing material. *Sound-absorbing panels (acoustic tiles) absorb part of the sound, and so reduce reflection back into the room and the echo effect. Koch* [179] recommends that workrooms should be lined with acoustic tiles if the following postulates can be made:

(a) if this operation will reduce the echo time in a factory by at least ¼ or in an office by ⅓;

(b) if the room is not more than 3 m high;

(c) if the room is more than 3 m high, but its volume is not much more than 5000 m³.

Up to the present, acoustic tiles have been used mainly in offices of more than 50 m² in area, in accounts offices, in cashiers' offices, and in offices giving counter services. Sound reductions of 5 dB, or even up to 10 dB can be expected in such places. On the other hand the effects of installing acoustic tiles in noisy workshops, machine rooms and factories are not always clear or easy to assess. It must be remembered that when an operative works close to a source of noise he is mainly affected by the sound reaching him direct, and reflected sound is of slight importance to him. Covering the walls and ceiling with acoustic tiles will therefore give no protection to such a worker. Ceiling tiles will be of slight benefit to him only if he is working several metres from the source of the noise.

Personal ear protectors

If an operative is forced to work in a noisy environment, and technological means have failed to reduce the noise to a safe level of below $L_{eq} = 85$ dB(A), then as a last resort there remains only personal ear protection.

The following possibilities are worth considering:

(a) *ear-plugs to block the outer ear passages;* these may be of cotton-wool, wax or synthetic material;

(b) *protective caps, which cover the entire external ear* (ear-muffs);

(c) a sound-proof helmet, enclosing the head, including the ears.

Ear plugs

A simple plug of cotton-wool or wax in each outer ear passage is an old device that is often used. *Properly used, ear-plugs can reduce the noise level by up to 30 dB.* Instead of cotton-wool, conical plugs of a synthetic material (Selectone) are available, which cut out more of the higher frequencies than the lower ones, and so interfere less with conversation. All types of ear-plugs have the drawback that they need to be used carefully and correctly if they are to give sufficient protection, and if used regularly they may set up irritations, such as aural eczema.

Protective caps (ear-muffs)

Caps which enclose the whole of the external ear (ear-muffs) give good protection, and if correctly fitted, can reduce noise levels by about 40–50 dB in the frequencies 1000–8000 Hz. The insulating pad must fit closely round the ear, on to the skull, so that it does not exert any pressure on the ear itself, otherwise it may set up pains and headaches. *Figure 182* shows the effects of using an ear-muff.

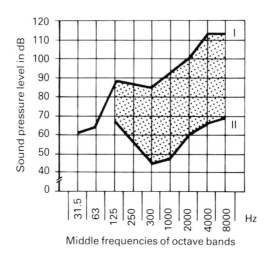

Middle frequencies of octave bands

Fig. 182   Protective effect of an ear-muff in the vicinity of a circular saw cutting light metal. I: noise spectrum at the unprotected ear. II: noise spectrum inside the ear-muff. After *"Bericht der Suva"* [285].

The graph shows that the noise of a circular saw, with frequencies between 500 and 8000 Hz is reduced at the eardrum by about 40–45 dB.

Unfortunately many workers object to any kind of ear protectors. Their main ground of complaint is "acoustic isolation"; the worker thinks that he is missing some vital information from his surroundings. He also thinks that his output, and hence his earnings, will be adversely affected. In spite of these objections, the wearing of ear protectors should be encouraged in the interests of the workers themselves.

As general guidelines we might suggest:
for noise levels L 85–100 dB(A): ear-plugs;
for noise levels above 100 dB(A): ear-muffs.

Medical research

Noise protection in its widest sense also includes those medical measures that are laid down for noisy places of employment in many countries, and which provide for periodic audiometric tests. The purpose of these is:

(a) to detect noise damage at its onset;

(b) to supply a basis for devising protective measures against noise, and for the introduction of personal protection into factories.

Music and work

Since from the point of view of physics music is a form of sound, it is appropriate to discuss it briefly in the present chapter.

Throughout the ages, music has been used more and more to lighten human labours and many working songs exist, some of the best-known being those of spinning women, soldiers' marching songs, and the famous Song of the Volga Boatmen. All songs of this kind are melodious, with a well-marked rhythm, and their effect is to rouse the singers and urge them on to greater effort.

Physiological effects

As we have already seen, an acoustic stimulus passes through the inner ear and along the auditory nerve into the interbrain, to the reticular formation of the activating system. From there a secondary nervous pathway brings the entire conscious sphere of the cerebral cortex into a state of advanced readiness for action. Hence noise can have a stimulating effect, especially in boring situations. Music that is strongly rhythmical, with marked variations in loudness, affects the brain in similar ways, bringing the whole organism to "action stations". Noise also has a distracting effect, however, so that activities that call for thought and alertness are disturbed. We might expect the same to be true of stirring, rhythmical music. In fact we do find that music is particularly welcome as a background to dull, repetitive work, but its effect on intellectual work is debatable.

The extent of the distraction and disturbance provoked by music depends a good deal on its nature. Up to a point the distraction can be minimised by choosing suitable music.

Studies in industry

As part of the campaign to improve working conditions, for more than 20 years music has been used, here and there, to relieve the boredom of certain jobs. Thus English research showed that the introduction of music into a factory making ready-made clothing improved the output by female workers. From their experience, the investigators recommended that the music should be restricted to a period in the mornings, between 10.00 and 11.15 am. *Kerr* [174] questioned 666 workers in an American factory, whose jobs included spool-winding, operating presses, and assembling radio tubes, about how they would like music distributed throughout the day. The overwhelming majority wanted continuous music all day long; if it had to be restricted, then most of them preferred a series of between 10 and 16 periods, equally distributed throughout the day. Mid-morning and mid-afternoon were periods when music was particularly welcome. It appeared that the younger employees and the women liked music more than the others.

Smith [293] studied the effect of music on 1000 American

employees who were engaged upon a variety of repetitive manual tasks. Before starting the investigation, which lasted 12 weeks, 98 % of them expressed a preference for music while they worked. The older workers preferred classical and quiet instrumental music, without singing, while the young and middle-aged were decidedly in favour of light music of all kinds.

In an assembly shop the introduction of light music resulted in an average increase in production of 7 % by day and 17 % by night. The increase was greatest when music was played for 12 % of the day shift and 50 % of the night shift. When classical music was played, production was less than with light music: this effect was most marked at those times of day when production is usually at its lowest.

A particularly interesting observation of *Smith* was that music increased the accident rate by day, but decreased it at night.

Similar research was carried out in French factories, which by and large confirmed these results. In Switzerland music was introduced in textile mills, shoe factories and the like, with similar results.

Background music

Originally, music at work was rhythmical, with a clear melody. The workers listened to it consciously, and sometimes hummed the tune. More recently, starting in America, administrative offices, business houses, salerooms, railway stations, waiting rooms, restaurants and even residential rooms have been provided with a different kind of music, persistent, but very quiet, unobtrusive, hardly impinging on the consciousness. This is "background music" which is supposed to surround one with a pleasant "envelope" of agreeable sound, which should have the advantage of not being distracting, and therefore being suitable for work which demands concentration, such as designing or planning.

Recommendations

As far as present knowledge goes, the question whether music at work is a good thing or not may be answered as follows.

Music at work helps to create a pleasant atmosphere, which stimulates the worker. This is particularly so if the work is boring, or repetitive, or makes few demands on thought or alertness. Music is less helpful in big, noisy workshops, or in jobs where mental alertness is essential.

When the music chosen is not quiet, background music, but assertive music that needs to be listened to, it should be played only for part of the working day. A short period of rousing music should begin the day, which should end with more festive tunes, while the rest of the day should have four periods of 30 minutes each of light music.

The tempo should be neither too slow and soporific, nor too fast, causing irritation and haste.

Vibrations

Vibrations are mechanical oscillations, produced by either regular or irregular periodic movements of a body about its resting position. They are termed mechanical oscillations because in the last analysis it is the small changes of position that are important.

The nature of vibrations, and their effects on people have been written about in detail by *Dupuis* [73] [74], *Coermann* [55], *Wisner* [331], *Grether* [118], as well as in the *VDI Guidelines* [315], in the *ISO Publication* [157], and in the *Guidelines of the British Standards Institute* [35].

A little physics

The following five physical quantities are important for the understanding of what follows.
1. *Point of application to the body.*
2. *Frequency of oscillations.*
3. *Acceleration of oscillations.*
4. *Duration of effect.*
5. *Individual frequency and resonance.*

1. Two points at which vibrations enter the body are significant ergonomically: feet/buttocks (when driving or riding in a vehicle) and hands (when operating vibrating tools or a machine). The direction of oscillation is important. Most often this lies in the vertical plane (head to foot) or approximately along the line of hand and arm.

2. The extent of the physiological and pathological effects of vibrations is strongly frequency-dependent. Particularly important frequencies are those which fall into the range of natural frequencies of the body, and so cause resonance (see item no. 5). Often a low and high range of frequencies are distinguished. The threshold lies between 30 and 50 Hz. Vibrations of motor-vehicles belong to the low, those of motor-driven tools to the high ranges of frequencies.

3. Within the frequency range that is physiologically important, *the acceleration of the oscillations is usually taken as a measure of the vibrational load.*[1] *The unit is the acceleration due to gravity (g) = 9.8 m per sec².*

4. The effect of vibrations depends greatly on their duration. Their ill-effects increase very rapidly as time goes on.

5. Every mechanical system which possesses the elementary properties of mass and elasticity is capable of being set in oscillation. The force which sets the system in motion is known

---

[1] In Germany the strength of perception of vibrations is calculated as the K value, derived from the frequency and various other components of the oscillation (see *VDI Guideline 2057* [315]).

as the *exciting force*, and the resultant oscillations are called the *forced vibrations*.

Each system has its own natural frequency, and the nearer the frequency of the exciting force comes to this, the greater will be the amplitude of the forced vibrations. When the amplitude of the forced vibrations exceeds that of the exciting force the system is said to be *in resonance*.

On the other hand the oscillations of any system are subject to *damping*, which reduces their amplitude. Thus, for example, when we are standing up any vertical vibrations set up in the legs are quickly damped.

Frequencies above 30 Hz are particularly heavily damped by the tissues of the body; thus for an exciting frequency of 35 Hz the amplitude of oscillation is reduced to ½ in the hands, to ⅓ in the elbows and to 1/10 in the shoulders.

Oscillatory characteristics of the human body

The human body does not vibrate as a simple mass, with its own natural frequency. Studies by *Dupuis* [74], *Coermann* [55], and others have shown that the natural frequencies are different in different parts. The body of a sitting person reacts to vertical vibrations as follows:

3–4 Hz: strong resonance in the cervical vertebrae;

4 Hz: peak of resonance in the lumbar vertebrae;

5 Hz: very strong resonance in the shoulder girdle (up to 2-fold increase);

20–30 Hz: resonance between head and shoulders;

60–90 Hz: resonance in the eyeballs;

100–200 Hz: resonance in the lower jaw.

In general, the most effective exciting frequency for vertical vibrations lies between 4 and 8 Hz. In more detail:

(a) *vibrations between 2.5 and 5 Hz generate strong resonance in the vertebrae of the neck and the lumbar region* (oscillations amplified up to 240%);

(b) *between 4 and 6 Hz resonances are set up in the trunk, shoulders and neck* (amplification up to 200%);

(c) *between 20 and 30 Hz sets up the strongest resonances between head and shoulders* (amplification up to 350%).

There has been little study of other points of application, and oscillations in other directions. It can only be said that the natural frequencies of the smaller components of the body, such as groups of muscles, eyes, and so on, lie in higher ranges of frequency. Hence the operation of machines with frequencies above 30 Hz are likely to set up resonances in the fingers, hands and arms. On the other hand the damping effect of bodily tissues is greater for these higher frequencies, and tends to confine them to the vicinity of the point of application.

Vibrations at the work place

Up to the present, vibration experienced at work has been measured mainly in construction machinery, in tractors, and in heavy vehicles (trucks). The results of *Dupuis* [74], shown in *Figure 183*, may be taken as an example.

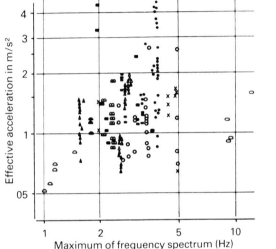

◻	Motor car
◼	Ambulance
•	Mini-car
▲	Omnibus
▲	Heavy lorry (truck)
■	Earth-moving machine
x	Self-propelled working machine
○	Agricultural tractor
•	Tractor drawing trailer
◙	Tractor with spiked wheels
○	Caterpillar tractor

Fig. 183 Vertical accelerations during vibrations between the driver and his seat in various motorised vehicles. Ordinate: effective value of acceleration = root mean square of the amplitude. After *Dupuis* [73] [74].

It is clear from this diagram that *the acceleration of vertical oscillations in various vehicles lies between 0.5–5 m per sec²*, with the highest values being recorded for earth-moving machines and tractors. Operating motor-powered tools involves high levels of vibration in the hands and wrists, some examples of which are summarised in *Table 51*.

Table 51 Vibrations in motorised hand-tools. Effective acceleration = root mean square of the accelerations at various amplitudes. These figures apply to oscillations along the direction of the arm; oscillations perpendicular to this are often higher. After *Dupuis* [73].

| Type of tool | Effective acceleration (m/s²) | | Ground |
	On fingers	On wrist	frequency in Hz
Power saw	17.5	1.1	120
Soil borer	21.0	3.5	110
Pneumatic compass saw	–	9.9	–
Two-wheeled cultivator	3	2.8	82

Physiological effects	Vibration affects the musculature, circulation and respiratory system to a lesser extent, and visual perception and psychomotor performance more seriously.
Muscular reflexes	Vibration seems to generate muscular reflexes which have a protective function, causing the extended muscle to shorten. According to *Hettinger* [143], after a long period of work with a pneumatic hammer, increasing fatigue causes these reflexes to diminish, or to disappear. The reflex activity of the muscles also explains the often-observed increase of energy consumption, heart rate and respiratory rate when a person is exposed to strong vibrations. These vibrational effects on metabolism, circulation and respiration are small and have little significance.
Visual powers	The adverse effect of vibration on eyesight is most important because it impairs the efficiency of drivers of tractors, trucks, constructional machines and other vehicles, and increases the risk of accidents. *Visual acuity is poorer* [125], *and the image in the visual field becomes blurred and unsteady* [121]. Visual powers are not affected by vibrations of less than 2 Hz. Measurable optical aberrations appear from 4 Hz upwards, and are greatest in the range 10–30 Hz. With a vibration of 50 Hz and an oscillatory acceleration of 2 m per sec², visual acuity is reduced by half, according to *Guignard* [125].
Skill	*Strong vibration impairs performance in various psychomotor tests*, of which that of *Kaminsky* [170] may be quoted as an example: a marked decline in manual skill was noted after using a power saw. Simplifying somewhat, we may say that strong vibration impairs visual perception, mental processing of information, and the carrying out of skilled movements [118] [125].
Driving tests	These psychophysiological effects of vibration are particularly evident in all kinds of simulated driving tests: (a) over the range 2–16 Hz (especially around 4 Hz) driving efficiency is impaired, and the effects increase with increasing acceleration of the oscillations; (b) driving errors increase when the seat is subjected to accelerations of the order of 0.5 m per sec²; (c) when the accelerations reach 2.5 m per sec², the number of errors becomes so great that such vibrations must be rated as positively dangerous. This consensus of physiological effects of vibration point to the conclusion that *strong mechanical oscillations reduce efficiency, and in many situations may lead to the risk of errors and accidents.*

Vibration as a nuisance

Vibration is subjectively felt as an imposition and a burden, impressions ranging from a minor annoyance to an unbearable nuisance. The extent of the nuisance depends in the first instance on the exciting frequency, on the rate of acceleration of the oscillations, and on the length of time they continue. The source of the nuisance lies in the physiological effects, and in the resonances set up in various parts of the body. *Figure 184* shows the results of an investigation by *Chaney* [48] on seated test subjects; curves are given for subjective feelings of equal intensity, in relation to frequency and acceleration of oscillation.

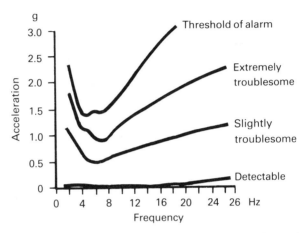

Fig. 184 Curves of equal subjective impressions of seated test subjects in relation to the exciting frequency and the acceleration of the vibrations. After *Chaney* [48].

From these results the following conclusions may be drawn for vertical oscillations applied to seated persons:

(a) *the most intense subjective sensitivity lies in the frequency range 4–8 Hz;*

(b) *the average threshold of "very severe" intensity comes at an acceleration of 1 g (i.e. about 10 m per sec²);*

(c) *at accelerations of 1.5 g (i.e. about 15 m per sec²) the vibrations became dangerous and intolerable.*

The same author carried out similar tests on standing persons. The curves of equal subjective intensity were higher than before, because of the damping effects in the legs, to which we referred earlier. The threshold of "very severe" intensity came 0.2–0.3 g higher than for seated persons.

Complaints

The complaints, which are suffered in addition to the annoyance of vibration, vary greatly, but some of them are frequency-dependent. According to *Magid* and his colleagues [210], the following are the most common complaints:

(a) interference with breathing, especially severe at vibrations of 1–4 Hz;

(b) pains in chest and abdomen, muscular reactions, rattling of the jaws, and severe discomfort, chiefly from 4–10 Hz;

(c) backache, particularly from 8–12 Hz;

(d) muscular tension, headaches, eyestrain, pains in the throat, disturbance of speech, irritation in intestines and in the bladder, at frequencies 10–20 Hz.

In addition we may mention sea and travel sickness, with nausea and vomiting, brought about by oscillations of 0.2–0.7 Hz, with the greatest effect at 0.3 Hz.

Damage to health   Exposure to vibration at one's place of work can, if repeated daily, lead to morbid changes in the organs affected. The effects are different in the two parts of the body most commonly subjected to vibration. Vertical oscillations when either standing up or sitting down, caused by vibration from underneath (e.g. in a vehicle) can cause degenerative changes in the spine, whereas the vibrations of power tools affect mainly the hands and arms.

Spinal ailments   Tractor drivers in various countries have been recorded as suffering an accumulation of disc troubles and arthritic complaints in the

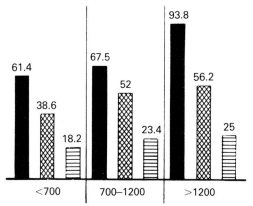

Hours of tractor driving per year

■ Unhealthy signs in X-ray of spine

▨ Aches in spine

▥ Stomach aches

Fig. 185 Unhealthy signs in X-ray pictures of vertebral column, and aches and pains, among 100 tractor drivers, in relation to their annual total of tractor-driving hours. The vertical columns express the percentage of the 100 test subjects who show symptoms at that number of driving hours. After *Dupuis* and *Christ* [75].

spine, as well as an above-average incidence of intestinal ailments, prostate troubles and haemorrhoids [74] [125]. In three successive studies of the same group of tractor drivers, at intervals of five years, *Dupuis* and *Christ* [75] found that X-ray pictures of the spine showed increasing spinal damage. They were able to relate the extent of the damage to the total time spent driving the tractor annually. Their results are shown in *Figure 185.*

The cumulative appearance of spinal damage among workers who were subjected to a high level of vertical oscillation lead one to suppose that heavy and prolonged vibration causes excessive wear and tear on the intervertebral discs and the joints. This is still hypothetical, however, since the causal link has not yet been firmly established.

Hand and
arm troubles

Workers who use motorised tools for years on end (e.g. power saws or pneumatic hammers) may suffer various ailments of the hands and arms, and, according to *Wisner* [331], the frequency of vibration is a decisive factor.

Arthritis

Tools with a frequency of vibration below 40 Hz, e.g. a heavy pneumatic hammer, can cause degenerative symptoms in the bones, joints, and tendons of hands and arms, leading to arthritis in the wrist, elbow and occasionally in the shoulder.

Atrophy

The effects on the bones may lead to atrophy, which in rare cases may involve so much loss of calcium that the risk of fracture is substantially increased. In a few countries these possible consequences of using a pneumatic hammer are classified as an industrial disease.

"Dead fingers"

Power tools with frequencies between 40 and 300 Hz usually have a very small amplitude of oscillation (from 0.2–5 mm), and their vibrations are quickly damped in the tissues. *Such vibrations may have ill-effects on the blood vessels and nerves of the hands,* resulting in one or more of the fingers going "dead". Usually the middle finger is most affected, becoming white or bluish, cold and numb. After a little while the finger turns pink again and is painful. The cause is a cramp-like condition of the blood vessels, known as Raynaud's Disease.

These symptoms were also observed among miners using a pneumatic drill with higher frequency of vibration, as well as among forestry workers using power saws with frequencies between 50 and 200 Hz.

"Dead fingers" make their appearance, at the earliest, six months after the beginning of the work using the vibrating tool, and cold is an important factor in the onset of this condition. Raynaud's

Disease is commoner in northern countries than in warmer latitudes. It must be assumed that cold makes the blood vessels more sensitive to vibration, and more liable to vascular cramp.

Users of power tools with even higher frequencies have also been known to suffer from ailments affecting the circulation, and loss of sensation. One example of such tools is polishing machines operating at 300 to 1000 Hz, which caused painful swellings and loss of sensation in the hands, which often did not disappear when the work was finished.

Limiting values

No limiting values have yet been evaluated for hand-operated power tools, which would enable a clear line to be drawn between safe and unsafe limits of exposure. On the other hand attention has been given in many countries to the question of fixing limits for what may or may not be expected in the way of vertical oscillations [73] [74] [125] [347].

From these studies it may be concluded that vibration becomes intolerable in the following circumstances:

below 2 Hz:	at accelerations of 3–4 g
between 4 and 14 Hz:	at accelerations of 1.2–3.2 g
above 14 Hz:	at accelerations of 5–9 g

The *International Standard Organisatio (ISO)* has been concerned with this problem for several years; its guidelines for the assessment of vertical oscillations are shown in simplified form in *Figure 186*.

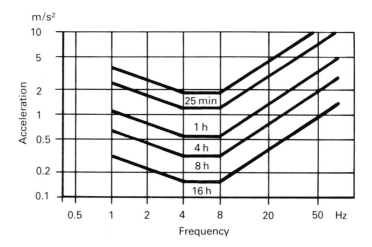

Fig. 186 Threshold values for vertical vibrations before efficiency is affected. ("Fatigue-decreased proficiency boundary"), for exposure times ranging from 25 minutes to 16 hours. For the calculation of threshold values "safety" and "comfort", see text. After *ISO* [157].

This *ISO Norm* distinguishes between three criteria, and envisages three levels of limiting values:

(a) *the criterion of comfort* (reduced comfort boundary). This applies mostly to vehicles, and to the automobile industry;

(b) *the criterion of the maintenance of efficiency* (fatigue-decreased proficiency boundary). A decisive factor in this criterion is working efficiency (proficiency). It applies to tractors, construction machinery and heavy vehicles. These limiting values are shown in *Figure 186*;

(c) *the criterion of safety* (exposure limits). Protection against damage to health is the criterion here.

Starting from the "efficiency" of the second criterion (*Figure 186*), the other two can be derived as follows:

the first, the criterion of comfort, can be derived by dividing the acceleration by 3.15;

the third, the criterion of safety, can be derived by multiplying the acceleration by 2.

Recommendations	From the standpoint of ergonomics, tractors, heavy vehicles and construction machinery, with their frequencies most often between 2 and 5 Hz and operating for an 8-hr day, require a limit of oscillational acceleration of *0.3–0.45 m per sec²*. Comparison with *Figure 183* will show that these limits are often exceeded, and the vibrations encountered are permissible only for a few hours at a time.
Driving seats	It is clear from this that *the ergonomic construction of driving seats requires them to be so damped that vertical oscillations are reduced within these safety limits.* It is already possible to do this by special types of suspension. By 1956, *Simon* and others [290] had already developed driving seats in which the impact of vibrations in the frequency range 3–5 Hz had been reduced by more than 50 %. *Dupuis* [73] obtained similar results with special seats which reduced the effects of vibration by 50–60 %.
Damping of hand-operated power tools	The same applies in principal to hand-operated power tools. Damping elements within the tool itself, and even more between the tool and the hand-grip, can reduce vibration considerably. The grip itself can be damped by making it of flexible material. Further improvement can be obtained by wearing thick gloves and avoiding working in very cold conditions.

17
Indoor climate

The indoor climate, as understood in this chapter, means the physical conditions under which work is carried on. The principal components are:
(a) the temperature of the air;
(b) the temperature of the surrounding surfaces;
(c) the humidity of the air;
(d) air movements.

Two important questions are *the degree of comfort*, and *the oppressiveness of working under very hot conditions*. These topics have been discussed in detail by *Newburgh* [247], *Winslow* [329], *Fanger* [81], *Wenzel* [326], *Leithead* and *Lind* [196], *Metz* [223], *Frank* [84], *Dukes-Dobos* [72], and finally in *ASHRAE handbook of fundamentals* [5], as well as in the *Report of the Symposium "Occupational exposures to hot environments"* [148].

Thermal regulation in man

Body temperature

The temperature of the human body is not, as often assumed, uniform throughout. A constant temperature, which fluctuates a little around 37°C, is found only in the interior of the brain, in the heart, and in the abdominal organs (core temperature). A constant core temperature is a prerequisite for the normal functioning of the most important vital functions, and wide or prolonged variations are incompatible with the life of a warm-blooded animal. In contrast to the core temperature, that in the muscles, limbs, and above all in the skin (the shell temperature) shows certain variations. Physiological studies have shown that when the surrounding air is cool there is a steep temperature gradient in the skin, from inside outwards. For example, in cool air, the temperature even 2 cm beneath the surface of the skin may have fallen to

35°C, whereas in warmer surroundings it may still be 35–36°C only a few millimetres below the surface. This capacity for adaptation allows the human body to tolerate a temporary heat deficit which may amount to several hundred kilocalories for the whole body. The muscles too, show considerable fluctuations in temperature, being several degrees warmer during strenuous effort than when they are at rest.

Control processes   The control mechanisms throughout the body, which are necessary to maintain a constant core temperature, are shown diagrammatically in *Figure 187*. At the centre of the control system is the heat control centre located in the interbrain. This is the overall director of the thermal regulation of the body, and so is comparable to a thermoregulator.

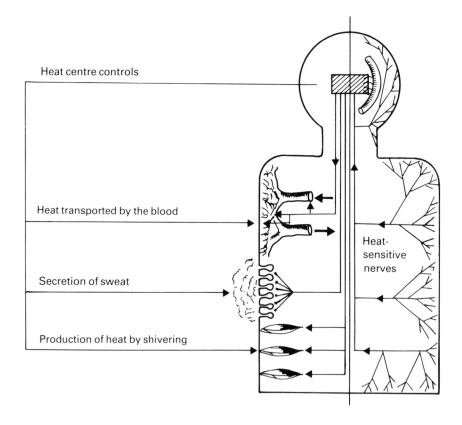

Heat centre controls

Heat transported by the blood

Secretion of sweat

Production of heat by shivering

Heat-sensitive nerves

Fig. 187   Diagram of the physiological control of the heat balance of the body. The heat centre located in the brain stem regulates the flow of blood through the capillaries of the skin, as well as the secretion of sweat. These two mechanisms, between them, adjust the heat balance of the body according to external and internal conditions.

The nerve cells of the heat control centre receive information about temperatures throughout the body, sometimes directly and sometimes from heat-sensitive nerves in the skin. The heat control centre in turn sends out the impulses that are necessary to direct and control the regulatory mechanism which keeps the core temperature constant. In this way are controlled the heat production of the body, its diffusion by way of the circulatory system, and the heat loss by secretion of sweat in the skin, thus enabling the process of thermal regulation to be carried on.

Heat transport by the blood

The most important item in thermal regulation is the heat transport function of the blood, whereby the blood vessels, especially the capillaries, act as distributors of heat, picking up heat from warm tissues and giving it out again to cooler tissues. In this way the blood can transport heat from the interior of the body to areas of skin that are cooled by the external temperature; conversely, if the exterior is artificially heated, heat can be transported away into the interior. *The key to this mechanism is the control of blood circulation in the skin.*

Secretion of sweat

The second regulatory mechanism directed by the heat control centre is the secretion of sweat in the skin. This, too, is under nervous control.

Shivering

The third regulatory mechanism is by raising the rate at which heat is produced by the body, a process that is set in motion whenever the body is subjected to cooling. This arises by an increase in metabolic heat in the muscles and other organs, a special manifestation of which in the rapid muscular movements known as "shivering".

Heat exchange

As explained earlier in this book, the body converts chemical energy into mechanical energy and heat. The body uses this heat to maintain a constant core temperature and dissipates any excess heat into its surroundings.

There is thus a constant exchange of heat between the body and its surroundings which is regulated in part by physiological control, and partly by the ordinary laws of physics. The latter involve four different processes:

(a) conduction;
(b) convection;
(c) evaporation;
(d) radiation.

Conduction of heat

Heat exchange by conduction depends first and foremost on the conductivity of objects and materials in contact with the skin. Anyone who sits down in winter, first on a stone and then on a tree-trunk can discover this for himself. The stone feels very cold because it conducts heat away from the body; the tree-trunk feels much less cold because its conductivity is less. Conductivity of heat is of practical importance in the choice of floorings, furniture and the parts of machinery that have to be touched (control handles, etc.). Loss of heat by contact with such objects is to be avoided, either from the feet or from other parts of the body; it is decidedly unpleasant, and is liable to cause ailments such as rheumatism and arthritis. Hence *work places should have well-insulated floorings (e.g. cork, linoleum, or wood), while bench tops, machine parts (particularly handles and control), and tools should be protected with felt, leather, wood or other material of low thermal conductivity.*

Convection

Heat exchange by convection depends primarily on the difference in temperature between the skin and the surrounding air, and on the extent of air movement. Under normal circumstances convection accounts for about 25–30 % of the total heat exchange of the body.

Evaporation of sweat

Loss of heat by sweating occurs because the sweat on the skin evaporates, consuming heat. This latent heat of vaporisation amounts to 0.58 kcals per gram of water evaporated. Under normal conditions each person will evaporate about one litre per day (insensible perspiration), and lose 600 kcal or more of heat in the process: about one quarter of the total daily loss of heat.

If, however, the temperature of the surroundings exceeds the limits of comfort, then the hot skin begins reflex sweating, with a sharp increase in the rate of loss of heat.

The extent of heat loss by the evaporation of water depends on the area of skin from which sweat can evaporate, and upon the difference in water vapour pressure between the air next the skin and that further away. Thus *the relative humidity of the surrounding air is an important factor.* A lesser factor is that of air movement, which on the one hand increases the gradient of water vapour pressure, but on the other hand cools the skin by convection, and thereby reduces the amount of sweating.

At environmental temperatures (of walls and of the air) above 25°C the clothed human body is able to lose hardly any heat by either convection or radiation, and sweating is the only compensatory mechanism left. Hence the loss of heat by evaporation of sweat rises steeply after a particular critical temperature has been reached.

Radiation of heat

Warm bodies radiate electromagnetic waves of relatively long wavelength, which are absorbed by other bodies (objects and surfaces) and converted into heat. This is called infra-red radiation, or radiant heat. It does not depend upon any material medium for its transmission in contrast to conducted or converted heat. Such heat exchange by radiation takes place between the human body and its surroundings (walls, inanimate objects, other persons), in both directions, and all the time. In contrast to conduction or convection, heat radiation is hardly affected by the temperature, humidity or movement of the air; it depends mainly on the temperature difference between the skin and adjacent surfaces. In temperate countries the surrounding objects and surfaces are usually cooler than the human skin, and so the human body loses a considerable amount of radiant heat in the course of a day.

Loss of heat by radiation is not noticeable as long as the amount is not excessive, but it may become uncomfortable when standing close to a cold wall or a large window, even though the air temperature is high enough. In such circumstances the loss of heat may be considerable, because the decisive factor is not the air temperature, but the temperature difference between the skin and the adjacent cold surface.

The amount of radiant heat lost in a day by a fully clothed person varies very greatly, according to circumstances. Excluding high summer, the average daily loss in temperate climates is about 1000–1500 kcal, accounting for some 40–60 % of the total heat lost from the body.

Overall heat balance

Figure 188 shows three of the principal ways in which the human body exchanges heat with its surroundings.

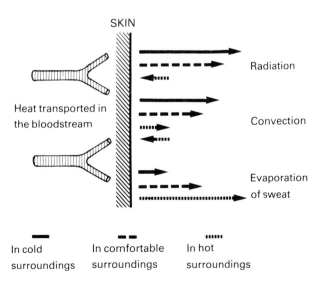

Fig. 188 Diagram of heat exchange between the human body and its surroundings. The length of the arrows gives a rough indication of the heat transferred by each of the three processes, under different conditions.

To summarise, it can be said that the following four physical factors are decisive:

(a) air temperature (for exchange of heat by convection);
(b) air movement (also for convection);
(c) temperatures of adjacent surfaces: walls, ceiling, floor, machinery, etc. (for exchange of heat by radiation);
(d) relative humidity of the air (for loss of heat by evaporation of sweat).

Comfort

Physiological basis of comfort

One hardly notices the internal climate of a room, as long as it is comfortable, but the more it deviates from a comfortable standard, the more it attracts attention.

The sensation of discomfort can increase from mere annoyance to one of pain, according to the extent to which the heat balance is disturbed. This discomfort is a practical biological device in all warm-blooded animals, which stimulates them to take the necessary steps to restore a correct heat balance. An animal can only react by seeking out another place that is neither too hot, nor too cold, but man has the use of clothing, as well as being able to modify his environment by technological means.

Side-effects of discomfort

Discomfort brings about functional changes that may affect the entire body. Overheating leads to weariness and sleepiness, loss of performance and increased liability to errors. This damping down of activity causes the body to produce less heat internally.

Conversely, overcooling induces restlessness, which in turn reduces alertness and concentration, particularly on mental tasks. In this case stimulation of the body into greater activity causes it to produce more internal heat.

Thus the maintenance of a comfortable climate indoors is essential for well-being and maximum efficiency.

Temperature zones in physiological terms

If a test subject is placed in a climatic chamber and exposed to different temperatures, a range can be found in which the heat exchanges of his body are in a state of balance. This is called the *zone of vasomotor regulation*, because within this range the heat balance is maintained chiefly by regulating the flow of blood to different parts of the body. This is shown in *Figure 189*, where it is termed the *comfort zone*. For a clothed and resting person in winter this zone lies between 20–23°C.

If there is a slight excess of heat above this comfort level, this is dealt with by warming the peripheral parts of the body and by increasing perspiration. This is the *zone of evaporative control*. If,

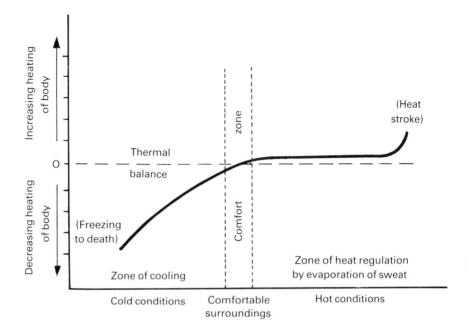

Fig. 189 Heat balance of the body between exposure to extremes of heat and cold.

however, the heat continues to increase and exceeds a certain level (*the limit of tolerance*), the core temperature rises quickly and steeply, and in quite a short time leads to death by heat stroke. Temperatures below the zone of vasomotor regulation are characterised by a negative heat balance for the body, since more heat is being lost than is being generated internally. This is the *zone of bodily cooling*. At first the cooling is confined to the peripheral parts of the body, which can tolerate a heat deficit for a time. The heat balance in these three zones is shown diagrammatically in *Figure 189*.

Comfortable ranges of temperature

If test subjects are asked to say when they feel really comfortable, the range is a comparatively narrow one, perhaps only 2 or 3°C. Obviously people feel comfortable only when the vasomotor regulation system is not heavily stressed, that is when the blood circulation to the skin is subject to no more than normal fluctuations. On the other hand either a negative or a positive heat balance (i.e. either a deficit or an accumulation of heat in the body shell) is felt as discomfort.

The temperature range within which a person feels comfortable is very variable. It depends firstly upon the amount of clothing being worn, and then upon how much physical effort is being performed. Whether other factors such as food, time of year, time of day, age and sex are equally important is arguable.

Four climatic factors and comfort

People's impressions of comfort are influenced by the same four climatic factors that determine the heat exchange, so we may repeat them here:

(a) *the temperature of the air;*
(b) *the temperatures of adjacent surfaces;*
(c) *the humidity of the air;*
(d) *air movements.*

Thus each factor is involved in its own "balance", and several research workers have tried to find a unit of measurement that would take account of them all. An example is the so-called "Kata-value" which attempts to use the rate of cooling of an artificial body as an index of comfort, but this method has not been put into practice. Research workers have come to rely more and more on the subjective impressions of test persons as a measure of the degree of comfort, resulting in the concept of *effective temperatures.*

Effective temperatures

Houghton and *Yaglou* [149] were the first to investigate the relation between temperature and humidity of the air that would result in the same effective temperature. The test subjects were placed in a climatic chamber in still air, at a given temperature, and with 100 % relative humidity. They were required to note their impressions of the temperature, and to remember them. Then the relative humidity was reduced, and the temperature varied until the test subjects felt the same sensation of warmth as before.

This is the *effective (or perceived) temperature.* These investigations were followed by others, in which the effects of temperatures of adjacent surfaces and of air movements on the effective temperature were analysed. These led to the formulation of "indices of comfort", and "zones of comfort", which could be read off from nomograms in relation to three or four of the climatic factors listed above. We may refer to the experiments of the Danish worker *Fanger* [81] who compiled an equation of comfort out of the ratio between the four climatic factors, the amount of clothing, and the extent of physical activity. This equation is very complex, however, and results can be worked out only by using electronic data-processing equipment. *Fanger* [81] set out the most important results in 28 diagrams, showing curves of degree of comfort in relation to several factors. An account of these would be beyond the scope of this chapter, so we must confine ourselves to considering a few relationships between the climatic factors already mentioned which are important in the evaluation of the climate indoors. The leading question is what is the effect on effective temperatures of various combinations of:

(a) air temperature and temperature of surrounding surfaces;
(b) air temperature and relative humidity;
(c) air temperature and air movement.

Air temperature and that of adjacent surfaces

Physiological research has shown that the effective (perceived) temperature is essentially a mean between that of the air and of adjoining surfaces. Expressed as a formula:

$$\text{Effective temperature} = \frac{T_A + T_S}{2}$$

where T_A = mean air temperature, and T_S = mean temperature of adjacent surfaces.

It is imporant for comfort that the difference between T_A and T_S should be small. Large areas of cold walls or windows are particularly uncomfortable, even if the air temperature is adequate. *It is a good rule of thumb that the mean temperature of adjacent areas should not differ from that of the air by more than 2 or 3°C, either up or down.*

Temperature and humidity of the air

The effect of atmospheric humidity was prominent in the older research [149], but more recent work [180] [246] has shown that after a prolonged stay in the same room, the impression of temperature was little affected by the humidity of the air. The following combinations of atmospheric humidity (%) and air temperature will produce equal effective temperatures:

70 % RH and 20°C
50 % RH and 20.5°C
30 % RH and 21°C.

So it is clear that within the range 30–70 %, relative humidity has little influence on effective temperature. According to *Frank* [84], it can be assumed that between 18–24°C the relative humidity can fluctuate between 30–70 % without creating thermal discomfort. The threshold at which the room begins to feel stuffy lies between the following pairs of values:

80 % RH at 18°C
60 % RH and 24°C.

If the relative humidity falls below 30 % there is a danger of too dry an atmosphere, about which we shall speak briefly later on.

Temperature and movement of the air

The third factor to influence effective temperature is air movement. *Yaglou* [340] carried out experiments to determine the ratios of air temperature and free movement of the air, which combined to give the same effective temperature of 20°C. His results are set out in *Table 52*.

Table 52 Air movement and effective temperature. After *Yaglou* [340].
The two columns show the dry-bulb and wet-bulb temperatures of a
psychrometer (wet-and dry-bulb hygrometer).

Air movement, metres per sec	Dry bulb °C	Wet bulb °C:100 % RH
0	20	20.3
0.5	21	21.3
1.0	22	22.2
1.5	22.8	23
2.0	23.5	23.8

More recent research has shown [81] [264] that air movements in
excess of 0.5 metres per second are unpleasant even when the air
is warm, and that the discomfort depends on the direction in
which the air is flowing, and on the parts of the body exposed to it.
(a) *air currents from behind are more unpleasant than those
     coming from in front;*
(b) *neck and feet are particularly sensitive to draughts;*
(c) *a cool draught is more unpleasant than a warm current.*

Both from the literature [84] and from general experience it can be
said that seated persons find draughts of more than 0.2 metres per
second to be unpleasant. Occasionally, when very precise work is
being carried on, which keeps the operator motionless for long
periods, even draughts of 0.1 metres per second can be unpleasant. In contrast, standing work, especially if it involves
strenuous physical effort, can be carried on without ill effects in
draughts of up to 0.5 metres per second.

Dryness of the air during heating periods

Dryness of the air   As we have seen, dryness of the air is a winter problem both at
work and at home, hygienically and medically. For many years
now, the tendency has been to prefer higher and higher temperature indoors during the period of the year when buildings are
heated, and this has led to lower and lower values of the relative
humidity.
In one of our investigations we carried out random tests in a
number of offices, both in winter and in summer, to measure the
relative humidity of the air. The results are shown in *Figure 190*.

Medical effects   Most ear, nose and throat specialists think that the present-day
tendency to excessive dryness of the air in heated rooms causes
an increased incidence of catarrhal ailments, and chronic irritation
of the nasal and bronchial passages. They often note a desiccation

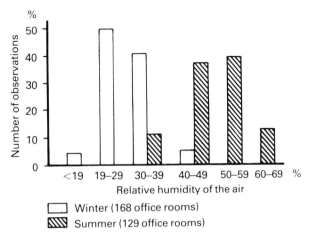

Fig. 190 **Relative humidity of the air in summer and in winter.** Winter samples taken in 168 offices without air conditioning. Summer samples taken in 60 offices with and 60 without air conditioning.

of the mucous membranes of the air ducts, which they think obstructs the flow of mucus over the ciliary tracts, resulting in a diminished resistance to infection. These views of the effects of dry air are shown diagrammatically in *Figure 191*.

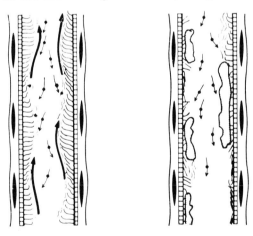

Normal air-passage
Dust particles adhere to the
mucus and are carried away
with it by the ciliated epithelium

Desiccated air-passage
Mucus has coagulated into
clumps to which dust particles
do not adhere

Fig. 191 **Diagram showing the effects of dry air on the self-cleaning capacity of the mucous membranes in the throat.**
Left: normal mucous membrane with a ciliated epithelium which rids the membrane of any contaminating particles.
Right: the ciliated epithelium is desiccated, and parts of it are no longer visible: the mucus forms into clumps, and does not get rid of the dust particles.

These medical interpretations have been confirmed by several other studies [117] [261], but two comparative studies in workrooms [123] [286] have shown no effect of atmospheric dryness on absenteeism.

These observations may be summarised by saying that relative humidities of 40–50% in heated rooms are desirable for comfort and below 30% become unhygienic, because they adversely affect the mucous membranes of nose and throat.

It may be mentioned here that most of the humidifiers sold commercially are inadequate for their purpose. As a rough guide, a workroom of 100 m³ volume requires a minimum output of 1 litre of water per hour.

Field studies

We have already seen that there is considerable individual variation in what people consider to be a comfortable temperature, and field studies emphasise this. *Figure 192* shows the results of an old piece of research by *McConnell* and *Spiegelman* [219], which recorded the views of 745 office workers about temperatures in New York during the summer of 1940.

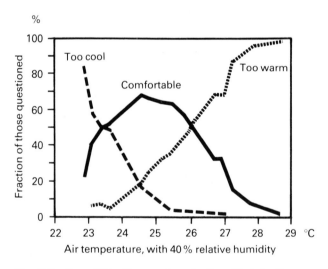

Fig. 192 How 745 office workers in New York in the summer of 1940 reacted to the air temperature. After *McConnell* and *Spiegelman* [219].

The most striking result was the wide range of individual responses. At best (24°C) only 65% of those questioned found this temperature comfortable, all the rest finding it either too hot or too cold. In one of our own studies [102], carried out in the winter

1964/65, we took measurements in 168 offices without air conditioning and at the same time questioned 410 employees (140 men and 270 women) about their impressions of temperature. The results are shown graphically in *Figure 193*.

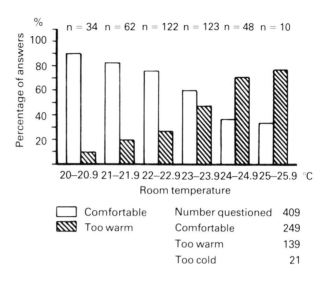

Fig. 193   **Reactions of 409 office workers to temperature in the winter 1964/5.** n= number of people questioned and also of temperature records. The question asked: "Do you find the indoor climate in the office today comfortable, too warm or too cold?" After *Grandjean* [102].

Air temperatures were high, varying between 22°C and 24°C most of the time. The reply "the room is pleasant" became less frequent as the temperature rose, while "too warm" took its place. *There were many replies of "pleasant" when the temperature was only 21°C.*
Similar studies were caried out in the summers of 1966 and 1967. This survey comprised 311 offices, 122 of them air-conditioned, and 189 not. A total of 1191 employees were questioned, and the results are summarised in *Figure 194*.

In the offices without air conditioning, the air temperatures mostly varied between 20°C and 27°C; in the air-conditioned offices the upper limit of temperature was usually 24°C. As appears from the diagram, temperatures above 24°C were assessed as "too warm" by most people, and a similar observation was made in the American Survey of *McConnell* and *Spiegelman* [219]. So we may conclude that *in summer room temperatures are comfortable as long as they do not exceed 24°C.*

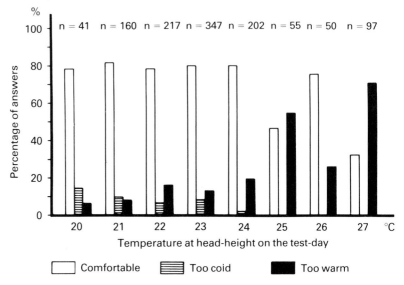

Fig. 194 **Reaction to temperatures in the summer.** n= number of people questioned, and also of temperature readings. The question asked: "Do you find the indoor climate in the office today comfortable, too warm or too cold?" After *Grandjean* [101].

Recommendations for comfort indoors

Sedentary office work

As far as present knowledge goes, the following guidelines may be applied to sedentary work which does not involve manual effort:

(a) *the air temperature* in winter should be 21°C,[1] and in summer temperatures between 20°C and 24°C are comfortable;

(b) *surface temperatures of adjacent objects* should be at roughly the same temperature as the air, or at least not more than 2–3°C different. No single surface (e.g. the outside wall of the room) should be more than 4°C colder than the air in the room;

(c) *the relative humidity of the air in the room* should not fall below 30% in winter, otherwise there will be a danger of desiccation problems in the air passages. In summer the natural relative humidity usually fluctuates between 40% and 60% and is considered to be comfortable;

(d) *draughts* at the levels of head and knees should not exceed 0.2 metres per second.

---

[1] In view of the current need to conserve energy, winter temperatures of 20 C may be considered: the office workers should wear suitable clothing.

With bodily activities the internal production of heat rises steeply. The relevant considerations are set out in *Figure 195*.

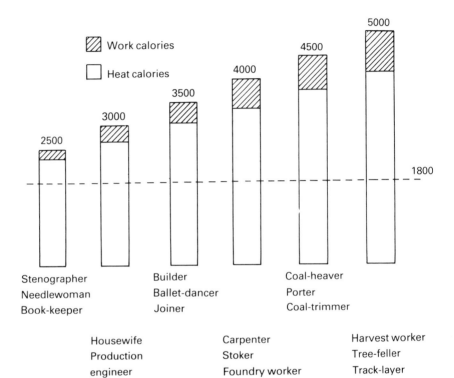

Fig. 195 Heat production during various occupations. The height of the columns, and the adjacent numbers, indicate the overall calorie consumption per 24 hours; the black portion of each column indicates the work calories, and the white portion the heat production. After *Lehmann* [194].

The more active people are, the more heat they generate. If they are to remain comfortable, the air temperature within the room must be lowered, so that it is easier to get rid of the surplus heat. At a relative humidity of 50%, various kinds of activity require the room temperatures listed in *Table 53*.

Table 53 Recommended room temperatures for various activities.

Type of work	Room temperature, °C
Sedentary mental	21
Sedentary light manual	19
Standing light manual	18
Standing heavy manual	17
Severe work	15–16

The temperatures quoted are valid both for winter and summer, but when the temperature outdoors is very high it is often difficult to achieve the lower temperatures appropriate for some kinds of work.

Air pollution and ventilation

Deterioration
of air

If a room has people in it, the air undergoes deterioration in various ways, changing its character:

(a) *release of odours;*
(b) *formation of water vapour;*
(c) *release of heat;*
(d) *production of carbon dioxide;*
(e) *air pollution,* either entering from outside, or generated by activities within the room.

The first four of these arise mainly from the human body itself. The last, air pollution, depends on the situation of the building and on what activities are carried on indoors.

Among the air changes that are of human origin, the odours given off by the skin are most important, since in very low concentrations they give rise to feelings of unpleasantness, disgust, distaste and revulsion. These odours are a mixture of organic gases and vapours, which are not toxic in the concentrations usually encountered, but which are most undesirable on account of their subjective annoyance. When the air pollution in a room is primarily human in origin, it is the personal odours that are most

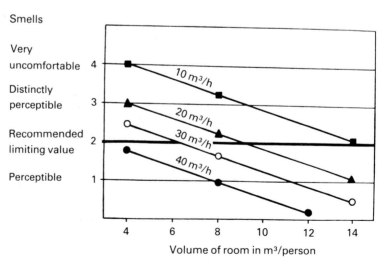

Fig. 196 **Guidelines for the fresh-air requirements of sedentary workers in relation to the air space available to each person.** After *Yaglou* and others [341].

offensive, far more than the changes in carbon dioxide or water vapour. A general guideline for the quantity of fresh air needed in a room with people in it is:

30 m³ of fresh air per person per hour.

But this is only a very rough figure. *Figure 196* summarises the results obtained by *Yaglou* and his colleagues [341] in a long series of experiments.

From these results we can make the following recommendations:

Volume per person in m³	Fresh air per person in m³/h	
	Minimal	Desirable
5	35	50
10	20	40
15	10	30

Smoking is a major cause of air pollution in place of employment, and the following recommendations appear in the literature:

rooms where smoking is prohibited: 30 m³/person/hour.
rooms where smoking is permitted: 40 m³/person/hour.

Natural or forced ventilation

After the extent of internal pollution, the situation of the building and the window area are decisive factors in deciding whether it is necessary to have any form of forced ventilation, or air conditioning. If it is impossible to open the windows in summer, either because of traffic noise or because of the pollution of the urban air, then artificial ventilation of some kind is essential.

Windows and climate indoors

Modern buildings tend to have lower ceilings, and more window area than buildings erected 50 years ago. These changes have important effects on the climate indoors. Lower ceilings make fresh air even more necessary. Huge window areas act as cooling surfaces in winter, and from spring to autumn they allow a great deal of heat to penetrate into the room. This becomes particularly obvious if K values (thermal conductivity) are used as a measure of the heat-balance of a building. The windows of older buildings occupied 15–30 % of the external surface, whereas they are often more than 50 % of the surface of modern buildings. If we assume a K value of a window as being 3.5, and that of a wall as 0.8, then if the windows occupy 50 % of the outer walls, it can be calculated that 82 % of the total heat loss occurs from the windows, and only 18 % from the walls and roof.

This means that in such a building four times as much heat is lost through the windows as through the walls and roof. Improving the thermal insulation of the window is four times as effective as insulating the roof and walls, and is greatly to be recommended.

Modern buildings admit more light and bring the occupants into closer contact with the outside world, with the sky, and with nature; at the same time they raise serious problems of controlling the climate indoors and lead to a greater dissipation of energy.

Heat in industry

As the temperature rises above the optimum for comfort, problems arise: first of a subjective nature, subsequently physical problems which impair the efficiency of the workers. Some of these problems and their symptoms, in the range between a comfortable temperature and the highest tolerable limit, are listed in *Table 54*.

Table 54   Effects of deviations from a comfortable working temperature.

20°C	1. Comfortable temperature	Maximum efficiency
	2. Discomfort; increased irritability; loss of concentration; loss of efficiency in mental tasks	Psychical problems
	3. Increase of errors; loss of efficiency in skilled tasks; more accidents	Psycho-physiological problems
	4. Loss of performance of heavy work; disturbed water- and salt-balances; heavy stresses on heart and circulation; intense fatigue and threat of exhaustion	Physiological problems
35–40°C	5. Limit of tolerance of high temperature	

Importance of sweating

We have seen above (*Figure 189*) that the range of temperatures between a comfortable level and the upper limit of tolerance, some 10–15°C, is a zone in which heat regulation is achieved by the evaporation of sweat from the skin. As the external temperature rises, the body can lose less and less heat by convection and by radiation (because of the lower temperature gradient), so that *sweating becomes the only way in which excess heat can be lost.* In fact a point is soon reached at which convection and radiation convey heat *into* the body, and this heat, together with that produced internally, must be lost by the evaporation of sweat. *When working in high temperatures, therefore, secretion and evaporation of sweat are of paramount importance for the preservation of a correct heat balance.*

Mechanisms for physiological adaptation

If the ambient temperature rises, the following physiological effects may be produced:

(a) increased fatigue, with accompanying loss of efficiency for both physical and mental tasks;

(b) rise in heart rate;

(c) rise in blood pressure;

(d) reduced activity of the digestive organs;

(e) a slight increase in core temperature and sharp rise in shell temperature (temperature of the skin may rise from 32°C to 36–37°C);

(f) massive increase in blood flow through the skin (from a few ml per cm^3 of skin tissue per min, to 20–30 ml);

(g) increased production of sweat, which becomes copious if the skin temperature reaches 34°C or more.

The effect of these adaptive changes is clearly *to transport more heat to the skin, by means of an increased flow of blood.* This increased flow is at the expense of the blood supply to the musculature (hence the reduced performance and efficiency), and to the digestive organs (which also reduce their activity). Since thermal regulation is now the overriding problem, the other systems must take second place; the muscles work less effectively and the stomach refuses food (nausea).

Similarly the heart and circulatory system adapt themselves. The rise in blood pressure, coupled with the dilatation of the blood vessels of the skin (and the simultaneous constriction of the blood vessels to the internal organs), the increased pumping action of the heart, all contribute to the increased flow of blood and transport of heat to the skin. If the skin temperature should reach 34°C, reflex action of the heat centre produces a copious flow of sweat, which is poured out from about 2½ million sweat glands in the human skin.

Effects of overheating

If these control measures are inadequate, then the core temperature of the body begins to rise, leading to hyperpyrexia, which can quickly be fatal. Clinical surveys have shown that during military exercises, core temperatures of around 39°C have resulted in heat stroke, followed by death, although *Robinson* and *Gerking* [263] reported that with rectal temperatures of 39–40°C heat collapse was not necessarily fatal.

Heat stroke

As the bodily heat accumulates, the first alarming symptoms include a general feeling of listlessness, loss of performance in spite of every effort, bright red skin, and increased heart rate, with feeble pulse. These are followed by severe headache, giddiness, shortness of breath, perhaps vomiting, and muscular cramps as a result of loss of salt. The final stage is unconsciousness, which may end in death within 24 hours, in spite of every medical attention, including rapid cooling of the body. Death from sun-stroke is really a special case of heat stroke, in which direct heat from the sun on the head may be the decisive factor.

Liability to heat stroke is an individual matter, and varies greatly from one person to another. The risk is much greater for a fat person than for a slim one, and may be increased to six times normal if a person is 25 kg overweight. Other factors involved include capacity for heat adaptation, age, intake of food, and particularly the amount of physical exercise taken. The work of *Wyndham* and colleagues [339] may be mentioned here. They recorded deaths from heat stroke among South African miners working in temperatures of 30°C and up, with 100% relative humidity; at 34.5°C the mortality was one in a thousand. Similar results were recorded for non-fatal heat collapse. Even with lower relative humidities, heat stroke sometimes occurs among the general population if the temperature outdoors rises above 35°C. The actual cause of death from heat stroke is not known. Autopsies have revealed swelling and minute haemorrhages, as well as dead nerve tissue in the brain. Specialists think that perhaps over-heating of the brain may lead to some kind of irreversible poison-ing, or to death of vital nerve cells. If the extent of this brain damage exceeds a certain level, no physical treatment by cooling and medical attention is of any use.

Fevers and sport

It is interesting to note that the core temperature may rise very high (41°C) during a fever, and up to 39.5°C in strenuous sporting activities, without provoking collapse through heat stroke. Under these conditions the heat centre is able to cope with the rise in core temperature, preserving the vital functions during a fever, and maintaining a high performance during sport. Heat stroke is a consequence of excessive heat from outside the body, where it is passively accumulated. The brain seems to have some sort of protective mechanism against active heat accumulation generated within the body itself, but this does not operate against heat from outside.

Heat tolerance

The important question in any problem of working in excessive heat is the tolerable heat load, or heat tolerance.

From a physiological point of view, most authors agree that *rectal temperature should not exceed an upper limit of 38°C* [72] [196], and several authors have compiled indices of climatic factors which would ensure that this physiological limit is not exceeded; examples are *Belding* and *Hatch* [22], and more recently *Dukes-Dobos* [72]. Since these indices of heat load are very complicated, taking into account air temperature, humidity and radiant heat, we may adopt a simpler system, after *Wenzel* [327]. These limits of heat tolerance are shown as shaded bands in *Figure 197*.

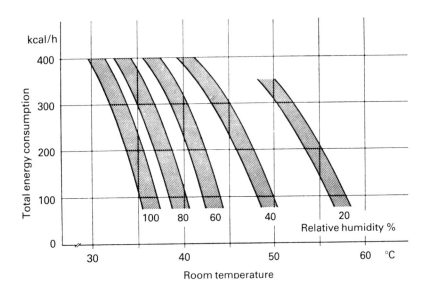

Fig. 197 Limiting values for heat load, in relation to physical effort (energy consumption), the relative humidity of the air, and the air-temperature. Air temperature approximately the same as radiant temperature; air movements between 0.1 and 0.9 metres per second. Working time 3–6 hours. Modified from *Wenzel* [327].

Practical limits

These limits are valid for unclothed young men, who are highly efficient and well accustomed to heat. Hence they cannot be applied directly to conditions in industry! There have been no detailed investigations under working conditions, but it can be assumed that the curved bands in *Figure 197* should be moved at least 5–10°C to the left to make them applicable to working

Table 55 Proposed temperature limits for acceptable heat loads during daytime work.

Overall consumption of energy kcal/h	Examples	Upper limit of temperature (°C)	
		Effective temperature	Temp. with 50% RH
400	Heavy work; walking, with 30 kg load	26–28	30.5–33
250	Moderately heavy work; walking at 4 km/h	29–31	34–37
100	Light sedentary work	33–35	40–44

conditions in industry, and even further to the left for older workers, or for jobs which require the use of protective clothing or breathing apparatus. *Table 55* sets out estimated figures for acceptable temperature ranges for work in the daytime, which are applicable to capable, healthy men, wearing clothing appropriate to the job.

If the heat load is greater than those quoted in *Table 55*, and cannot be significantly reduced by technical means, then *the working time in the heat must be shortened*. Suggestions are given in *Table 56*.

Table 56 **Permissible working times in hot, humid conditions, at heavy work of 450 kcal/h.** After *McConnell* and *Yaglou* [218].

Wet-bulb temperature (C)[1]	Permissible working time (min)
30	140
32	90
34	65
36	50
38	39
40	30
42	22

[1] The wet-bulb temperature is the reading of a thermometer with its bulb covered with a wick dipping into water. Its reading is lower, the drier the surrounding air, because of increased evaporation.

Working under radiant heat

The limiting values quoted in *Figure 197* and *Tables 55* and *56* apply to conditions where the surrounding surfaces are at approximately the same temperature as the air. They cannot be applied without modification to work places that are exposed to great radiant heat, such as, for example, in front of furnaces. Radiant heat is often measured by means of a *globe thermometer*. This usually consists of a hollow sphere of copper, about 15 cm in diameter, into which a normal mercury thermometer is inserted. The outer surface of the copper sphere is painted black, to absorb the radiation and to convert it into heat. *The globe thermometer takes up an average value between the temperature of the air and that of surrounding surfaces.*

Various methods have been devised of integrating the various temperature factors into one. One of these is to combine the wet-bulb and globe thermometer readings into the WBGT (wet-bulb–globe temperature). (WBGT) = 0.7 wet bulb reading + 0.3 globe thermometer reading.) Using this formula, the following values of WBGT for continuous work were proposed in the U.S.A.:

500 kcal/h	WBGT of 25°C
350 kcal/h	WBGT of 27°C
200 kcal/h	WBGT of 30°C.

Physiological limits

In practice it is usually very difficult to express the heat load and the degree of physical work exactly in kcalories, so physiological means have to be sought to assess these. As we have already seen, heart rate, core temperature (rectal temperature) and perspiration can all be used for this purpose. Upper limits of these parameters for working in heat for an entire working day are:

(a) heart rate (daily average):   100–110/min
(b) rectal temperature: 38°C
(c) evaporation of sweat: 0.5 litres/h.

Acclimatisation to heat

Experience shows that workers take time to become accustomed to working under very hot conditions, and only after several weeks is their performance equal to that of workers who are already "heat-adapted". This is a genuine acclimatisation of the body to heat, which proceeds by the following steps:

(a) the body gradually increases its perspiration, losing more and more heat in the process. A worker who is heat-adapted can lose two litres of sweat per hour, and up to six litres per working day;

(b) as part of the acclimatisation process the sweat becomes more dilute, with a lower concentration of salts. The sweat glands "learn" to conserve salts, so that larger amounts of sweat can be produced without creating a salt deficit in the body. Such a deficit would lead to muscular cramps and eventually to exhaustion, and possibly death;

(c) during acclimatisation there is a loss of weight, which helps heat loss by reducing the amount of insulating fat, and also reduces energy consumption;

(d) as acclimatisation proceeds, the worker drinks more fluids, to compensate for the greater amount of water lost by sweating;

(e) the blood system and heart also adapt themselves to provide for the improved performance after acclimatisation.

When a worker has become acclimatised to heat he feels thirsty whenever his body needs more liquid, and so he tends to drink small amounts frequently. The question whether he should take additional salt is debatable. In America good results have been obtained from giving either salt tablets or salt-rich foods (meat broth, etc.) both to factory workers and to troops exposed to excessive heat. In Europe heat stresses are not usually as great as they are in the U.S.A., so that up to the present the provision of additional salt has not been necessary.

Recommendations The following guidelines may be recommended for workers in very hot conditions:

(a) a worker should be acclimatised to heat by stages. He should begin by spending only 50% of his working time in the heat and increase this by 10% each day. The same procedure should follow when a worker returns from illness or holiday;

(b) the greater the heat load and the greater the physical effort performed under heat stress, the longer and more frequent should be the pauses (cooling periods). If the limit of heat tolerance is being exceeded, the working day must be shortened;

(c) the worker should drink small amounts of fluid at frequent intervals; never more than 0.25 litres at a time, and a cupful every 10–15 minutes is recommended;

(d) as drinks we recommend lightly sweetened tea or coffee, varied with an occasional drink of soup. If large quantities of fluid are needed it is best to drink plain water, with an occasional drink of tea or coffee. The beverages should be lukewarm or warm, so that they are the more quickly absorbed by the digestive system;

(e) iced drinks, fruit juices and alcoholic drinks are not to be recommended. Milk drinks, too, are unsuitable for working in hot conditions, because they put more stress on the digestive organs;

(f) the drinks should be available close to the worker, so that he can drink whenever he feels the need;

(g) where the radiant heat is excessive (e.g. blast furnaces) the worker must be protected by special goggles, screens and protective clothing against the risk of burning his eyes and hands.

In addition to these personal precautions, every attempt must be made to reduce the impact of the heat on the worker. Some possible measures include improving ventilation, both natural and forced, perhaps artificial drying of the air, and screens to protect him against radiant heat.

18
Checklist for the analysis of work places

Purpose

This questionnaire is designed to ensure that work places are physiologically well designed, by checking the details in a logical order.

It may be compared to the "checklist" used by an airline pilot before take-off, to ensure that everything necessary has been done and no small detail has been overlooked.

Similarly, this questionnaire may be used to check that no important aspect of work places has been forgotten.

Preliminary
questions

1 *Assessing the work load*
   11 Main occupation:
      Secondary occupation:
   12 Is the work physically arduous?
   13 Does the work make heavy demands on skill, vigilance, or perception?
   14 Is the work made harder by external factors such as climate, noise, or lighting?
   15 Is the work made harder by the way it is organised: shift work, lack of rest pauses, working against the clock?

Checklist

2 *Questions relating to physical stress*

   21 *Bodily posture:*
      (sitting, standing, stooping)
      211 Does this posture involve much static muscular effort?
      212 Is the working height correct?
      213 Is the range of movement of grips and handles anatomically correct?
      214 Is there enough room to move about?

215 Can the work be seen clearly and any instruments read with the body in a natural position?

216 Does the body have to take up an unnatural posture when pedals are operated?

Questions relating to sedentary work:

217 Is the seat correctly adjusted to the working height?

218 Does the seat cause aches or pains?

219 Is a foot rest necessary?

22 *Muscular work*

Is the muscular effort predominantly static or dynamic?

221 Is any form of strenuous static work involved?

222 If so, can this be avoided by providing clamps or supports for the work?

223 Can it be made easier by supports for hands or elbows?

224 Must loads be lifted?

225 Are the weights of these loads acceptable?

226 Is there some suitable way of lifting and transporting these loads?

227 Is strenuous dynamic work required?

228 Are the working pulses below the limits?

229 Is the work performed with a high enough degree of efficiency?

3 *Questions relating to perception, vigilance and skill*

31 *Demands on perception*

311 Is the lighting good?

312 Are the instruments well arranged and appropriate to the task?

313 Are numbers, words, symbols and scale divisions of a size to suit the reading distance?

314 Are instruments, components and labels in full view, so as to avoid mistakes?

315 Are any magnifying devices necessary?

316 Are instruments and controls properly placed in relation to each other?

317 Can any audible signals be heard clearly and without danger of missing one?

32 *Demands on vigilance*

321 Is vigilance disturbed by noise?

322 Is vigilance impaired by other people's activities?

323 Is vigilance impaired by what is going on at the same work place?

33 *Demands on skill*

331 Is the skilled work performed under visual control?

332 Does it require a long training period?

333 Has every facility been given to acquire an automatic skill?

334 Do the directions and sequence of movements follow a stereotyped pattern?

335 Are the controls set out so that bodily postures are natural?

336 Is much effort needed to operate manual controls?

337 Are the controls adequate for their purpose?

4 *Questions relating to the working environment*

41 *Light and colour*

411 Is the lighting bright enough during daytime?

412 Is the artificial lighting bright enough?

413 Are excessive contrasts present in the working field?

414 Must the worker keep looking from a bright to a dark area, and vice versa?

415 Are there reflective surfaces in the working field?

416 Are the light sources properly arranged?

417 Is the lighting steady? (No flickering fluorescent tubes; tubes out of phase with each other; no stroboscopic effects from moving machinery.)

418 Is there excessive brightness contrast between different colours?

419 Are "eye-catchers" sensibly used?

41.10 Is the colour scheme in the room restful and friendly?

42 *Indoor climate*

421 Is the air temperature comfortable?

422 Are the surrounding surfaces at approximately the same temperature as the air?

423 Are there any perceptible draughts?

424 Is the relative humidity physiologically suitable?

425 Are the eating appliances placed correctly?

426 Is the air changed often enough?

Questions relating to work under hot conditions

427 Is the heat load acceptable?

428 Are the workers suitably clothed?

429 Is the supply of liquids sufficient?

42.10 Can the heat load be reduced by protective devices?

43 *Protection against noise*

431 Does the noise disturb vigilance or mental effort?
432 Does the noise interfere with conversation?
433 Is the noise level so high that there is a danger of damage to hearing?
434 Can the noise level be reduced?

44 *Health safeguards*

441 Does the air in the room contain any toxic substances?
442 Can the spread of toxic substances be stopped at source?
443 Can ventilation equipment be installed?
444 Is there contact with any substance that might cause dermatitis?
445 Does the layout of the work place make accidents possible?
446 Does the performance of the work involve risk of accidents?
447 Might accidents occur through the action of a third person?
448 Is there any risk of burns or explosion?

5 *Questions relating to the organisation of the work*

51 *Boredom*

511 How much time is allowed for each task?
512 Does the repetitive work create a bad posture of the body?
513 Is it possible to make the work less monotonous by broadening the tasks or by rotation of jobs?
514 Does the organisation of the work permit social contacts?

52 *Shift work*

521 Are there day and night shifts?
522 Does the existing shift system permit only short periods on night shift?
523 How many free weekends are there per year?
524 Are conditions favourable for daytime sleep?

53 *Pauses (breaks) and eating habits*

531 Are there official breaks?
532 Are the timing and lengths of these breaks reasonable?

533 Would additional short breaks be desirable?
534 Would flexible hours be advantageous?
535 Is there adequate provision for food (snacks) during breaks?
536 Is the midday break long enough?

Literature

1   *Aanonsen, A.:* Shiftwork and health. Norwegian Monographs on Medical Science, Universitets forlaget, Oslo (1964).

2   *Abramson, N.:* Information theory and coding. McGraw Hill, New York (1963).

3   *Akerblom, B.:* Standing and sitting posture. Nordiska Bokhandeln Stockholm (1948).

4   *Allensbacher Berichte:* Wer will die Viertagewoche? Institut für Demoskopie Allensbach, D–7753 Allensbach am Bodensee, Nr 16 (1971).

5   *American Society of Heating, Refrigerating and Air-Conditioning Engineers:* Handbook of fundamentals. ASHRAE, New York (1967) und (1972).

6   *Andersen, E. J.:* The main results of the Danish medico–psycho–social investigations of shiftworkers. Proceedings, Int. Congr. Occup. Health, III, p. 135, Helsinki (1957).

7   *Andersson, B. J. G. and Ortengren, R.:* Lumbar disc pressure and myoelectric back muscle activity during sitting. 1. Studies on an office chair. Scan. J. Rehabilitation Medicine *3*, 115–121 (1974). The same author with colleagues also in Scan. J. Rehabilitation Medicine *3*, 104–114 (1974), *3*, 122–127 (1974), *3*, 128–135 (1974).

8   *Andlauer, P., Carpentier, J. and Cazamian, P.:* Ergonomie du travail de nuit et des horaires alternants. Editions Cujas, Paris (1977).

9   *Andlauer, P. and Fourré, L.:* Aspects ergonomique du travail en équipes alternantes. Rapport du Centre d'études de Physiologie appliquée au travail, Strasbourg (1962) and in: La Revue française du travail, 35–50 (1965).

10 *Astrand, I.:* Aerobic work capacity in men and women with special reference to age. Acta physiol. scand. *49*, Suppl. 169 (1960).

11 *Astrand, P. O.:* Experimental studies of physical working capacity in relation to sex and age. Munksgaard, Kopenhagen (1952).

12 *Attneave, F.:* Applications of information theory to psychology. Holt, Rinchart and Winston, New York (1959).

13 *Barmack, J. E.:* The length of the work period and the work curve. J. Exper. Psychol. *25*, 109—115 (1939)

14 *Barnes, R. M.:* An investigation of some hand motions used in factory work. University of Iowa, Iowa City, Studies in Engineering, Bulletin 6 (1936).

15 *Barnes, R. M.:* Motion and time study. 3rd edition, Wiley, New York (1949).

16 *Barnes, R. M., Hardaway, H. and Podalsky, O.:* Which pedal is best? Factory Management and Maintenance, *100,* 98–99 (1942).

17 *Bartenwerfer, H. G.:* Über die Auswirkungen einförmiger Arbeitsvorgänge. Marburger Sitzungsberichte *80,* 1–70 (1957).

18 *Baschera, P. and Grandjean, E.:* Effects of repetitive tasks with different degrees of complexity on CFF and subjective state. (To be published.)

19 *Gierer, R., Martin, E., Baschera, P. and Grandjean, E.:* Ein neues Gerät zur Bestimmung der Flimmerverschmelzungsfrequenz. (To be published in *European Journal of Applied Physiology and Occupational Physiology* 1979).

20 *van Beck, H. G.:* The influence of assembly line organisation on output, quality and morale. Occup. Psychol. *38,* 161–172 (1964).

21 *Behrens, R.:* Überstunden und Krankenstand. Zbl. Arbeitsmed. *7,* 15–21 (1957).

22 *Belding, H. S. and Hatch, T. F.:* Index for evaluating heat stress in terms of resulting physiological strains. Heating, Piping and Air Conditioning *27,* 129–135 (1955).

23 *Berger, C.:* Stroke-width, form and horizontal spacing of numerals as determinants of the threshold of recognition. J. of Applied Psychology *28,* 208–231 and 336–346 (1944).

24 *Bills, A. G.:* Blocking: a new principle of mental fatigue. Am. J. of Psychology *43,* 230–239 (1931).

25 *Bjerner, B., Holm, A. and Swensson, A.:* Diurnal variation in mental performance. Brit. J. Industrial Medicine *12,* 103–110 (1955).

26 *Blackwell, H. R. and Blackwell, O. M.:* The effect of illumination quantity upon the performance of different visual tasks. Illuminating Engineering *63,* 143–152 (1968).

27 *Blum, M. L. and Naylor, J. C.:* Industrial psychology. Harper and Row, New York (1968).

28 *Bonjer, F.:* Physiological aspects of shiftwork. Proceedings XIII Int. Congress of Industrial Medicine, 848–858, New York (1958).

29 *Bonvallet, M., Dell, P. and Hiebel, G.:* Tonus sympathique et activité électrique corticale. J. Electroenceph. and clin. Neurophysiology *6,* 119–125 (1954).

30 *Bouisset, S.:* Postures et mouvements. In Physiologie du Travail Tome 1, J. Scherrer, Masson, Paris (1967).

31 *Bouisset, S., Laville, A. and Monod, H.:* Recherches physiologiques sur l'économie des mouvements. Proceedings of 2nd I.E.A. Congress, Ergonomics *7,* 61–67 (1964).

32 *Bouisset, S. and Monod, H.:* Un essai de determination de caractéristiques anthropométriques en vue de l'aménagement de postes de travail. Travail Humain *24,* 35–50 (1961).

33 *Bouisset, S. and Monod, H.:* Etude d'un travail musculaire léger. I. Zone de moindre dépense energétique. Arch. internat. Physiol. Biochem. *70,* 259–272 (1962).

34 *Branton, P. and Grayson, G.:* An evaluation of train seats by observation of sitting behavior. Ergonomics *10,* 35–51 (1967).

35 *British Standards Institute:* Guide to the evaluation of human exposure to whole-body vibration. No. DD 32, London (1974).

36 *Broadbent, B. E.:* Perception and communication. Pergamon Press, London/New York (1958).

37 *Broadbent, D. E.:* Effects of noise on behaviour. In Handbook of noise control, ed. by C. M. Harris, McGraw Hill, New York (1957).

38 *Broadbent, D. E.:* Effect of noise on an intellectual task. Journal of the Acoustical Society of America *30,* 824–827 (1958).

39 *Brouha, L.:* Physiology in industry. 2nd edition, Pergamon Press, Oxford (1967).

40 *Brown, H. G.:* Some effects of shift work on social and domestic life.

Yorkshire Bull. of Economic and Social Research (1959).

41 Brown, J. S., Knauft, E. B. and Rosenbaum, G.: The accuracy of positioning reactions as a function of their direction and extent. Am. J. Psychology 61, 167–182 (1948).

42 Brown, J. S. and Slater-Hammel, A. T.: Discrete movements in the horizontal plane. J. Experimental Psychology 39, 84–95 (1949).

43 Brundke, M.: Langzeitmessungen der Pulsfrequenz und Möglichkeiten der Aussage über die Arbeitsbeanspruchung. Beitrag in: Pulsfrequenz und Arbeitsuntersuchungen, Schriftenreihe Arbeitswissenschaft und Praxis, Band 28, Beuth-Vertrieb, Berlin, Köln and Frankfurt/M (1973).

44 Busch, G. and Wachholder, K.: Der Einfluss ermüdender geistiger Beanspruchung auf die Flimmerverschmelzungsfrequenz. Arbeitsphysiologie 15, 149–156 (1953).

45 Caldwell, L. S.. The effect of the special position of a control on the strength of six linear hand movements. U.S. Army Medical Research Laboratory, Fort Knox, Kentucky, Report Nr 411 (1959).

46 Caplan, R. D., Cobb, S., French, J. R. P., van Harrison, R. and Pinneau, S. R.: Job demands and worker health. H E W publication No. (NIOSH) 75–160 (1975).

47 Carpentier, J. and Wisner, A.: L'Aménagement des conditions du travail par équipes successives. Ministère du travail; Agence nationale pour l'amélioration des conditions de travail, Paris (1976).

48 Chaney, R. E.: Subjective reaction to wholebody vibration. Boeing Company, Human Factors Tech. Report D3, 64–74, Wichita, Kansas (1964).

49 Chapanis, A. and Gropper, B. A.: The effect of the operator's handedness on some directional stereotypes in control-display relationships. Human Factors, 10, 303–320 (1968).

50 Chapanis, A. and Lindenbaum, L.: A reaction time study of four control display linkages. Human Factors, 1, 1–7 (1959).

51 Chavasse, P., Pilon, J. and Wisner, A.: Manuel de bruits et vibrations. Institut National de Sécurité, Paris (1955).

52 Chazalette, A.: Etude sur les conséquences du travail en équipes alternantes et leus facteurs explicatifs. Groupe de Sociologie Urbaine, Lyon (1973). Quoted from (8).

53 Christensen, E. H.: L'Homme au travail. Serie "Sécurité, hygiène et médecine du travail" No. 4, Bureau International du Travail, Genève (1964).

54 Clarke, H. H., Elkins, E. C., Martin, G. M. and Wakim, K. G.: Relationship between body position and the application of muscle power to movements of joints. Archive Physic. Med. 31, 81–89 (1950).

55 Coermann, R.: The mechanical impedance of the human body in sitting and standing position at low frequencies. In Human Vibration Research, Pergamon Press, New York (1963).

56 Colquhoun, W. P. (Ed.): Biological rhythms and human performance. Academic Press, London New York (1971).

57 Conroy, R. T. W. L. and Mills, J. N.: Human Circadian Rhythms. Churchill, London (1970).

58 Corlett, E. N. and Bishop, R. P.: A technique for assessing postural discomfort. Ergonomics 19, 175–182 (1976).

59 Damon, F. A.: The use of biomechanics in manufacturing operations. The Western Electric Engineer 9, October (1965), quoted from E. J. McCormick (220).

60 Damon, A., Stoudt, H. W. and McFarland, R. A.: The human body in equipment design. Harvard University Press, Cambridge, Mass. (1966).

61 Davies, D. R. and Tune, G. S.: Human Vigilance Performance. Staples Press, London (1970).

62 Davis, P. R.: Variations of the human intra-abdominal pressure during weight-lifting in different postures. J. Anat. London 90, 601–610 (1956).

63 *Davis, P. R.:* The causation of hernia by weight lifting. Lancet *2*, 155–157 (1959).

64 *Davis, P. R. and Stubbs, D. A.:* Safe levels of manual forces for young males. Applied Ergonomics *8*, 141–150 (1977).

65 *Davis, P. R. and Stubbs, D. A.:* A method of establishing safe handling forces in working situations. Report of the Internat. Symp. on safety in manual materials handling. NIOSH (1977).

66 *Diffrient, N., Tilley, A. R. and Bardagjy, J. C.,* with *Henry Dreyfuss Associates:* Humanscale 1/2/3. MIT Press, Cambridge, Mass. (1974).

67 *DIN Norm No. 4552:* Drehstuhl mit in der Höhe nicht verstellbarer Rückenlehne mit oder ohne Armstützen. Beuth-Vertrieb, Berlin and Köln (October 1975).

68 *DIN Normen No. 4844 und No. 5381:* Sicherheitskennzeichnung, II. Sicherheitsfarben und Kennfarben. Beuth-Vertrieb, Berlin and Köln (1977) and (1976).

69 *DIN Norm No. 5035:* Blatt 1 und Blatt 2: Innenraumbeleuchtung mit künstlichem Licht. Beuth-Vertrieb, Berlin and Köln (1972).

70 *DIN Norm No. 33401:* Stellteile. Entwurf 1974. Beuth-Vertrieb, Berlin and Köln (1974).

71 *Dubois-Poulsen, A.:* Notions de physiologie ergonomique de l'appareil visuel. In "Physiologie du Travail" 114–183 by J. Scherrer, Tome 2, Masson, Paris (1967).

72 *Dukes-Dobos, F. N.:* Rational and provisions of the work practices standard for work in hot environments as recommended by NIOSH. In St. M. Horvath and R. C. Jensen (Ed.): Standards for occupational exposures to hot environments. U.S. Dep. Health, Education and Welfare. Nat. Institute Occup. Safety and Health, No. 76–100, Cincinnati, Ohio (1976).

73 *Dupuis, H.:* Mechanische Schwingungen, sowie: Messung und Bewertung von Schwingungen und Stössen. In H. Schmidtke (Ed.): Ergonomie, Band 2, S. 211–236, Carl Hanser Verlag, München (1974).

74 *Dupuis, H.:* Zur physiologischen Beanspruchung des Menschen durch mechanische Schwingungen. Fortschr.-Ber. VDI Zeitschrift, Reihe 11, Nr. 7, 1–168 (1969).

75 *Dupuis, H. und Christ W.:* Untersuchung der Möglichkeit von Gesundheitschädigungen im Bereich der Wirbelsäule bei Schlepperfahrern. Arbeiten aus dem Max-Planck-Institut für Landarbeit und Landtechnik, Bad Kreuznach Heft A. 72/2 (1972).

76 *Durnin, J. V. G. A. and Passmore, R.:* Energy, work and leisure. Heinemann Educational Books Ltd., London (1967).

77 *Eastman, M. C. and Kamon, E.:* Posture and subjective evaluation at flat and slanted desks. Human Factors *18,* 15–26 (1976).

78 *Egli, R., Grandjean, E. and Rhiner, A.:* Arbeitsphysiologischer Beitrag zur Verbesserung der Sägetechnik. Int. Z. ang. Physiol. einschl. Arbeitsphysiologie *16,* 325–334 (1956).
Egli, R., Grandjean, E. and Turrian, H.: Arbeitsphysiologische Untersuchungen an Hackgeräten. Arbeitsphysiologie, *15,* 231–234 (1943).

79 *Ellis, D. S.:* Speed of manipulative performance as a function of work-surface height. J. appl. Psychol. *35,* 289–296 (1951).

80 *Endo, T. and Kogi, K.:* Monotony effects of the work of motormen during high-speed train operation. J. Human Ergol. *4,* 129–140 (1975).

81 *Fanger, P. O.:* Thermal comfort. McGraw Hill Book Comp., New York (1972). Thermal comfort – Analysis and application in environmental engineering. Copenhagen: Danish Technical Press (1970).

82 *Fletcher and Munson, W. A.:* Loudness, its definition, measurement and calculation. J. Acoustical Soc. of America *5,* 82—108 (1933).

83 *Fortuin, G. C.:* Visual power and visibility. Philips Research Report (1957).

84 *Frank, W.:* Raumklima und Thermische Behaglichkeit. Berichte aus

der Bauforschung, Heft 104, Wilhelm Ernst u. Sohn, Berlin-München-Düsseldorf (1975).

85 *Frankenhaeuser, M.:* Man in technological society: stress, adaptation and tolerance limits. Reports from the Psychological Laboratories, University of Stockholm, Suppl. 26, Dec. (1974).

86 *Frankenhaeuser, M. and Johansson, G.:* On the psycho-physiological consequences of understimulation and overstimulation. Reports from the Psychological Laboratories, University of Stockholm, Suppl. 25 (1974).

87 *Frankenhaeuser, M., Nordheden, B., Myrsten, A. L. and Post, B.:* Psychophysiological reactions to understimulation and overstimulation. Acta Psychologica *35,* 298–308 (1971).

88 *Friedmann, G.:* Grenzen der Arbeitsteilung. Europäische Verlagsanstalt, Frankfurt/M (1959).

89 *Fröberg, J., Karlsson, C. G. and Levi, L.:* Shiftwork, a study of catecholamine excretion, selfratings and attitudes. Studia Laboris et Salutis Report No. 11, 10–20, Stockholm (1972).

90 *Furrer, W. und Lauber, A.:* Raum- und Bauakustik, Lärmabwehr. Birkhäuser, Basel (1972).

91 *Ganong, W. F.:* Lehrbuch der Medizinischen Physiologie. Springer Verlag, Berlin, Heidelberg and New York (1974).

92 *General Electric Co.:* See better work better. Lamp Division: Bulletin No. 1, Cleveland (1953).

93 *Geyer, B. H. and Johnson, C. W.:* Memory in man and machines. General Electric Review *60,* 29—33 (1957).

94 *Graf, O.:* Studien über Fliessarbeitsprobleme an einer praxisnahen Experimentieranlage. Forschungsbericht d. Wirtschafts- und Verkehrsministerium Nordrhein-Westfalen. Nr 114 und Nr 115, Westdeutscher Verlag, Köln and Opladen (1954).

95 *Graf, O.:* Vorschläge einer Neuregelung. In E. Blume, K. Doese, H. Fischer, O. Graf, J. Höffner and H. Kraut: Gutachten über die kontinuierliche Arbeitsweise. Quoted from (274) and (176).

96 *Graf, O., Pirtkien, R., Rutenfranz, J. and Ulich, E.:* Nervöse Belastung im Betrieb. I. Nachtarbeit und nervöse Belastung. Westdeutscher Verlag, Köln and Opladen (1958).

97 *Graf, P., Müller, R. and Meier, H.-P.:* Sozio-psychologische Fluglärmuntersuchung im Gebiet der drei Schweizer Flughäfen Zürich, Genf, Basel. Ed.: Arbeitsgemeinschaft für sozio-psychologische Fluglärmuntersuchungen. Bern (Juni 1973). Orders: Eidg Luftamt, Bundeshaus, Bern.

98 *Grandjean, E.:* Wohnphysiologie. Verlag für Architektur Artemis Zürich (1973).

99 *Grandjean, E.:* Physiologische Untersuchungen über die nervöse Ermüdung bei Telephonistinnen und Büroangestellten. Int. Z. Angew. Physiol. *17,* 400–418 (1959).

100 *Grandjean, E.:* Fatigue. Yant Memorial Lecture 1970. Am. Ind. Hyg. Ass. J. *31,* 401–411 (1970).

101 *Grandjean, E.:* Raumklimatische Untersuchungen in Büros während der warmen Jahreszeit. Heizung–Lüftung–Haustechnik *19,* 118–123 (1968).

102 *Grandjean, E.:* Raumklimatische Wirkungen verschiedener Heizsysteme in Büros. Schw. Blätter f. Heizung u. Lüftung, *33,* 3–6 (1966).

103 *Grandjean, E. und Bättig, K.:* Das Verhalten der subjektiven Verschmelzungsschwellen des Auges unter verschiedenen Arbeits- und Versuchsbedingungen. Helv. physiol. pharmacol. Acta *13,* 178–190 (1955).

104 *Grandjean, E., Bloch, W., Egli, R. and Gfeller, H.:* Arbeitswissenschaftliche Untersuchungen über die Produktionsleistung und die physiologische Beanspruchung von Nadelrichterinnen. Ind. Organisation *24,* 2–11 (1955).

362

105 *Grandjean, E., Böni, A. and Kretschmar, H.:* Entwicklung eines Ruhesesselprofils für gesunde und rückenkranke Menschen. Wohnungsmedizin 5, 51–56 (1967).

106 *Grandjean, E. and Burandt, H. U.:* Körpermasse der Belegschaft eines schweizerischen Industriebetriebes. Ind. Organisation 31, 239–242 (1962).

107 *Grandjean, E. and Burandt, H. U.:* Das Sitzverhalten von Büroangestellten. Ind. Organisation 31, 243–250 (1962).

108 *Grandjean, E., Egli, R., Rhiner, A. and Steinlin, H.:* Der menschliche Energieverbrauch der gebräuchlichsten Waldsägen. Helv. physiol. pharmacol. Acta 10, 342–248 (1952).

109 *Grandjean, E., Horisberger, B., Havas, L. and Abt, K.:* Arbeitsphysiologische Untersuchungen mit verschiedenen Beleuchtungssystemen an einer Feinarbeit. Ind. Organisation 28, 231–239 (1959).

110 *Grandjean, E., Hünting, W., Wotzka, G. and Schärer, R.:* An ergonomic investigation of multipurpose chairs. Human Factors 15, 247–255 (1973).

111 *Grandjean, E., Kahlcke, H. and Wotzka, G.:* Ergonomische Untersuchung von Schul-Zeichentischen. Werk 57, 53–56 (1970).

112 *Grandjean, E., Kretschmar, H. and Wotzka, G.:* Arbeitsanalysen beim Verkaufspersonal eines Warenhauses. Z. Präventivmed. 13, 1–9 (1968).

113 *Grandjean, E., Kretschmar, H. and Wey, K.:* Erhebungen über die Ermüdung und den Gesundheitszustand beim Verkaufspersonal eines Warenhauses. Z. Präventivmed. 13, 10–21 (1968).

114 *Grandjean, E., Streit, K. and Perret, E.:* Arbeitsphysiologische Untersuchungen über die Ermüdung bei Tätigkeiten mit geringen und hohen Anforderungen an die Aufmerksamkeit. Arbeitsmedizin, Sozialmedizin und Arbeitshygiene 1, 2–11 (1966).

115 *Grandjean, E. and Wotzka, G.:* Correlations between subjective and physiological criteria of fatigue. Proceedings, 17th Int. Congress, Int. Ass. of applied Psychology pp. 375—380, Editest, Bruxelles (1971).

116 *Grandjean, E., Wotzka, G., Schaad, R. and Gilden, A.:* Fatigue and Stress in Air Traffic Controllers. Ergonomics 14, 159–165 (1971).

117 *Green, G. H.:* Indoor humidity and respiratory health. Respiratory Technology 11, 18–22 (1975).

118 *Grether, W. F.:* Vibration and human performance. Human Factors 13, 203–216 (1971).

119 *Grether, W. F. and Williams, A. C.:* Speed and accuracy of dial reading as a function of dial diameter and angular separation of scale divisions. In P. M. Fitts, "Psychological research in equipment design". Army Air Force, Aviation Psychology Program, Report 19 (1947).

120 *Griefahn, Barbara, Jansen, G. and Klosterkötter, W.:* Zur Problematik lärmbedingter Schlafstörungen – eine Auswertung von Schlafliteratur. Umwelt Bundesamt, Bericht No. 4, Bismarkplatz 1, D-1000 Berlin 3 (1976).

121 *Griffin, M. J.:* Whole-body vibration levels affecting visual acuity. In Human reaction to vibration. Proceedings of the Conference, Salford University (1973).

122 *Groll, Edith and Haider, M.:* Belastungsunterschiede bei Arbeiterinnen in Früh- und Spätschicht. Int. Z. angew. Physiol. einschl. Arbeitsphysiol, 21, 305–309 (1965).

123 *Gubéran, E., Dang, V. B. and Sweetnam, P. M.:* L'Humidification de l'air des locaux prévient-elle les maladies respiratoires pendant l'hiver? Schweiz. med Wschr. 108, 827–831 (1978).

124 *Gubser, A.:* Monotonie im Industriebetrieb. Huber, Bern and Stuttgart (1968).

125 *Guignard, J. C. and King, P. F.:* Aeromedical aspects of vibration and noise. NATO Advisory groupe for Aerospace Research and Development, A D 754631, (1972).

126 *Guth, S. K.:* Light and comfort. Industr. Medicine and Surgery *27,* 570–574 (1958).

127 *Haggard, H. W. and Greenberg, L. A.:* Diet and physical efficiency. Yale University Press, New Haven (1935).

128 *Haider, M.:* Direkte und indirekte Selbsteinstufungen der Auswirkungen von freier Arbeit und von Fliessbandarbeit. Psychologie und Praxis *5,* 1–11 (1961).

129 *Haider, M.:* Ermüdung, Beanspruchung und Leistung. Franz Deuticke, Wien (1962).

130 *Haider, M.:* Experimentelle Untersuchungen über Daueraufmerksamkeit und Cerebrale Vigilanz bei einförmigen Tätigkeiten. Zschr. Exp. und Angew. Psychologie *10,* 1–18 (1963).

131 *Handbuch für Beleuchtung.* Verlag W. Girardet, Essen (1975).

132 *Hanhart, A.:* Die Arbeitspause im Betrieb. Emil Oesch-Verlag, Thalwil/Zürich (1954).

133 *Harris, W., Mackie, R. R. et al.:* A study of the relationship among fatigue, hours of service and safety of operations of truck and bus drivers. Human Factor Research Inc. Santa Barbara, Goleta, Cal. 93017, Report No. 1727–2 (1972).

134 *Hashimoto, K.:* Physiological features of monotony manifested under high speed driving situations. Railway Labour Science Institute, Japan National Railways. In Proceedings 16th Int. Congress of Occupational Health, Tokyo, 85–88 (1969).

135 *Hashimoto, K.:* Estimation of the driver's work load in high-speed electric car operation on the new Tokaido line in Japan. Proc. 2nd International Congress on Ergonomics, Dortmund, Ergonomics, Suppl. 463–469 (1964).

136 *Hashimoto, K., Kogi, K., Endo, T. and 10 other collaborators:* Physiological strain of the electric-car driving of the exclusive new Tokaido line and fatigue by the driving schedules. Bull. Railw. Labour Sci. Res. Institute *19,* 1–31 (1966), *20* 1–9 (1967) and *26,* 1–16 (1972).

137 *Hawel, W.:* Untersuchungen eines Bezugsystems für die psychologische Schallbewertung. Arbeitswissenschaft *6,* 123–127 (1967).

138 *den Hertog, F. J. and Kerkhoff, W. H. C.:* Vom Fliessband zur selbständigen Gruppe. Ind. Organisation *43,* 21–24 (1974).

139 *Hertzberg, H. T. E., Daniels, G. S. and Churchill, E.:* Anthropometry of flying personnel – 1950. USAF Wright Air Development Center, Tech. Report, 52–321 (1954).

140 *Hess, W. R.:* Die funktionelle Organisation des vegetativen Nervensystems. Benno Schwabe, Basel (1948).

141 *Hettinger, Th.:* Muskelkraft bei Männern und Frauen. Zbl. Arb. Wiss. *14,* 79–84 (1960).

142 *Hettinger, Th.:* Angewandte Ergonomie. Bartmann-Verlag, Frechen BRD (1970).

143 *Hettinger, Th., Kaminsky, G. and Schmale, H.:* Ergonomie am Arbeitsplatz. Friedrich Kiehl Verlag, Ludwigshaben (1976).

144 *Hettinger, Th. and Müller, E. A.:* Der Einfluss des Schuhgewichtes auf den Energieumsatz beim Gehen und Lastentragen. Arbeitsphysiologie *15,* 33–40 (1953).

145 *Hilgendorf, L.:* Information input and response time. Ergonomics *9,* 31–37 (1966).

146 *Hill, J. H. and Chernikoff, R.:* Altimeter display evaluation: final report. USN, NEL Report 6242, Jan. 26 (1965). Quoted from McCormick (220).

147 *Hopkinson, R. G. and Collins, J.:* The ergonomics of lighting. McDonald, London (1970).

148 *Horvath, St. M. and Jensen, R. C.* (ed.): Standards for occupational exposures to hot environments. Proceedings of Symposium Febr. 1973, Nat. Institute Occup. Safety and Health, No. 76–100, Cincinnati, Ohio (1976).

149 *Houghten, F. C. and Yaglou, C. P.:* Determining lines of equal comfort. ASHVE Transactions *29,* 163–171 (1923).

150 *Hoyos, C.:* Kompatibilität in: Ergonomie, 2. Band von H. Schmidtke Carl Hanser Verlag, München (1974).

151 *Hünting, W. and Grandjean, E.:* Die Wirkungen verschiedener Sitzflächenneigungen auf die Körperhaltung und das subjektive Komfortempfinden. Int. Z. angew. Physiol. *31,* 1–9 (1972).

152 *Hünting, W. and Grandjean, E.:* Sitzverhalten und subjektives Wohlbefinden auf schwenkbaren und fixierten Formsitzen. Z. Arbeitswissenschaft *30,* 161–164 (1976).

153 *Hünting, W., Nemecek, J. and Grandjean, E.:* Die physische Belastung von Arbeitern an der Gesenkschmiede – eine Fallstudie. Sozial- und Präv. med. *19,* 275–278 (1974).

154 *IES lighting handbook.* 5th ed. Illuminating Engineering Society, New York (1972)

155 *Iskander, A.:* Ueber den Einfluss von Pausen auf das Anlernen sensumotorischer Fertigkeiten. Arbeitswissenschaft und Praxis, Band 6, Beuth-Vertrieb, Berlin Köln Frankfurt/M (1968).

156 *ISO* (International Organisation for Standardisation): TC 43, Assessment of noise-exposure during work for hearing conversation purposes. Geneva (1971).

157 *ISO* (International Organisation for Standardisation): 2631: Guide for the evaluation of human exposure to whole-body vibration. Geneva (1974).

158 *Jacob, E. and Scholz, H.:* Beleuchtung im Betrieb. Beuth-Verlag, Berlin, Köln and Frankfurt (1962).

159 *Jansen, G.:* Zur nervösen Belastung durch Lärm. Zbl. Arb. med. Arb. schutz, Beiheft 9, Dietrich Steinkopff, Darmstadt (1967).

160 *Jasper, H.:* Quoted from W. F. Ganong: Lehrbuch der medizinischen Physiologie. Deutsche Ausgabe. Springer Verlag, Berlin, Heidelberg, New York (1974).

161 *Jasper, H.:* Diffuse projection systems; the integrative action of the thalamic reticular system. Electroenceph. clin. Neuro-physiol. (Canada) *1,* 405–420 (1949).

162 *Jenkins, W. O.:* The tactual discrimination of shapes for coding aircraft type controls. In P. M. Fitts: Psychological Research on Equipment Design. Army Air Force, Aviation Psychology Program, Report 19 (1947).

163 *Jerison, H. J.:* Effects of noise on human performance. J. of Applied Psychology *43,* 96–101 (1959).

164 *Jerison, H. J. and Pickett, R. M.:* Vigilance: the importance of the elicited observing rate. Science, *143,* 970–971 (1964).

165 *Johansson, G., Aronsson, G. and Lindström, B. O.:* Social psychological and neuroendocrine stress reactions in highly mechanized work. Report from the Psychological Laboratories, University of Stockholm, No. 488 (1976).

166 *Jouvet, M.:* The states of sleep. Scientific American *216,* 62–66 (1967).

167 *Jürgens, H. W.:* Körpermasse in Ergonomie 1 von H. Schmidtke, Carl Hanser Verlag, München (1973).

168 *Kahn, H.:* Toward the year 2000. American Academy of Arts and Sciences (1967).

169 *Kalsbeck, J. W. H.:* Sinus Arrhythmia and the dual task method in measuring mental load. In W. T. Singleton *et al.*: Measurement of man at work. Taylor and Francis, London (1971).

170 *Kaminsky, G.:* Praxisnahe Untersuchungen über die Auswirkungen von Schwingungen über das Hand-Arm-Körper-System auf die Handgeschicklichkeit. Ergonomics *7,* 439–447 (1964)

171 *Kappauf, W. E., Smith, W. M. and Bray, C. W.:* A methodological study

of dial readings. Princeton University, Dep. of Psychology, Report 3 (August 1947).

172 *Karrasch, K. and Müller, E. A.:* Das Verhalten der Pulsfrequenz in der Erholungsperiode nach körperlicher Arbeit. Arbeitsphysiologie *14,* 369–382 (1951).

173 *Keegan, J. J.:* Alterations of the lumbar curve related to posture and seating. J. Bone Joint Surgery *35,* 567–589 (1953).

174 *Kerr, W. A.:* Worker attitudes toward scheduling of industrial music. J. appl. Psychology *30,* 575–578 (1946).

175 *Klosterkötter, W.:* Larmeinwirkung auf den Menschen. In VDI Bericht Nr 134: Beurteilung und Minderung von Arbeitslärm. VDI-Verlag, Düsseldorf (1969).

176 *Knauth, P., Rohmert, W. and Rutenfranz, J.:* Arbeitsphysiologische Kriterien zur Schichtplangestaltung bei kontinuierlicher Arbeitsweise. Z. Arb. wiss. *30,* 240–244 (1976).

177 *Knauth, P. and Rutenfranz, J.:* Untersuchungen zum Problem des Schlafverhaltens bei experimenteller Schichtarbeit. Int. Arch. Arbeits-med. *30,* 1–22 (1972).

178 *Knauth, P. and Rutenfranz, J.:* Das Verhalten der Körpertemperatur in verschiedenen Schichtsystemen bei experimenteller Schichtarbeit. Z. Arb. wiss. *31,* 18–21 (1977).

179 *Koch, A.:* Betriebslärm. Bundesinstitut für Arbeitsschutz, Soest Westfalen (1956)

180 *Koch, K. W., Jennings, B. H. and Humphreys, C. H.:* Is humidity important in the temperature comfort range? ASHRAE Transactions *66,* 63–68 (1960).

181 *Koella, W. P. and Czicman, J. S.:* Mechanism of the EEG-synchronizing action of serotonin. Am. J. Physiol. *211,* 926–931 (1966).

182 *Kornhauser, A.:* Mental health of the industrial worker. A Detroit study. J. Wiley, New York, London, Sidney (1965).

183 *Krämer, J.:* Biomechanische Veränderungen im lumbalen Bewe-gungssegment. Hippokrates, Stuttgart (1973).

184 *Kroemer, K. H. E.:* Heute zutreffende Körpermasse. Arbeitswissen-schaft *3,* 42–45 (1964).

185 *Kroemer, K. H. E.:* Foot operation of controls. Ergonomics *14,* 333–339 (1971).

186 *Kroemer, K. H. E.:* Was man von Schaltern, Kurbeln und Pedalen wissen muss. Sonderheft der REFA-Nachrichten, Verband für Arbeits-studien, REFA c.V., Darmstadt (1967).

187 *Kryter, K. D.:* The effects of noise on man. Academic Press Inc., New York (1970).

188 *Kryter, K. D.:* Damage risk criterion and contours based on permanent and temporary hearing loss data. Am. Ind. Hyg. Ass. J. *26,* 34–44 (1965).

189 *Kurtze, G.:* Physik und Technik der Lärmbekämpfung. G. Braun, Karlsruhe (1964).

190 *Laville, A., Teiger, C. and Duraffourg, J.:* Conséquences du travail répétitif sous cadence sur la santé des travailleurs et les accidents. Rapport No. 29, Lab. Physiologie du Travail, Conservatoire National des Arts et Métiers, Paris (1973).

191 *Lecret, F.:* La Fatigue du conducteur. Cahier d'étude de l'Organisme National de Sécurité Routière (ONSER) Bull. No. 38, Paris (1976).

192 *Lecret, F., Pin, M. C., Cura, J. B. and Pottier, M.:* Les Variations de la vigilance au cours de la conduite sur autoroute. 6 ème Congrès de la SELF, Paris (1968). Quoted from F. Lecret (191).

193 *Legendre, R. and Piéron, H.:* Quoted from Pappenheimer (252) Z. allg. Physiologie *14,* 235 (1913).

194 *Lehmann, G.:* Praktische Arbeitsphysiologie. 2. Auflage. Thieme Verlag, Stuttgart (1962).

195 *Lehmann, G. and Stier, F.:* Mensch und Gerät. Handbuch der

gesamten Arbeitsmedizin, Band 1, 718–788. Urban und Schwarzenberg, Berlin (1961).

196 *Leithead, C. S. and Lind, A. R.:* Heat stress and heat disorders. Cassel, London (1964).

197 *Leplat, J.:* Attention et incertitude dans les travaux de surveillance et d'inspection. Sciences du Comportement No. 6, Dunod, Paris (1968).

198 *Levi, L.* (Ed.): Society, Stress and Disease. I: The Psychosocial Environment and Psychosomatic Diseases. Oxford University Press (1971). II: Childhood and Adolescence. Oxford University Press (1975).

199 *Levi, L.:* Emotions – their parameters and measurement. Raven Press, New York (1975).

200 *Lille, F.:* Le Sommeil de jour d'un groupe de travailleurs de nuit. Le Travail Humain *30,* 85–97 (1967).

201 *Lind, A. R. and McNicol, G. W.:* Cardiovascular responses to holding and carrying weight by hand and by shoulder harness. J. applied Physiology *25,* 261–267 (1968).

202 *van Loon, J. H.:* Diurnal body temperature curves in shift workers. Ergonomics *6,* 267–273 (1963).

203 *Luckiesh, H. and Moss, F. K.:* The science of seeing. Van Nostrand, New York (1937).

204 *Lundervold, A.:* Electromyographic investigations during typewriting. Ergonomics *1,* 226–233 (1958).

205 *Lundervold, A.:* Electromyographic investigations of position and manner of working in typewriting. Acta Physiol. Scand., suppl. *84,* 171–183 (1951).

206 *Lundgren, N. P. V.:* Menschengerechte Gestaltung der Schwerarbeit. Ind. Organisation *29,* 401–412 (1960).

207 *Lukas, J. S.:* Noise and sleep; a literature review and a proposed criterion for assessing effect. J. Acoust. Soc. Am. *58,* 1232–1242 (1975).

208 *Mackworth, J. F.:* Vigilance and habituation. Penguin Books, Harmondsworth, England (1969).

209 *Mackworth, N. H.:* Researches on the measurement of human performance. H. M. Stationery Office, London (1950).

210 *Magid, E. B., Coermann, B. R. and Ziegenruecker, G. H.:* Human tolerance to whole body sinusoidal vibration. Aerospace Medicine *31,* 915—924 (1960).

211 *Magoun, H. W.:* The working brain. Thomas, Springfield, Illinois (1965).

212 *Magoun, H. W. and Rhines, R.:* An inhibitory mechanism in the bulbar reticular formation. J. Neurophysiol., U.S.A. *9,* 165–171 (1964).

213 *Malhotra, M. S. and Sengupta, J.:* Carrying of school bags by children. Ergonomics *8,* 55–60 (1965).

214 *Maric, D.:* L'Aménagement du temps de travail. Bureau International du Travail, 1211 Genève (1977).

215 *Martin, E.:* Auswirkungen einer einförmigen Tätigkeit auf das subjektive Befinden und die Flimmerverschmelzungsfrequenz. Dissertation, Eidg. Technische Hochschule (1977).

216 *Martin. and Weber, A.:* Wirkungen eintönig-repetitiver Tätigkeiten auf das subjektive Befinden und die Flimmerverschmelzungsfrequenz. Z. Arb. wiss. *30,* 183–187 (1976).

217 *Maurice, M. and Monteil, C.:* Vie quotidienne et horaires de travail. Université de Paris, Institut des Sciences Sociales du Travail, Paris (1965).

218 *McConnell, W. J. and Yaglou, C. P.:* Work tests conducted in atmospheres of high temperatures and various humidities in still and moving air. J. Am. Soc. Heating Ventilating Eng. *31,* 217—221 (1925).

219 *McConnell, W. J. and Spiegelman, M.:* Reactions of 745 clerks to

summer air conditioning. Heating, Piping, Air Conditioning *12,* 317–322 (1940).

220 *McCormick, E. J.:* Human Factors in Engineering. 3rd edition. McGraw Hill Book Company, New York (1970).

221 *McFarland, R. A.:* Human Factors in air transport design. McGraw Hill, New York, Toronto, London (1946).

222 *McFarland, R. A. and Stoudt, H. W.:* Human body size and passenger car design. SP-142 A Soc. Automotive Engineers, New York (1960).

223 *Metz, B.:* Ambiances thermiques. In J. Scherrer: Physiologie du Travail, Tome 2. Masson, Paris (1967).

224 *Michaels, H.:* Raubbau in 4 Tagen. Die Zeit, Nr 34, S. 25 (1971).

225 *Monnier, M. and Schoenenberger, G. A.:* Erzeugung. Isolierung und Charakterisierung eines physiologischen Schlaffaktors ''delta''. Schweiz. med. Wschr. *103,* 1733–1743 (1973).

226 *Monod, H.:* Contributions à l'étude du travail statique. Thèse, Faculté de Médecine, Paris (1956).

227 *Monod, H.:* La dépense energétique chez l'homme. In Physiologie du Travail, edited by J. Scherrer. Masson, Paris (1967).

228 *Morgan, C. T., Chapanis, A., Cook, J. S. and Lund, M. W.:* Human engineering guide to equipment design. McGraw Hill, New York (1963).

229 *Moruzzi, G.:* Physiologie des Schlafes. Endeavour *22,* 31–36 (1963).

230 *Mott, P. E., Mann, C., McLoughlin and Warwick, P.:* Shiftwork: the social, psychological and physical consequences. University of Michigan Press, Ann Arbor (1965).

231 *Müller, E. A.:* Die günstigste Anordnung im Sitzen betätigter Fusshebel. Arbeitsphysiol. *9,* 125–137 (1936).

232 *Müller, E. A.:* Die Pulszahl als Kennzeichen für Stoffumtausch und Ermüdbarkeit des arbeitenden Muskels. Arbeitsphysiologie *12,* 92–106 (1942).

233 *Müller, E. A.:* Die physische Ermüdung. In Handbuch der gesamten Arbeitsmedizin, Band 1. Urban und Schwarzenberg, Berlin (1961).

234 *Müller, E. A. and Spitzer, H.:* Arbeit recht verstanden. Oldenbourg Verlag, München (1952).

235 *Müller-Limmroth, W.:* Sinnesorgane. in: Ergonomie, 1. Band von H. Schmidtke. Carl Hanser Verlag, München (1973).

236 *Münchinger, R.:* Gewichtheben und Bandscheibenbelastung. Schweiz. Zschr. Sportmedizin *8,* 65–78 (1960).

237 *Murrell, K. F. H.:* Ergonomics; man in his working environment. Chapman and Hall, London (1965).

238 *Murrell, K. F. H.:* Human performance in industry. Reinhold Publishing Corp. New York (1965).

239 *Murrell, K. F. H., Laurie, W. D. and McCarthy, C.:* The relationship between dial size, reading distance and reading accuracy. Ergonomics *1,* 182–190 (1958).

240 *Nachemson, A.:* Lumbal intradiscal pressure. Results from in vitro and in vivo experiments with some clinical implications. 7. Wiss. Konf. Deutscher Naturforscher and Aerzte. Springer, Berlin, Heidelberg, New York (1974).

241 *Nachemson, A. and Elfström, G.:* Intravital Dynamic Pressure Measurements in Lumbar Discs. Scand. J. Rehabilitation Medicine, Supplement 1. Almqvist and Wiksell, Stockholm (1970).

242 *Nemecek, J. and Grandjean, E.:* Etude ergonomique d'un travail pénible dans l'industrie textile. Le Travail Humain *38,* 167–174 (1975).

243 *Nemecek, J. and Grandjean, E.:* Das Grossraumbüro in arbeits-physiologischer Sicht. Ind. Organisation *40,* 233—243 (1971).

244 *Nemecek, J. and Turrian, Verena:* Der Bürolärm und seine Wirkungen. Kampf dem Lärm *25,* 50–57 (1978).

245 *Neumann, J. and Timpe, K. P.:* Arbeitsgestaltung. Psychophysiolo-

gische Probleme bei Ueberwachungs- und Steuerungstätigkeiten. V.E.B. Deutscher Verlag der Wissenschaften, Berlin (1970).

246 *Nevins, R. G., Rohles, F. H., Springer, W. and Feyerherm, A. M.:* A temperature-humidity chart of thermal comfort of seated persons. ASHRAE Journal *8,* 55–61 (1966).

247 *Newburgh, L. H.:* Physiology of heat regulation and the science of clothing. W. B. Saunders, Philadelphia (1949).

248 *Nitsch, J.:* Theorie und Skalierung der Ermüdung. Dissertation, Köln (1970).

249 *Northrup, H. R.:* Hours of work. Harper and Row, New York (1965).

250 *O'Hanlon, J. F.:* Heart rate variability: A new index of drivers alertness/fatigue. Human Factors Research Inc. Santa Barbara, Goleta, Cal. Report No. 1812–1 (1971).

251 *O'Hanlon, J. F., Royal, J. W. and Beatty, J.:* EEG theta regulation and radar monitoring performance in a controlled field experiment. Human Factors Research Inc. Santa Barbara, Cal. Techn. Rep. 1738-F (1975).

252 *Pappenheimer, J. R., Miller, T. V. and Goodrich, C. A.:* Sleep promoting effects of cerebrospinal fluid from sleep-deprived goats. Proc. Nat. Acad. Sci., U.S.A. *58,* 513–518 (1967).

253 *Pearson, R. G. and Byars, G. E.:* The development and validation of a checklist for measuring subjective fatigue. Report 56–115. School of Aviation Medicine, USAF, Randolph AFB, Texas, Dec. (1956).

254 *Pierce, J. R. and Karlin, J. E.:* Reading rates and the information rate of a human channel. Bell Telephone Technical Journal *36,* 497–516 (1957).

255 *Pin, M. C.:* Application de techniques electrophysiologiques à l'étude de la conduite automobile. Cahiers d'étude de l'ONSER, Bull. No. 15, Paris (1966).

256 *Proceedings of the International Conference on Enhancing the Quality of Working Life.* Arden House, Harriman, New York (Sept. 1972).

257 *Prokop, O. and Prokop, L.:* Ermüdung und Einschlafen am Steuer. Deutsche Zschr. f. ger. Med. *44,* 343–350 (1955).

258 *Pternitis, C.:* Etudes electrophysiologiques au cours du sommeil chez l'ouvrier posté. In Ergonomie du travail de nuit et des horaires alternants, P. Andlauer and collab. (Ed.) Editions Cujas, Paris (1977).

259 *Rey, Paule and Rey, J. P.:* Effect of an intermittent light stimulation on the critical fusion frequency. Ergonomics *8,* 173–180 (1965).

260 *Ridder, C.:* Basic design measurements for sitting. Agricultural Experiment Station, University of Arkansas, Fayetteville, Bull. 616 (1959).

261 *Ritzel, G.:* Sozialmedizinische Erhebungen zur Pathogenese und Prophylaxe von Erkältungskrankheiten. Z. Präventivmed. *11,* 9–16 (1966).

262 *Robinson, D. W. and Dadson, R. S.:* Threshold of hearing and equalloudness relations for pure tones and the loudness function. J. Acoustical Soc. of America *29,* 1284–1288 (1957).

263 *Robinson, G. and Gerking, S. D.:* The thermal balance of men working in severe heat. Am. J. Physiology *149,* 102–108 (1947).

264 *Roedler, F.:* Wärmephysiologische und hygienische Grundlagen. In Rietschel und Raiss (Hrsg.): Heiz- und Klima Technik. 15th edition, Springer, Berlin (1968).

265 *Rohmert, W.:* Die Grundlagen der Beurteilung statischer Arbeit. Forschungsberichte des Landes Nordrhein-Westfalen Nr 938. Westdeutscher Verlag, Köln and Opladen (1960).

266 *Rohmert, W.:* Maximalkräfte von Männern im Bewegungsraum der Arme und Beine. Forschungsberichte des Landes Nordrhein-Westfalen Nr. 1616, Westdeutscher Verlag, Köln and Opladen (1966).

267 *Rohmert, W.:* Statische Haltearbeit des Menschen. Sonderheft der REFA-Nachrichten. Beuth-Vertrieb, Berlin, Köln and Frankfurt/M (1960).

268 *Rohmert, W. and Jenik, P.:* Maximalkräfte von Frauen im Bewegungsraum der Arme und Beine.
Schriftenreihe "Arbeitswissenschaft und Praxis". Beuth-Vertrieb, Berlin, Köln and Frankfurt/M (1972).

269 *Rohmert, W. and Hettinger, Th.:* Körperkräfte im Bewegungsraum. RKW-Schrifteenreihe Arbeitsphysiologie. Beuth-Vertrieb, Berlin, Köln, Frankfurt/M (1963).

270 *Rohmert, W. and Hettinger, Th.:* Ergebnisse achtstündiger Untersuchungen am Kurbel- und Fahrradergometer. Quoted in (142).

271 *Rohmert, W., Rutenfranz, J. and Ulich, E.:* Das Anlernen sensumotorischer Fertigkeiten. Europäische Verlagsanstalt, Frankfurt/M (1971).

272 *Rosemeyer, B.:* Elektromyographische Untersuchungen der Rücken- und Schultermuskulatur im Stehen und Sitzen unter Berücksichtigung der Haltung des Autofahrers. Arch. orthop. Unfall-Chir. *69,* 59–70 (1971).

273 *Rosenkranz, R.:* Die Viertagewoche und die kritische Untersuchung des Ist-Zustandes. Rationelles Büro *22,* Nr 7 (1971).

274 *Rutenfranz, J. and Knauth, P.:* Rhythmusphysiologie und Schichtarbeit. in: Schicht- und Nachtarbeit. Institut für Gesellschaftspolitik (pub.) Sensenverlag, 1090-Wien (1976).

275 *Ryan, A. H. and Warner, M.:* The effect of automobile driving on the reactions of the driver. Amer. J. Physiol. *48,* 403–409 (1936).

276 *Saito, H., Kishida, K., Endo, Y. and Saito, M.:* Studies on bottle inspection task. J. of Science of Labour *48,* 475–525 (1972).

277 *Scherrer, J.:* Physiologie Musculaire. In Physiologie du Travail. Edited by J. Scherrer. Masson, Paris (1967).

278 *Schmidtke, H.:* Bedienungs- und Steuerarmaturen. In H. Schmidtke, Ergonomie, 2. Band. Carl Hanser Verlag, München (1974).

279 *Schmidtke, H.:* Wachsamkeitsprobleme. In H. Schmidtke, Ergonomie, 1. Band. Carl Hanser Verlag, München (1973).

280 *Schmidtke, H.:* Die Ermüdung — Symptome, Theorien, Messversuche. Hans Huber, Bern and Stuttgart (1965).

281 *Schmidtke, H. and Stier, F.:* Der Aufbau komplexer Bewegungsabläufe aus Elementarbewegungen. Forschungsbericht des Landes Nordrhein-Westfalen, Nr 822, Westdeutscher Verlag, Köln and Opladen (1960).

282 *Schoberth, H.:* Sitzhaltung – Sitzschaden – Sitzmöbel. Springer, Berlin (1962).

283 *Scholz, H.:* Die physische Arbeitsbelastung der Giessereiarbeiter. Forschungsbericht des Landes Nordrhein-Westfalen Nr 1185, Westdeutscher Verlag, Köln and Opladen (1963).

284 *Scholz, H. and Sieber, W.:* Quoted from W. Rohmert and Th. Hettinger: Arbeitsgestaltung und Muskelermüdung. R.K.W.-Reihe Arbeitsphysiologie – Arbeitspsychologie, Beuth-Vertrieb Berlin, Köln, Frankfurt (1963).

285 *Schweiz. Unfallversichungsanstalt:* Gehörschädigender Lärm am Arbeitsplatz. Schweiz. Blätter für Arbeitssicherheit Nr 113, Luzern (1974).

286 *Serati, A. and Wüthrich, M.:* Luftfeuchtigkeit und Saisonkrankheiten. Sch weiz. Med. Wschr. *99,* 48–50 (1969).

287 *Shackel, B.:* Applied Ergonomics Handbook. Reprints from Applied Ergonomics Vol. 1 and Vol. 2. I.P.C. Science and Technology Press, Guildford, U.K. (1974).

288 *Shannon, C. E. and Weaver, W.:* The mathematical theory of communication. The University of Illinois Press, Urbana (1949).

289 *Shepherd, R. D. and Walker, J.:* Absence and the physical conditions of work. Brit. J. Industrial Med. *14*, 266–274 (1957).

290 *Simons, A. K., Radke, A. O. and Oswald, W. C.:* A study of truck ride characteristics in military vehicles. Bostrom Research Laboratories, Milwaukee, Report 118 (1956).

291 *Singleton, W. T.:* Introduction à l'ergonomie. Organisation Mondiale de la Santé, Genève (1974).

292 *Sleight, R. B.:* The effect of instrument dial shape on legibility. J. of Applied Psychology *32*, 170–188 (1948).

293 *Smith, H. C.:* Applied Psychology Monographs, Nr. 14 (1947).

294 *Spitzer, H.:* Physiologische Grundlagen für den Erholungszuschlag bei Schwerarbeit. REFA-Nachrichten, Heft 2, Darmstadt (1951).

295 *Steinbuch, K.:* Information processing in man. Quoted from McCormick (220).

296 *Steinicke, G.:* Die Wirkung von Lärm auf den Schlaf des Menschen. Forschungsberichte des Wirtschafts- und Verkehrsministerium des Landes Nordrhein-Westfalen Nr 416 (1957).

297 *Stier, F. and Meyer, H. O.:* Physiologische Grundlagen der Arbeits-gestaltung. Verband für Arbeitsstudien (REFA) Darmstadt.

298 *Stubbs, D. A.:* Manual handling in the construction industry. Report, Constructing Industry Training Board (1973).

299 *Swensson, A. Ed.:* Night and shiftwork. Proceedings of an International Symposium, Nat. Institute of Occupational Health, Stockholm, Sweden (1969), and a second symposium: Studia Laboris et Salutis, Rep. No. 11, Stockholm (1972).

300 *Swink, J. R.:* Intersensory comparisons of reaction time using an electropulse tactile stimulus. Human factors *8*, 143–145 (1966).

301 *Taylor, C. L.:* The biomechanics of the normal and of the amputated upper extremity. In Klopsteg and Wilson: Human limbs and their substitutes. McGraw Hill, New York, S. 169–221 (1954).

302 *Teeple, J. B.:* Work of carrying loads. J. Perceptual and Motor Skills *7*, 60–68 (1957).

303 *Thiberg, S.:* Anatomy for planners, Parts I–IV. Statens Institut för Byggnadsforskning (1965–1970).

304 *Thiis-Evensen, E.:* Shiftwork and Health. Industrial Medicine, Chicago, *27*, 493–497 (1958).

305 Thomas, D. R.: Exposure time as a variable in dial reading experiments. J. of Applied Psychology *41*, 150–152 (1957).

306 *Tichauer, E. R.:* Biomechanics sustains occupational safety and health. Industrial Engineering *27*, 46–56 (1976).

307 *Tichauer, E. R.:* Potential of Biomechanics for solving specific hazard problems. Conference. Am. Soc. of Safety Engineers, 149—187, Park Ridge, Illinois (1968).

308 *Tichauer, E. R.:* Occupational Biomechanics. Rehabilitation Monograph No. 51, New York University, Center for Safety (1975).

309 *Ulich, E.:* Arbeitswechsel und Aufgabenerweiterung. REFA-Nachrichten, *25*, 265–275 (1972).

310 *Ulich, E.:* Some experiments on the function of mental training in the acquisition of motor skills. Ergonomics *10*, 411–419 (1967).

311 *Ulich, E.:* Schicht- und Nachtarbeit im Betrieb. RKW Publikation, Westdeutscher Verlag Köln and Opladen (1964).

312 *Ulich, E., Groskurth, P. and Bruggemann, Agnes:* Neue Formen der Arbeitsgestaltung. Europäische Verlagsanstalt, Frankfurt/M (1973).

313 *U.S. Dep. of Health, Education and Welfare:* Weight, Height and Selected Body Dimensions of Adults: United States 1960–62. Vital and Health Statistics, Series 11, no. 8 Washington: Government Printing Office (1966).

314 *Vernon, M. H.:* Industrial fatigue and efficiency. Dutton, New York (1921).

315 *VDI Richtlinie 2057:* Beurteilung mechanischer Schwingungen auf den Menschen. VDI Verlag, Düsseldorf (October 1963).

316 *VDI Richtlinie 2058:* Beurteilung von Arbeitslärm am Arbeitsplatz hinsichtlich Gehörschäden. Blatt 2. VDI Verlag, Düsseldorf (1970).

317 *Wagenhäuser, F. J.:* Die Rheumamorbidität: Eine klinisch epidemiologische Untersuchung. Hans Huber, Bern (1969).

318 *Wakim, K. G., Gersten, J. W., Elkins, E. C. and Martin, G. M.:* Objective recording of muscle strength. Arch. Physic. Med. *31*, 90–99 (1950).

319 *Ward, W. D.:* Noise-induced hearing damage. In Otolaryngology, ed. by M. M. Paparella and D. A. Shumrick, Vol. 2. Saunders, Philadelphia (1973).

320 *Wargo, M. J.:* Human operator response speed, frequency and flexibility: a review and analysis. Human Factors *9*, 221—238 (1967).

321 *Warrick, M. J., Kibler, A. W. and Topmiller D. A.:* Response time to unexpected stimuli. Human Factors *7*, 81–86 (1965).

322 *Weber, A., Jermini, C. and Grandjean, E.:* Beziehung zwischen objektiven und subjektiven Messmethoden bei experimentell erzeugter Ermüdung. Z. Präventivmed. *18*, 279–283 (1973).

323 *Weber, A., Martin, E., Udris, I., Jermini, C. and Grandjean, E.:* Objektive und subjektive Ermüdung in experimentellen Monotonie-Situationen. Soz. u. Präventivmed. *19*, 285–287 (1974).

324 *Welford, A. T.:* Fundamentals of Skill. Methnen, London (1968).

325 *van Wely, P.:* Design and disease. Applied Ergonomics *1*, 262–269 (1970).

326 *Wenzel, G.:* Klima, sowie Klimamessung und Klimabewertung. In H. Schmidtke (Ed.): Ergonomie, Band 2. Carl Hanser Verlag, München (1974).

327 *Wenzel, H. G.:* Möglichkeiten und Probleme der Beurteilung von Hitzebelastungen des Menschen. Arbeitswissenschaft *3*, 73–83 (1964).

328 *Weston, H. C.:* Sight, light and efficiency. Lewis, London (1949).

329 *Winslow, C. E. A. and Herrington, L. P.:* Temperature and human life. Princeton University Press, Princeton, N.J. (1949).

330 *Wisner, A.:* Audition et Bruits. In J. Scherrer (Ed.): Physiologie du travail, Tome 2. Masson, Paris (1967).

331 *Wisner, A.:* Effets des vibrations sur l'homme. In J. Scherrer (Ed.): Physiologie du Travail, Tome 2. Masson, Paris (1967).

332 *Wisner, A., Laville, A. and Richard, M. E.:* Conditions de travail des femmes O.S. de la construction électronique. Rapport No. 2 Lab. Physiologie du Travail, Conservatoire National des Arts et Métiers, Paris (1967).

333 *Wisner, A. and Rebiffé, R.:* L'Utilisation des données anthropométriques dans la conception du poste de travail. Travail Humain *26*, 193–217 (1963).

334 *Wirths, W.:* Ist eine Zwischenverpflegung während der Arbeitszeit ernährungsphysiologisch notwendig? Published in: Ernährungspädagogisches Colloquium, Mühlenstelle, Bonn, Bericht 10 (January 1976).

335 *Woodson, W. E. and Conover, D. W.:* Human Engineering Guide for Equipment Designers. Berkeley, California, University of California Press (1964).

336 *Wyatt, S.:* A study of variations in output. Medical Res. Council Emergency Rept. 5, 1–16 (1944).

337 *Wyatt, S. and Marriott, R.:* A study of attitudes to factory work. Med. Res. Council Special Report Series 292, London (1956).

338 *Wyatt, S. and Marriott, R.:* Night work and shift changes. Brit. J. Industrial Med. *10*, 164–170 (1953).

339 *Wyndham, C. H., et al.:* Examination of heat stress indices. Arch. industr. Hyg. Occup. Med. *7*, 221–233 (1953).

340 *Yaglou, C. P., Riley, E. C. and Coggins, D. I.:* Ventilation requirements. and the science of clothing. Saunders, Philadelphia (1949).

341 *Yaglou, C. P., Riley, E. C. and Coggins, D. I.L* Ventilation requirements. ASHVE Transactions *42,* 133–158 (1936).

342 *Yamaguchi, Y.:* Quoted from E. Grandjean: Wohnphysiologie. Artemis, Zürich (1973).

343 *Yamaguchi, Y., Umezawa, F. and Jshinada, Y.:* Sitting posture: an electromyographic study on healthy and notalgic people. J. Jap. Orthop. Ass.: *46,* 51–56 (1972).

344 *Yllö, A.:* The biotechnology of card-punching. Ergonomics *5,* 75–79 (1962).

345 *Yoshitake, H.:* Relations between the symptoms and the feeling of fatigue. In Methodology in human fatigue assessment, edited by K. Hashimoto, K. Kogi and E. Grandjean. Taylor and Francis, London (1971).

346 *Zeier, H. and Bättig, K.:* Psychovegetative Belastung und Aufmerksamkeitsspannung von Fahrzeuglenkern auf Autobahnabschnitten mit und ohne Geschwindigkeitsbegrenzung. Zschr. f. Verkehrssicherheit *23* (1977) 1.

347 *Ziegenruecker, G. H. and Magid, E. B.:* Human tolerance to whole-body sinusoidal vibration. Short-time, one-minute and three-minutes studies. Aerospace Med. *31,* 915–924 (1960).

Abbreviations and units

1 l = 1 litre = 1 dm³ = 1 decimetre cubed
1 ml = 1 millilitre = 1 cm³ = 1 cubic centimetre
1 dl = 1 decilitre = 100 cm³

1 kcal = 1 kilocalorie = 1000 calories = 4.2 kilojoules

s = second
min = minute
h = hour

g = gram; kg = kilogram; mg = milligram = one thousandth of a
  gram
ng = nanogram = 10^{-9} gram = one millionth of a milligram

kp = kilopond (an old unit of force, = 1 kg force)

μ (mu). Prefix indicating 10^{-6} (1 millionth)
1 μm = 1 micron = 10^{-6} m

lx = lux (unit of illumination). 1 lx = 1 lm/m²
lm = lumen (intensity of illumination)
cd = candela (unit of luminous intensity)
sb = stilb (unit of luminance). 1 sb = 10 000 cd/m²
asb = apostilb (unit of luminance). 1 asb = 0.32 cd/m²

b = bar (unit of pressure). 1 bar is approximately = 1 atmosphere
  = 760 mm Hg
μb = microbar = 10^{-6} b

dB = decibel (measure of sound pressure)

Index